SECOND EDITION

ENVIRONMENTAL ETHICS

K.S. Shrader-Frechette, Ph.D.
Professor of the Philosophy of Science
University of South Florida

THE BOXWOOD PRESS

2nd Printing, 1993

by
The Boxwood Press

Distributed by

The Boxwood Press
183 Ocean View Blvd.
Pacific Grove, CA 93950

Phone: 408—375-9110
Fax: 408—375-0430

ISBN: 0-940168-22-7

Library of Congress Cataloging in Publication Data:

Shrader-Frechette, Kristin S. 1944-
Environmental ethics.

Bibliography: p.
Includes index.

1. Human ecology—Moral and religious aspects. Introduction; readings.
2. Man-Influence on nature. Introduction; readings.
3. Environmental policy. Introduction; readings.
4. Environmental protection. Introduction; readings.

GF80.S47 304.2 80-21114

Printed in U.S.A.

For Eric and Danielle
And all children
For they shall inherit the earth

CONTENTS

PREFACE

IN 1950, there were 50 million cars worldwide, 75 percent of them in the USA. This number doubled by 1960 and quadrupled by 1970. By 1980 there were 300 million automobiles on the planet, and by 1990, the figure had risen to 400 million cars, only 33 percent of which are in the USA. Even clean-burning cars produce 20 pounds of carbon dioxide for every gallon of gasoline that they burn. "Greenpeace," the environmental organization, claims that automobiles are the largest single source of greenhouse gases that threaten to raise temperatures worldwide. What is worse, Stanford Professor Paul Ehrlich claims that carbon-dioxide-induced global warming could lead to the famine deaths of a billion persons before the year 2020.

Should we change our use of automobiles in order to stop global warming? Should we rethink our policies about threats such as toxic wastes, groundwater pollution, nuclear weapons, and fossil fuels in order to avoid further environmental destruction? If environmental degradation were purely, or even primarily a problem demanding scientific or technological solutions, then its resolution would probably have been accomplished by now. As it is, however, our crises of pollution and resource depletion reflect profound difficulties with some of the most basic principles in our accepted systems of values. They challenge us to assess the adequacy of those principles and, if need be, to discover a new framework for describing what it means to behave ethically.

Some of the basic lines of this new framework were suggested decades ago by writers such as Carson, Muir, Leopold, and Thoreau. The purpose of this book is to continue what they have begun. It organizes, philosophically develops, defends, rationally analyzes, and criticizes the body of ideas which are gradually coming to be known as "environmental ethics." Specifically I discuss (1) the metaphysical and methodological assumptions underlying alternative views of environmental ethics; (2) the ethical conflict between the attempt to maximize environmental, as opposed to other, values, e.g., economic benefits; and (4) the alternative means of resolving specific environmental issues, e.g., whether to use nuclear fission to generate electricity.

In a very practical sense, this book owes its origin to my frustration at the deficiences I discovered in typical USA technology assessments and environmental impact statements. While working, both as an individual researcher and as a member of an interdisciplinary assessment team, under grants from the US Environmental Protection Agency (1975-1977); and the US National Science Foundation (1982-1983; 1984-1987; 1988-1991); and the US Department of Energy (1991), I learned that most evaluations of the environment-related public policy were seriously incomplete. They often failed to investigate the methodological, ethical, and social assumptions implicit in the conclusions.

Instead they dealt only with easily quantifiable physical, biological, and chemical data. Such difficulties appeared to me to be, not a consequence or dishonesty or ill will, at least on the part of the teams of scientists with whom I worked, but rather a result of shortsightedness and the tendency to believe that all parameters relevant to an issue were scientific and quantifiable ones. All too often such research procedures have undercut the ultimate validity and the methodological justifiablity of numerous environment studies in which taxpayers have invested great amounts of money.

Since one of the deficiencies plaguing many current technology and environmental impact assessments is the failure of researchers to recognize and analyze social and ethical assumptions implicit in their work, one solution to such a difficulty is to clarify these presuppositions. I wrote this book to provide one step in that direction, by giving an elementary introduction to enviroment-related ethical problems. I hope that it is sophisticated enough to convince philosophers, assessors of environmental policy, and the general public that ethical and methodological naiavete is no more defensible in environmental assessment than is naivete regarding biological matters.

In a more profound sense, this book owes its origin to all I have learned from a number of scholars and environmental activists. It has been the compassion, humility, and sense of social justice that, years ago, let me first to admire them as persons and then to examine the quality of their ideas. I owe a great debt to my parents, Owen and Mildred Shrader, who were environmentalists long before it became fashionable, and to my husband, Maurice, whose brilliance, simplicity, and strength have made both this book and my life better than it might have been. I am also grateful to numerous environmentalists who have not only thought and written about ethics, technology, and the environment, but have had the courage to live out their beliefs. Both their reverence for the earth and for people, and the austerity of their lifestyles have shown me much of what is right and good.

My typists and proofreaders, Nacy Keen and Betty Myers Shrader, have been invaluable assistants. Without their patience, thoroughness, and generosity, this volume would never have been completed. Finally, I thank Dr. Ralph Buchsbaum, editor of Boxwood Press, and the referees who made numerous suggestions for improving the book. Whatever shortcomings that remain in it are my responsibility.

K.S. Shrader-Frechette

University of South Florida
May, 1991

FOREWORD

AS PAUL GOODMAN correctly observed, technology is a branch of moral philosophy as well as of science. It is directed toward the discovery of prudent and just means of obtaining goods for society, and it draws upon science to achieve these goods. For most technologies, however, there are no clear ethical directives for how they ought to be employed, and no explicit codes for determining when they have been used in a prudent and just way. To remedy this deficiency, thinkers such as Karl Jaspers and Robert Hutchins have argued that we must develop our understanding of technology and its implications. In the case of environment-related technologies, such as chemical pesticides and nuclear fission, we must work out a policy framework with at least the degree of sophistication attained by current systems of medical ethics and legal ethics. If we fail to do so, we might jeopardize the well-being of both humankind and the entire planet.

This book is intended as an introduction to environmental ethics. My object in writing it is to lay a foundation for bringing together the closely related disciplines of ecology and moral philosophy. Just as ecology provides guidelines for the way humankind ought to be integrated into the environment, so moral philosophy or ethics affords a framework for the way we ought to be integrated into society. As a combination of both disciplines, environmental ethics aims at enhancing our relationships both with others and with the world around us.

Because the field of environmental ethics is rooted in science and technology, this book addresses practical ecological problems of our contemporary technological society. Its earlier sections provide an ethical analysis of general and theoretical questions which are basic to all environmental decision-making. Later chapters are devoted to evaluation of policy options regarding particular environmental problems. As such, this volume includes material falling in the domain of applied as well as theoretical ethics.

Frequently, issues of environmental ethics have been treated in too simple a fashion. The result has been that they often are seen as involving either a conflict between "the good guys" (the environmentalists) and "the bad guys" (the polluters) or, on the other hand, a contest between uninformed idealists (environmentalists) and common-sensical realists (economists and industrialists). In collecting essays representative of many points of view, I have attempted both to avoid such oversimplification and to assemble readable selections. My goal has been to provide an interesting rational analysis of each issue treated, and to select essays representative of a wide variety of opinions on each topic, including those of persons not particularly sensitive to ecological matters. By gaining familiarity with opposing points of view, one obtains a more realistic idea of the sorts of problems that a well-thought-out system of environmental ethics must address.

The twelve chapters of the book are divided into four sections, each of which treats a different set of environmental issues. The earlier sections aim at developing a general body of principles and at formulating specific rights and duties relative to the environment. Later sections then employ these principles to resolve specific questions of ecological policy.

The first section provides introductory material for understanding the theoretical underpinnings of environmental decision-making. Chapter 1 addresses the questions of whether a new system of environmental ethics is needed in order to avoid the planetary crisis induced by pollution and resource depletion, and whether our traditional moral philosophies have helped us to ignore ecosystemic well-being.

Various ethics have been proposed in response to what is perceived either as a real, or only as an apparent, environmental crisis. Chapter 2 presents several models for environmental policy. These include "Frontier or Cowboy Ethics," Garrett Hardin's "Lifeboat Ethics," and the "Spaceship Ethics" of Lynton Caldwell and Buckminster Fuller. After analyzing these three systems, I argue that the third option presents the best available alternative for environmental ethics.

Section Two of the book is devoted to analysis of a number of issues involving rights and duties. These inquiries have been chosen because, regardless of the specific ethic or the particular planetary problem under discussion, certain questions are sure to arise when one attempts to consider any specific environmental policy. Chapter 3 is devoted to discussion of whether future generations may be said to have rights. Likewise, Chapters 4 and 5 provide alternative analyses of the question of whether natural objects and animals, respectively, may be said to have rights. The two remaining chapters in this section deal with proposals to extend or amend existing lists of human rights. Chapter 6 is addressed to whether persons may be said to have a right to a liveable environment, whereas Chapter 7 focuses on whether we may be said to have a duty to limit our consumption or possession of goods.

The topics treated in Section Three are similar in one respect to those discussed in the previous section; being basic issues, they inevitably arise in almost any discussion of either applied or theoretical environmental ethics. One of these central problems, addressed in Chapter 8, is how to resolve conflicts between environmental and economic goods. This problem arises in industry, for example, because pollution control is often said to be bought at the expense of economic expansion. Chapter 9 deals with a related and equally troubling question of choosing the greater good (viz., how to resolve conflicts between actions designed to help the poor and those directed at protecting the environment). Like the issue of the environment versus the economy, this problem also raises the difficult matter of of priorities. Does ethics require one to be concerned with basic questions of poverty and discrimination rather than with ecological issues? Should priority government funding be given to protection of the environment or to meeting the needs of the poor? These are only a few of the cluster of ethical problems discussed in Chapter 9.

Since a number of general questions, central to nearly all discussions of environmental ethics, are treated in the first three sections of the book, the final section is devoted to specific contemporary applications of these questions. Chapter 10 deals with what many people believe is the single most important factor affecting the environment: population. One of the ethical questions raised by this problem, and addressed in Chapter 10, is whether ecological considerations can be said to justify coercive methods of population control and, hence, limits on human freedom. Chapter 11 also deals with an issue which has been one of the main concerns of environmentalists. This is the pesticide problem, which focuses in part on the necessity of ethical evaluation of the assets of short-term protection from disease and crop destruction versus the liabilities of long-term environmental degradation. The short-term/long-term ethical analysis is also important to evaluation of nuclear energy, the topic of Chapter 12. On the one hand nuclear energy provides a way to meet immediate energy needs; on the other hand it threatens the well-being of hundreds of future generations because of the longevity of its lethal radioactive wastes.

Although many other contemporary problems could be treated in Section Four, I have limited discussion to the issues of population control, employment of pesticides, and utilization of nuclear power. Space restrictions have made it impossible to treat as many current problems as I would have liked, and these three appear to be among the most important, interesting, and controversial. Since Sections One through Three lay the foundations for discussing a wide variety of environmental issues, readers can easily apply considerations gleaned from this earlier framework to a variety of topics of their own choosing.

A second reason for limiting the number of specific problems treated near the end of the book is that many important areas of controversy (e.g., air pollution, water pollution, endangered species) have in part been addressed by recent federal legislation (e.g., the Clean Air Act of 1970, the Federal Water Pollution Control Act of 1970 and the Endangered Species Act of 1973). Because these problems arouse less interest now than they did a decade ago, many of them have not been included here. Several other important topics have not been considered in Section Four because the scientific underpinnings, necessary for developing defensible positions regarding them, have not yet become completely clear. The ozone controversy, for example, came into being nearly a decade ago, was later dismissed by many members of the scientific community, and now has re-emerged, but for quite different reasons than were originally thought to be important. Given the rate at which scientific information on this and other topics is changing, I decided not to include consideration of questions such as this one.

Each of the twelve chapters in the book addresses one specific topic and is divided into three parts. The first, the Introduction, provides a relatively unsophisticated overview of the problem under consideration and outlines broadly the alternative and/or opposed positions regarding it. The second part of each chapter contains two or three relevant readings, drawn from books, journals, and unpublished material. Finally, at the end of each chapter is a selected bibliography of some of the more important ethical/scientific/environmental readings on the question addressed in the chapter. For the most part, extremely difficult or very highly specialized scientific material has been omitted from this bibliography, since the purpose of the book is to introduce a wide range of undergraduates and general readers to some ethical issues raised by various environmental problems.

Although the chapters vary slightly in the level of scientific, legal, and philosophical understanding required to master them, this book is intended for the person with no previous background in any of these areas. As such, it is the only one available which provides various essays specifically on topics of theoretical and applied environmental ethics. More philosophically minded persons might want to spend additional time on certain readings, for example, those in Chapter 5, which exhibit a moderate degree of ethical sophistication. Scientists, environmentalists, and the general reader, on the other hand, might want to concentrate on the various concrete issues addressed by the three chapters in Section Four. In either case, the material can be covered in a semester, at a rate of approximately one chapter per week.

In writing this book, I have made every attempt to make the questions treated relevant to the concerns of today's students and intelligent laymen. Being an introductory book, this one must carry the double burden of (1) addressing important problems of environmental ethics, and (2) motivating its readers to begin more intensive analyses than many of those represented in the various readings. I am convinced that the topics treated here will do just that. They illustrate why environmental issues signal not merely a crisis of resources and pollution, but a crisis of the human spirit as well.

SECTION ONE

Theoretical Frameworks

1

DO WE NEED ENVIRONMENTAL ETHICS?

Introduction

EDWARD ABBEY shocked the world when he wrote, in *Desert Solitaire,* that he would sooner shoot a man than a snake. Snakes, like other parts of the natural community of interacting species, rarely cause a collapse of the ecosystems of which they are a part. They are kept in check by predators and competitors. Humans, however, are also part of a natural community of interacting species, but they often behave as if they are self-sufficient. Indeed, humans are destroying earth's life-support systems in a way that snakes and other nonhuman creatures are unable to do. We humans are responsible, for example, for 10,000 species extinctions per year, according to Harvard zoologist E.O. Wilson. Of course, extinction is a natural occurrence. We, however, have increased the rate of extinction more than a thousand times beyond what it would be without human intervention.

In response to widespread human destruction of the environment, many thinkers have argued that the worst pollution today is mind pollution. Our minds have been so polluted with human arrogance, insensitivity, greed, and ignorance about our environment that we have followed a purely human-centered (or anthropocentric) ethics. Indeed, traditional ethics or moral philosophy, especially in the West, has been anthropocentric. It takes insufficient account of the needs of nonhuman beings, except insofar as those beings affect humans. Many *anthropocentric ethics* teach that nonhuman beings have only *instrumental value,* as means to human ends, like food. Only humans, they say, have inherent value, value in themselves, apart from their worth to other persons.

One philosopher went so far as to claim that the typical human way of life is "a criminal act," a crime against nature and other persons. To avoid destruction of the earth and our life-support systems, thinkers such as Callicott, Leopold, and Taylor have urged us to abandon anthropocentric ethics and to embrace biocentric ethics. They claim that, within *biocentric ethics,* all beings have equal, *inherent value,* value in themselves, rather than merely instrumental value to humans.

Arguing for a biocentric environmental ethics, Taylor and others point out that all beings on the earth are interdependent and that, because of this interdependence, individualistic, anthropocentric ethics are not appropriate. Biocentric philosophers argue for a holistic approach that recognizes the inherent value of the entire planet. They criticize the human chauvinism, speciesism, and arrogance that many humans exhibit toward nature.

Responding to the biocentric arguments, many philosophers maintain that anthropocentric ethics can be an entironmental ethics. Thinkers like Frankena and Shrader-Frechette point out that all traditional anthropocentric ethics condemn the greed, arrogance, and insensitivity that causes environmental destruction. Moreover, they say, although all environmental degradation must stop, it is not possible to give all beings equal "rights," because not all beings are equal. They also warn that it would be impractical and impossible to give rights to all beings. Finally, they warn that, if we follow biocentric ethics and therefore claim that humans are equal to all other beings, then serious consequences to humans could follow. We could cause the deaths of millions of humans so as to protect certain species or ecosystems. (See the discussion of *Lifeboat Ethics* in Chapter 2.) Regan claims that, if we assume that humans are equal to nonhumans, then we will fall victim to "environmental fascism." Environmental fascism would make all human rights to life, liberty, and dignity subject to the welfare of the environment: we could kill humans to save snail darters.

The purpose of this chapter is to raise two questions that are central to the concerns discussed in the rest of the book: (1) What are the causes of environmental degradation? (2) Do we need some new, perhaps biocentric, ethics to correct this destruction?

The first reading is by Douglas Strong, Professor of History of San Diego State University, and Elizabeth Rosenfield, who holds a law degree from Stanford University. In their article, Strong and Rosenfield argue that adherence to the value systems espoused by most Americans today, viz., the Judea-Christian ethic and the Pragmatic-Utilitarian ethic, has contributed to environmental degradation. To correct this situation, Strong and Rosenfield propose some specific moral principles that they say ought to be adopted. These tenets focus on human obligations to the nonhuman world.

Although the position outlined by Strong and Rosenfield is part correct, the second reading argues that, insofar as *human injury* is concerned, conventional morality suffices to justify our demand for action against the polluter, the depleter of natural resources, and the destroyer of species and wildernesses. Insofar as environmental damage involves no human injury, however, the essay shows that a new ethic might be needed, since existing moral codes do not recognize the rights of nature, *per se*. On the whole, what the West needs is not a "new ethic" but better adherence to our current ethic.

Readings

1. Strong, D.H. and Rosenfield, E.S. 1976. Ethics or Expediency: An Environmental Question. From *Environmental Affairs,* 5 (Spring): 255-270. Copyright 1976 by *Environmental Affairs.* Reporinted by permission of the authors.

2. Shrader-Frechette, K.S. 1979. Environmental Responsibility and Classical Ethical Theories. Written especially for this volume.

ETHICS OR EXPEDIENCY: AN ENVIRONMENTAL QUESTION

Douglas H. Strong[1] *and Elizabeth S. Rosenfield*[2]

Introduction

MOST ETHICAL SYSTEMS existing in the world today fail to take into account the need for a healthful physical and social environment in which the human species can survive and the human spirit thrive. Economist Robert Heilbroner noted recently that "affluence does not buy morale, a sense of community, or even a quiescent conformity."[3] As the quality of life on earth declines, the Western, individualistic model of behavior based on the pursuit of self-interest and material improvement is losing credence. We struggle to make adjustments that will satisfy the poor and aspiring classes and nations while protecting the interests of the advantaged, but lack adequate ethical principles to guide us.

Americans as well as all other peoples of the world have two choices in response to the rising tide of social and environmental crises: begin now to seek guidelines for meeting such fundamental problems as how to achieve a better distribution of the world's wealth and how to protect the world's physical resources from ultimate exhaustion, or wait until drastic change is forced upon us by the severity of the problems which we have helped to create. We can adopt new social and environmental ethics now or wait until human degradation and environmental deterioration threaten our very existence.[4] Whichever path we elect to follow, we must recognize that the future depends upon our present decisions, and that neither as individuals nor as a society can we escape responsibility for them.

To draw a line between environmental ethics and social ethics may seem artificial and arbitrary, since every environmental problem is in some sense a social problem. For the purposes of this discussion, however, "environment" is used to mean physical environment as distinguished from social environment, and is limited still further to mean natural physical environment. A tenement house is certainly part of the physical environment of those who live in it, but whether the existence of such a dwelling is right or wrong is not a question of environmental ethics within the present definition. "Ethics," on the other hand, is used in a very broad sense to mean any value system by which to judge the rightness or wrongness and the desirability or wisdom of our actions and objectives.

Decisions on environmental issues are peculiarly difficult to make. As individuals we tend to be guided by personal considerations; as members of a society we tend to be guided by the society's cultural patterns; and as members of a species we must ultimately be guided by what is needful if the species is to survive. Thus the factors that influence our decisions range from the purely individualistic and hedonistic at one extreme to the purely biological at the other. Inevitably, personal inclination, cultural patterns, and concern for humans as a species will often be in conflict. For this reason alone we need ethical guidelines to help us make valid decisions.

How, for example, should we answer such questions as the following. Should couples be free to choose the number of children they will have, or should we, since there is a limit to the number of people the earth can support, begin to restrict the right to reproduce?[5] Is it justifiable to close down a factory that is causing a high degree of pollution, or to prohibit the manufacture of automobiles that use excessive quantities of fuel, even though these actions cause unemployment?[6] Should we stress the immediate needs of people alive today or the quality of the environment that will be our legacy to future generations? How should we balance our desire for individual freedom with the need for social controls to assure the survival and welfare of mankind?

In looking for criteria to help us answer these and similar questions, this article will start by examining two ethical systems that have contributed to our present difficulties, and will ask how they could be amended to serve us better.

The Judeo-Christian Ethic

Many recent commentators on environmental ethics begin with a discussion of Christian attitudes toward nature. Historian Lynn White Jr.'s controversial thesis, for example, is that the root of the ecological crisis lies in our Judeo-Christian heritage. White's belief is that the Christian teaching that nature exists apart from people and was created for their use has encouraged the unchecked development of science and technology at the expense of the environment and that as long as Western society maintains its basic Christian values—including the "Christian axiom that nature has no reason for existence save to serve man"—we shall suffer a "worsening ecological crisis."[7]

Critics of White's thesis argue that he overemphasizes the influence of religion on our beliefs about nature and on the way we have treated the environment. Biologist Richard Wright, for example, agrees with White that Christianity nurtured science and technology in its early growth, but insists that scientists themselves, not Christianity, are responsible for the destruction caused by their activities and discoveries.[8] Social scientist Lewis Moncrief, while he does not discount the influence of religion, stresses the impact of the democratic revolution and the scientific-technological revolutions of the eighteenth and nineteenth centuries as a cause of environmental degradation today.[9] Geographer Yi-Fu Tuan points out that the Christian West is not alone in its destructiveness. Even traditional China with its adaptive attitude toward nature has greatly altered and damaged its landscape through deforestation and consequent erosion.[10] Rene Dubos, even more critical of White, maintains that the ecological crisis today "has nothing to do with the Judeo-Christian tradition" but results from the pursuit of short-term economic self-interest.[11]

White has recently assured his many critics that he is well aware that religion is not the sole cause of the ecological crisis, that ecological damage was widespread prior to the rise of Christianity, and that it is widespread in non-Christian societies today.[12] He has said also that to understand a society's value system we sometimes need to rely "less on what that society says about itself than on what it actually does." Nevertheless, reaffirming his original thesis, he insists that in the last analysis our society is a product of individual beliefs about who we are, where we are going, and how we ought to behave toward nature and our fellow man—in other words, a product of our religion and our "value structures."

While White has concentrated on the effect of the Judeo-Christian tradition on the rise of science and technology in the past 1,000 years, David Crownfield has argued that our present irresponsible attitude toward the earth's resources can be traced to the pre-Christian seminomadic Hebrew sheepmen who were always able to move their flocks to greener pastures, leaving behind the land that had been over-grazed and eroded. Their profligate use of the land was encouraged by their belief in a divine shepherd who would intervene miraculously and deliver them from their earthly troubles. Meanwhile, during their secular existence, the earth and its resources were theirs to use as they saw fit. Calvinists would later argue that only by full use of all that earth had to offer did people show their readiness to migrate to heaven.[13]

Not only historians but a growing number of theologians are inquiring into the relationship between Christian ethics and the environmental crisis. Although Richard A. Baer, Jr., Frederick Elder, Joseph Sittler, and others disagree somewhat on the meaning of Scripture, they fear that some interpretations of the Bible have encouraged an arrogant attitude toward nature. They are calling for an interpretation of Genesis that sees human "dominion over the earth" as a kind of trusteeship. They believe that people were placed on earth to "tend God's garden," and that this being so, they have no absolute right to the land or its fruits.[14]

One recent interpreter of the Bible, David Crownfield, suggests that the Garden of Eden (before the fall) was a stable plant and animal community.[15] Humans were fruit-gatherers who cooperated with all other living things in a self-sustaining environment; the apple tree and all fruitful plants reseeded themselves and shared their abundance with other residents of the garden. Certainly not only Jesus but other early Christians lived simple lives close to the land and preached kinship with the earth and all living things. Also, speaking in this vein, theologian Sittler reflected the thought of many enlightened churchmen when he stated that the solution of the environmental crisis was to be found in "joyful and intelligent fellowship with our sister the earth—and *that* concept, utterly basic to the massive changes we must make if we are to survive as human beings, is a religious concept, a spiritual vision."[16]

The Pragmatic-Utilitarian Ethic

The second ethical system believed to have contributed to our environmental problems is the Pragmatic-Utilitarian ethic espoused by most Americns today. As popularly interpreted, this approach has become an ethic of expediency. Whatever is of the greatest immediate benefit to the greatest number of people at the lowest cost is judged desirable or "good," and therefore moral.

The primary problem is not with the theory that the greatest benefit at the lowest cost is good, but with the failure to take into account future good as well as present good. The shortsighted policies stemming from this philosophy have been ecologically disastrous. One outstanding example is the dust-bowl of the American midwestern plains, which resulted from over-grazing and over-ploughing the land. The indigenous buffalo-grass had held the soil against the periodic droughts and merciless winds. When it was ploughed under the preparation for crop-planting, the operation could have been said to be benefitting the greatest number at the lowest cost. The long-term effect was anything but beneficial, however, and everyone suffered. Over-grazing, strip-mining, dumping factory wastes into lakes and rivers, the indiscriminate use of pesticides—all of these practices have been justified as "practical" and therefore right.

A utilitarian ethic naturally makes for very flexible standards of conduct. Biologist Garrett Hardin argued, for instance, that "the morality of an act is a function of the state of the system at the time it is performed.[17] The frontiersman who polluted a nearby stream did not necessarily harm anyone if his closest neighbor was over 20 miles away. Nor was he culpable if he "wasted" resources that were greater than anyone's need for them. Today, in a crowded world with depleted resources, comparable behavior would be immoral. By the same token, freedom to reproduce according to individual election is no longer defensible in as much as it threatens to result in a population that the earth cannot support. Every human being in order to exist must consume a share of the earth's air and water and other essential resources, and none of these resources is inexhaustible. Hardin sees the unrestricted use of the land and its resources, coupled with the pursuit of individual self-interest, as ruin for all. The state of the system has so changed that what was once ethical is no longer so.[18]

Whether the Judeo-Christian Bible is reinterpreted, or the Pragmatic-Utilitarian ethic reformulated, we obviously need a more clearly defined ethical principle by which to judge whether any course of conduct that affects our environment is right or wrong. The hazards of our present unguided course are plain to be seen. A correlation clearly exists between an increased prevalence of degenerative diseases and nervous disorders, and an increase in pollution of air and water, in contamination of foods, in exposure to radiation, and in urban expansion at the expense of open country. Since those who have adapted best to the environmental conditions created by over-population and over-industrialization are most likely to survive, in generations to come we could conceivably propagate a physically and mentally inferior species, not only suited to an anthill existence but content with it.[19]

The Search for a Solution

Of many suggestions[20] that have been made, one of the most promising is that we begin our search for ethical guidelines by looking at our biological past. Our aim would be to determine, if possible, the sort of environment most conducive to mankind's health and well-being. People are unique among animals, but they are animals, and like all other life on earth, the product of evolution. They can reasonably assume, therefore, that the environment to which they are genetically best suited is the natural environment in which they evolved, that is, their global environment before they began to change it.[21]

In *The Tender Carnivore and the Sacred Game,*[22] ecologist Paul Shepard discusses primitive peoples who accepted the world as they found it and learned to live in harmony with it. Shepard says:

> We commonly hold that we want a society that operates economically and nearer to a steady state, a life that enables us to achieve personal fulfillment, enough small-group contact to let us have a consistent body of close friends, the means for according a respect for all of living things, a sense of connectedness of life, of atunement and alertness. Such is a description of those primitive people.[23]

Shepard goes on to point out that aboriginal peoples lived a remarkably good life in which the equivalent of only two or three days a week were spent in providing for their sustenance. These peoples spent most of their time in peaceful "socializing, sleeping, dancing, visiting, being hosts, telling stories, playing with children, making music." They lived in a "rich, stable, and diverse environment" and maintained a healthy and alert population that the land could comfortably support.[24]

This remarkably successful adaptation of humans to their environment began to break down some 10,000 years ago with the rise of agriculture. Shepard thinks that the objectives of the herders and farmers who appeared on the scene derived from the nature of their activities; "... they all shared the aim of completely humanizing the earth's surface, replacing wild with domestic, and creating landscapes from habitat."[25] Each person tried to maximize his or her share of space and goods.

In time the values and characteristics of the former huntergatherers changed. Diversity gave way to uniformity, life became geared to the drudgery of production, to competition, and to regimentation, and "civilized" people set themselves apart from nature.[26] When the demands of an increasing population exceeded the capacity of the land to provide for it, wars, epidemics, and famines resulted, followed by serfdom and slavery.[27] In brief, Shepard contends that the golden age of humans ended with this fundamental shift in their way of life and sense of relationship to the land, and that today we suffer the consequences of having remade our environment.[28]

While Shepard may be criticized for his glorification of our "primitive" past, much can be learned from a study of earlier values in relation to our present values and our hopes for the future. Precivilized people regarded the cosmos as a whole and made no distinction between personal, natural, and sacred worlds. They believed that humans and the nonhuman world were bound together in a single moral order, and that they had an obligation to care for the whole, of which they were a part. Many American Indians taught their children the concept of "immanent justice"—the idea that the physical universe is not indifferent to what one does, and that the land itself exacts retribution for its misuse.[29]

"Civilized" individuals, on the other hand, teach that the universe has no moral quality, that it is indifferent to them, and that they owe it nothing. And they teach that, far from being an integral part of the universe, they are beings of a different order, with the natural world and all of its resources there for them to exploit as they choose.[30] In the words of one author:

> The contemporary Western world, now imitated by the Orient, tends to regard the relationship of man to nature as a relation of man to physical matter in which application of physical science to man's material comfort is man's paramount assignment on earth.[31]

No matter how appealing Shepard's picture of this primitive way of life may be , we have come too far to return to those ways. Our lifestyles and the means by which we support ourselves, both dependent on industry, are too firmly entrenched. Not many of us would be willing to relinquish the bathtub with hot running water for an icy natural pool, or a job of so many hours a day and a fair certainty of dinner for the chances of the hunt. We can, however, learn an invaluable lesson from the concern of early humankind for the well-being of the earth, and from the fact that this concern was reflected in his ethics and governed his behavior. A regard for nature similar to that of primitive peoples, and conspicuously lacking in Judeo-Christian and Pragmatic-Utilitarian ethics, is the cornerstone of an environmental ethic in any era.

The environmental ethic formulated by Aldo Leopold over 25 years ago provides a useful starting point in our efforts to evolve an environmental ethic for our own era.[32] While a number of biologists and environmentalists have proposed ethical principles to guide us in making decisions that affect the environment, up to this time the principle proposed by Leopold is at once the most comprehensive and the simplest.

Leopold's thesis was that all ethical systems so far evolved rest on a single premise: that the individual is a member of a community of interdependent parts. Ethics first dealt with the relationships between individuals and then with the relationship between the individual and the social community. Leopold's belief was that ethics should be extended once more to include one's relationship to the natural world. This "land ethic," as he called it, simply enlarged the boundaries of the community to include soils, waters, plants, and all species of animals, each member of which is dependent on each of the other members for its own healthy existence. As a "plain member and citizen of this community," to use Leopold's designation, each person owes respect and a duty to each of his or her fellow members and to the community as a whole.[33]

Leopold's premises lead to the result that we are responsible for the "health" of the biotic community, which he defined as its "capacity for self renewal." He defined conservation as our effort to understand and preserve this capacity, and warned us that the development of a land ethic required us to "quit thinking about decent land use as solely an economic problem," and to think of it also in ethical terms. "A thing is right," he said, "when it tends to preserve the integrity, stability, and beauty of the biotic community. It is wrong when it tends otherwise." He believed an ethic of this kind to be an evolutionary possibility and an ecological necessity.[34]

Leopold's concept of people as part of a biotic community made up of soils, water, plants, and other animals, although not as all embracing, resembles the primitive concept of being at one with the universe. The environmental ethics of these earlier peoples, however, were rooted in instinct and religion, whereas Leopold's were based on his understanding of biology. Unfortunately, as Leopold himself has pointed out, a great deal in the field of biology and especially in the field of ecology is still not known.[35]

Toward the Development of an Environmental Ethic

Several steps are necessary in the development of an environmental ethic. The first is to formulate a basic principle. Leopold provided such a principle: a thing is right when it tends to preserve the integrity, stability, and beauty of the biotic community, and wrong when it tends to do otherwise. Less explicitly, but with intent similar to Leopold's, environmentalist Joseph W. Meeker suggested that we try to understand the enormously complex world in which we live and then to "learn how to shape ourselves to please the world."[36] In whatever way its fundamental principle is formulated, a responsible environmental ethic by its very nature imposes on people a duty to protect, preserve, and care for the natural physical world around them.

Since a principle of this kind can be implemented only to the extent that we know what care the earth requires, the next step is to put our best minds to the task of learning as much as possible about the intricate biological processes on which the health of the biotic community depends, and about which so little is fully understood. We know, of course, that once a species of plant or animal has been exterminated, there is no way to recreate it. Also, mankind's ultimate survival depends upon the ultimate survival of other species. Even were the prospect inviting, we could not solve our environmental problems by living alone under geodesic domes in a man-made environment. To do so would be a biological impossibility.

We do not know how many species of plants and animals we dare to exterminate, either intentionally or inadvertently—by air, water or soil pollution—before we risk toppling the biotic pyramid and ourselves with it. In the vast complex of nature, what are the biological functions of each constituent? Is there any species of plant or animal, no matter how apparently valueless, that does not serve some purpose not yet understood? How much modification can our environment stand and still support human life? These are very difficult, long-range questions that are not likely to be answered soon.

Meanwhile, innumerable questions of immediate concern need to be answered. These range from how best to dispose of industrial and atomic wastes, to how aerosol propellants affect the earth's protective shield against cosmic rays. Until such time as more money and trained personnel are available, research of the kind required to solve problems such as these will have to proceed at the expense of some other types of research now in progress. A great deal of time and energy is presently devoted to the development of new drugs and elaborate techniques and equipment for the treatment of chronic diseases, many of which, as noted earlier, are caused by adverse environmental conditions. With a new ethical principle to guide us, we would need to devote some of this energy and money to correcting as far as possible the conditions that cause the illnesses in the first place, and some to advancing our knowledge of ecology and biological processes in general.

Despite the vast amount that is still not known, we currently have sufficient scientific data to enable us to start drawing up guidelines for an environmental ethic. Many facts of environmental importance have been established and given wide publicity. The deleterious effects of chlorinated hydrocarbons in pesticides, of phosphates in soaps and detergents, and of pollutants in the exhaust fumes of automobiles, have become matters of common knowledge. Every day newspapers and journals report new scientific findings related to environmental problems. Each newly established fact would add another criterion of right and wrong to an environmental ethic, but development of such an ethic need not wait on new data or additional criteria.

Assuming that we have formulated an ethical principle and have taken steps to implement it, we are still faced with the very difficult problem of procuring general acceptance of its tenets. In primitive societies common moral conceptions held the members of a community together. Because a whole community accepted the same set of values, including values related to the environment, extraordinary pressures could be put on individuals to conform to environmental rules.[37] Modern society, however, is not in that enviable position.

Hardin, who saw individualistic self-interest leading us to disaster, believed that the only way to bring human conduct into line with environmental needs was by means of stringent, legal controls: "mutual coercion, mutually agreed upon by the majority of people affected."[38] Whatever the exact meaning of Hardin's formula, John J. McMahon took issue with the principle involved. McMahon believed that people were genetically programmed as much for cooperative, altruistic behavior as for competition and that education could, of itself, create a sense of responsibility and a willingness to sacrifice personal interest.[39]

Some argue in Hardin's defense that he simply had the courage to confront the reality of man's inherent egocentricity.[40] Freedom to live totally egocentric existence cannot be tolerated, and Hardin saw legal restrictions as offering the best chance for survival.[41] Hardin, however, was relying too heavily on regulation *per se.* The observation that we cannot legislate morals has been borne out time and time again.[42] As Leopold has noted, "the mechanism of operation is the same for any ethic: social approbation for right actions; social disapproval for wrong actions."[43] Once the ethical principles are widely accepted, approbation and disapprobation follow as a matter of course.

So far as the acceptance and implementation of an environmental ethic is concerned, a vision of the future and of a better, finer world is believed necessary to capture people's imaginations and inspire them to action.[44] The reason given for this view is that an environmental ethic looks not only to the present but to the future, a more distant future than any of us will live to see, and any appreciable number of individuals, unless motivated by a vision, will probably not concern themselves with what is so remote, nor make present personal sacrifices in the interest of generations yet to be born. A corollary of this view is that an environmental ethic will not succeed if it appeals only to self-interest.

We should, of course, nurture dreams and visions of a better future. Environmentalists, together with educators and sociologists, should set new goals such as adequate health care for everyone, improved housing in better designed communities, more equitable distribution of wealth, greater opportunity for creativity and self-fulfillment, and an environment that would maintain the complexity, diversity, and stability of all life on earth. Means of attaining these goals should be devised. A vision of such achievements does indeed capture the imagination.

So far as the *physical* environment is concerned, however, the authors very much doubt that a vision, no matter how inspiring to a sizable segment of society, is capable of engendering an ethic that will be accepted widely enough to be operable. Enough people cannot probably be made to feel a sense of moral right and wrong in respect to environmental practices through the power of a vision. The authors believe that too many people in any modern society are incapable of sharing a vision of the sort required.

More likely, the route to general acceptance of an environmental ethic will actually be by way of the much deplored concept of "self-interest." Some, perhaps most, ethical systems have succeeded because they appealed to self-interest. The Judeo-Christian ethic offers spiritual incentives to do God's bidding, backed up by the threat of punishment for transgressions—it "pays" to be good. The Utilitarian ethic that exalts the greatest good for the greatest number also appeals to self-interest—so long as the individual regards himself as one of the greatest number.

The appeal to self-interest in the case of an environmental ethic would be twofold. With respect to the present, an appeal would be made to our very natural desire to live out a healthy life. We cannot escape ingesting or breathing the deleterious substances which we are releasing into our waters, soils, and air. Whether affected by sulphur compounds that travel from auto to air, or mercury that travels from factory to water to fish, our health inevitably suffers, although just how seriously will only be known as the evidence gradually accumulates.

Looking to the distant future, the appeal would be to a natural interest in posterity. Most human beings, the authors believe, feel an identity with their species, and see in their descendants a kind of immortality for themselves. Consequently they have a personal interest in the continued existence of this earth and of humankind. The threat to the future is not, of course, that God will visit His wrath upon an erring world, but that the earth itself will take its vengeance. For every violation of its integrity, from the denuding of watersheds to the pollution of lakes and rivers, the earth exacts a price. The idea of "imminent justice" that was taught to American Indian children is beginning to appear as a simple but inescapable law of nature.[45]

For an appeal to self-interest on these grounds to succeed, we must be convinced of the danger to ourselves and to posterity. If the facts are presented clearly and simply, however, the layperson need not know a great deal about biology or ecology to understand the risks we run of causing irreparable damage or of starting irreversible and disastrous processes. How soon and how thoroughly he or she could be convinced would depend, if we can judge by the success of advertising methods, upon how thoroughly and repeatedly he or she was exposed to the facts.

To persuade enough people who cannot share a vision that an urgent need exists for a new approach to environmental problems, to persuade them that whatever we do that affects the ecology is morally either right or wrong, to educate them to the point where they are willing to make personal sacrifices (as envisaged by McMahon) is, to say the least, a formidable task. Educators at every teaching level, journalists of every order, the powers of radio and television, in fact all the powers of communication, would have to participate. To see it as an impossible task, however, is surely to lack another kind of vision.

Conclusion

Once an environmental ethic is adopted, what will be the answers to some of the questions posed earlier in this article? In the matter of personal freedom, the reconciliation of the individual's desire for independence on the one hand with the requirements of his neighbors on the other has been an issue since the beginning of community life. An environmental ethic will simply add a duty not to cause harm to the land. To the individual's present duty not to cause harm to other members of society, this duty toward the land could in fact be considered as simply another duty toward society, since whatever affects the environment adversely will eventually harm everyone. Whether the question involves the size of one's family or the cutting of trees on one's own property, the good of the earth will come first. What is ecologically sound will be right; what is ecologically unsound will be wrong.

With respect to any course of action that protects or improves the condition of the environment at the expense of someone's job, the duty to the environment will be, by its very nature, greater than the duty to the job-holder. Readjustments can be made. At least the chance for another job exists. But the child crippled by mercury poisoning, or whose lungs are damaged irretrievably by polluted air, has no chance for another life. As for the question, should we use the earth today as we see fit at the expense of those who come after us, the answer is implicit in the ethic itself. Care of the earth is its essence, the present and future health of the earth its objective.

To recapitulate, we suggest that enlightened self-interest can form the basis for an effective, generally acceptable environmental ethic.[46] As for those who seek a vision and can be guided by one, what could be more inspiring than the vision of people learning to understand the earth in all its complexity, the vision of people coming by reason and insight to see themselves as members of the biotic community, administors of its laws, insurers of its health, custodians of its beauty? For those who share this vision, an environmental ethic already exists.

Notes

1. Professor of History, San Diego State University, San Diego, California.

2. J.D., Stanford University. Resides in Atherton, California.

3. R. Heilbroner, *The Human Prospect,* 20 The New York Review of Books, Jan. 24, 1974, at 26.

4. Institute of Ecology, Man in the Living Environment (1972); B. Ward and R. Dubos, Only One Earth (1972); D. Meadows et al., The Limits to Growth (1972).

5. D. Callahan, *Ethics and Population Limitation,* 175 Science 487-94 (Feb. 4, 1972); P. Ehrlich & A. Ehrlich, Population, Resources, Environment (1970).

6. P. Drucker, *Saving the Crusade,* 244 Harper's Magazine, Jan. 1972, at 66-71; N. Faramelli, *Ecological Responsibility and Economic Justice,* in Western Man and Environmental Ethics 188-203 (I. Barbour ed. 1973).

7. L. White, Jr., *The Historical Roots of Our Ecological Crisis,* 155 Science 1203-07 (March 10, 1967).

8. R. Wright, *Responsibility for the Ecological Crisis,* 20 Bioscience 351-53 (Aug. 1, 1970).

9. L. Moncrief, *The Cultural Basis of Our Environmental Crisis,* 170 Science, 508-72 (Oct. 30, 1970).

10. Yi-Fu-Tuan, *Discrepancies Between Environmental Attitude and Behavior: Examples from Europe and China,* 12 Canadian Geographer 176-90 (1968); and Tuan, *Our Treatment of the Environment in Ideal and Actuality,* 58 American Scientist, May-June 1970, at 244-49.

11. R. Dubos, The Genius of the Place, Horace M. Albright Conservation Lectureship X (University of California School of Forestry, 1970).

12. L. White, Jr., *Continuing the Conversation,* in Western Man and Environmental Ethics, 55-64 (I. Barbour ed. 1973).

13. D. Crownfield, *The Curse of Abel: An Essay in Biblical Ecology,* 258 The North American Review, Summer 1973, at 58-63. *See also* P. Carroll, Puritanism and the Wilderness (1969).

14. For example, *see* F. Elder, Crisis in Eden: A Religious Study of Man and Environment (1970); F. Schaeffer, Pollution and the Death of Man: The Christian View of Ecology (1970); R. Baer, Jr., *Land Misuse: A Theological Concern,* 83 Christian Century 1239-41 (October 12, 1966); and R. Baer, *Conservation: An Arena for the Church's Concern,* 86 Christian Century 40-43 (Jan. 8, 1969); J. Sittler, Jr., *A Theology for Earth,* 37 The Christian Scholar 367-74 (June 1954); G. Fackre, *Ecology and Theology,* in Western Man and Environmental Ethics, 116-31 (Ian G. Barbour ed. 1973).

15. D. Crownfield, *supra* note 11, at 60-62.

16. J. Sittler, *Two Temptations—Two Corrections,* 45 National Parks and Conservation Magazine, December 1971, at 21.

17. G. Hardin, *The Tragedy of the Commons,* 162 Science 1245 (Dec. 13, 1968).

18. *Id.*

19. R. Dubos, Man, Medicine, and Environment 103-06 (1968); R. Dubos, So Human an Animal 16-17, 63-77, 144-60 (1968); J. Livingston, One Cosmic Instant 225 (1973).

20. Among the most useful proposals are the following: E. Goldsmith, Blueprint for Survival (1972); The California Tomorrow Plan (A. Heller ed. 1971).

21. H. Iltis, O. Loucks, and P. Andrews, *Criteria for an Optimum Human Environment,* 26 Bulletin of the Atomic Scientists 2-6 (Jan. 1970).

22. P. Shepard, The Tender Carnivore and the Sacred Game (1973).

23. P. Shepard, *Hunting for a Better Ecology,* 258 The North American Review, Summer 1973 at 15. For a brief account of the African Bushmen, *see* M. Sahlins, *The Original Affluent Society,* 4 Ecologist 181-89 (June 1974). *See also* Kenneth MacLeisch's account of the Tasaday, *Stone Age Cavemen of Mindanao,* 142 National Geographic Magazine, 218-49 (Aug. 1972).

24. P. Shepard, *Hunting for a Better Ecology,* 258 North American Review, Summer 1973, at 13.

25. P. Shepard, The Tender Carnivore and the Sacred Game 237 (1973).

26. *Id.* at 239.

27. *Id.* at 16-26.

28. *Id.* at 35-36.

29. R. Redfield, The Primitive World and Its Transformation 107-08 (1953).

30. John A. Livingston has traced the religious roots of the environmental crisis to paleolithic man who utilized the magic of cave paintings in an attempt to improve his chances for successful hunting. This, according to Livingston, was the beginning of man's conception of himself as separate from nature and of his efforts to exercise domination over it. J. Livingston, One Cosmic Instant 134-46 (1973).

31. Redfield, *supra* note 27, at 110.

32. A. Leopold, A Sand County Almanac (1949).

33. *Id.* at 204.

34. *Id.* at 221, 224-25.

35. *Id.* at 185-87.

36. J. Meeker, *Prologue to an Environmental Ethic,* 258 The North American Review, Summer 1973, at 22. *See also* J. Meeker, The Comedy of Survival: Studies in Literary Ecology (1974).

37. Conformity to community values ensured stability, but at the cost of flexibility and change. Modern societies experience rapid change, but suffer from a breakdown in sense of community and security. *See* W. Burch, Jr., Daydreams and Nightmares: A Sociological Essay on the American Environment 49 (1971).

38. G. Hardin, *The Tragedy of the Commons,* 162 Science 1247 (Dec. 13, 1968). *See also* G. Hardin, Exploring New Ethics for Survival (1972).

39. J. McMahon, *Garrett Hardin and the Search for Ecological Morality,* 37 The Living Wilderness, Spring 1973, at 23-29.

40. Margaret O. Strahl, Letter to the editor, in 37 The Living Wilderness, Summer 1973, at 41-42.

41. G. Hardin, *The Tragedy of the Commons,* 162 Science 1247 (Dec. 13, 1968).

42. *E.g.,* note the problems with laws to ban alcoholic beverages, certain books, and marijuana, or to restrict sexual activities.

43. A. Leopold, A Sand County Almanac, 225 (1949).

44. *E.g., see* L. Mumford, *Closing Statement,* in Future Environments of North America 718-29 (F. Darling & J. Milton eds. 1966).

45. R. Redfield, The Primitive World and Its Transformation, 107-08 (1953).

46. An environmental ethic that is accepted and complied with because it appeals to one's self interest is not, admittedly, on an altogether sound footing. Such a view continues in part the erroneous concept of one's separateness from nature. The authors believe, however, that an appeal to self-interest is the best if not the only hope for the adoption of an environmental ethic at the present time, when such an ethic is critically needed. Our hope is that ultimately people will recognize and accept the right of other species to exist simply for their own sake and not because people need them.

ENVIRONMENTAL RESPONSIBILITY AND CLASSICAL ETHICAL THEORIES

K.S. Shrader-Frechette

AMERICANS are very good at solving problems which have technological solutions. We have put transitors to work keeping records and computing, created more rubber than nature could, and made electricity our slave. The crux of the environmental crisis, however, is not technological but ethical.[1] Environmental questions, such as whether we should limit consumption, or adopt a zero tolerance level of man-made carcinogenic pollutants, are essentially ethical issues. Despite this fact, however, many "environmental problems have been approached by setting scientific standards and relying on innovation to meet them, while the behavior that caused the problem continues."[2]

No imaginable technology can restore extinct species or purify radiation-contaminated food chains. Increasing pollution and resource depletion cannot be sustained in a finite world with finite, nonrenewable resources. What is necessary, along with continued reliance on technology to help solve environmental problems, is the recognition that our behavior patterns and value systems need to change.

If we assume, with most environmental writers, that environmental degradation is severe and warrants at least some change in human behavior regarding nature, then numerous questions confront us. Two important ones are (1) whether we have the necessary technological skills for helping to remedy problems such as pollution and resource depletion, and (2) whether our current ethical systems provide a theoretical basis adequate for justifying a change in behavior toward the environment.

In the following pages I will investigate the second of these questions and attempt to determine whether, as most people seem to have argued, we need a "new ethic" in order to guard planetary welfare. First, I answer this question by distinguishing two different types of environmental ethics, one which is new, the other which is not. The second, and major, portion of the discussion is devoted to a brief analysis of several classical ethical philosophies and religious traditions. In examining them, I hope to show why they should provide a clear direction for environmental protection, even though in most cases this path has not been followed by humankind.

Do We Need a New Ethic to Protect the Environment?

Because of the relatively recent emergence of serious environmental problems, many writers believe that traditional ethical theories do not take sufficient account of them. Hence they conclude that a new ethic, an "environmental ethic," is needed to provide a theoretical basis adequate for justifying a change in behavior toward nature.[3]

The more precise answer to the question of whether a new ethic is needed, however, is both "yes" and "no." Yes, we do need a new ethic if by it one wishes to claim, for example, that natural objects and animals themselves have certain rights;[4] traditional ethical theories have accorded rights neither to animals (other than humans) nor to natural objects. No, we do not need a new ethic if by it one wants to affirm, for example, that one has a moral obligation not to infringe on the property rights of those who have an interest in protecting a given environment. Classical ethical theories have usually defended one's right to his property.

Environmental Ethics in a Primary and Secondary Sense

Another way of saying, both that we do and we do not need a new ethic, is that, on the one hand, if *prima facie* duties regarding the environment ought to be based on the needs and interests of nature alone, then we need a new ethic. On the other hand, if *prima facie* duties regarding the environment ought to be based on the needs and interests of the human community, then we do not need a new ethic, since this is already provided for in current normative theories.

As most nature writers have recognized, classical ethical theories describe environmental values as *secondary,* as a matter of what will promote human well-being. To the extent that environmental values are viewed as *primary* or intrinsic, or apart from their effect on human well-being, an environmental ethic might be viewed as new.[5] In the latter case, for example, certain actions might be said to be wrong, even if they allegedly harmed no human interests, but only those of nature.

The practical or empirical problem of distinguishing between an environmental ethic which is primary and one which is secondary, however, is that it is difficult to think of an action which would do irreparable damage to the environment or ecosystem, but which would not also threaten human well-being. In other words, all or nearly all actions which would do irreparable damage appear to be wrong under the secondary sense of environmental ethics, as well as under its primary sense. If people kill all the blue whales, for example, this action might be said to be wrong, not only in a primary sense, because it violates the "interests" of nature, but also in a secondary sense, because it deprives humans of the aesthetic pleasure that enjoyment of a variety of animal species could bring to members of future generations. Also, all persons might not now have the same right to enjoy all creatures of the earth, as they might once have had. Likewise, if a polluter dumps toxic wastes into a river, this action could be said to be wrong, not only in a primary sense, because the "interests" of the river are violated, but also in a secondary sense, because there are human interests in having clean water (e.g., for recreation or for drinking). As John Passmore points out, all ethical systems have taught that we should—or must—not injure our neighbor.[6]

What this comes down to is that, although there may well be a difference between an environmental ethic in a primary and a secondary sense, it is difficult to think of a viable norm for behavior which would hold in the primary, but not the secondary, sense of ethical justification. None of the authors who sanction this distinction, in fact, gives an example of such a norm.[7] What is clearly the case is that both a primary and a secondary type of environmental ethic together would provide greater protection to nature than would a secondary ethic alone. This benefit, however, is not without its own costs, since if both human and natural interests were said to be primary or intrinsic, then it is conceivable that in certain cases human welfare might wrongly be made subservient to environmental well-being. In fact, some authors have charged that this could be a dangerous consequence of according primary value to the environment. For example, it is conceivable that, if both trees and humans were said to have primary and intrinsic rights, then a person in a forest fire, faced with the choice of saving either the last specimens of a highly valued type of tree or a human life, could easily decide it was ethically desirable to save the trees.[8]

One might also argue that a primary and secondary type of environmental ethic is morally superior to, or more enlightened than, a seconday ethic. As J.A. Keller points out, when humans view nature only as a tool for their well-being, they lose something aesthetic or spiritual and fall victim to *hubris,* human pride and self-exaltation.[9] An environmental ethic in the primary sense may also be superior to one based solely on human interests since, as Holmes Rolston expressed it, a person may practice an ethic in an environmentally secondary sense out of fear; one can follow it in a primary sense only out of love.[10]

It is difficult to say, however, beyond these sketchy observations, whether one ought to distinguish environmental ethics in a primary and secondary sense, and if one does so, whether this will add anything to the validity of conclusions based on ethics of one or both types. Part of the presenting problem is that, without knowing what specific principles are incorporated into an ethic in the primary sense, it is difficult to determine whether environmental ethics in this primary sense are desirable. For example, on the one hand, if one said that we ought never to kill any plant, because plants (even poison ivy) have as much value as humans, then this principle of an allegedly primary ethic might be shown to be undesirable. On the other hand, if one claimed that we ought never to hurt unnecessarily any living, feeling thing, (even a mosquito) then this precept of an allegedly primary ethic might be said to be desirable. What is clear, however, is that, if the distinction between primary and secondary types of environmental ethics is meaningful (in the sense that the former finds its justification in looking at the rights of nature, whereas the latter does so by examining the rights of our own species), then ethics in the primary sense is indeed "new."

Environmental Values in Traditional Ethical Theories

Apart from discussing the legitimacy, need, and desirability of distinguishing between environmental ethics in primary and secondary senses, there is another way of determining whether we need a "new ethic" to cope with environmental problems. We can simply examine some of the major philosophies and religious traditions within which appropriate human behavior toward nature has been said to be prescribed or condemned.

Environmental Ethics and Judaeo-Christian Traditions

Perhaps more than any other cultural influence, Western Judaeo-Christianity has been blamed for mankind's mistreatment of the environment. In his classic essay, "The Historical Roots of Our Ecologic Crisis," Lynn White argues that in its dominant Western form, Judaeo-Christianity is anthropocentric and does not provide the basis for an appropriate use of technology. This is because Jews and Christians believe that God transcends the world and that humans, who bear God's image, therefore have an absolute right to dominate nature. As a consequence of these two beliefs, says White, they have sanctioned the development of a technology which aims at "the conquest of nature for the relief of man's estate." Because Judaeo-Christianity destroyed pagan animism, he says, it made exploitation of natural objects possible.[11]

White's account has enjoyed wide popularity, not only among environmentalists and historians, but among philosophers as well.[12] Alan Donagan, Professor of Philosophy at the University of Chicago, for example, agrees with White's view that, because humans have been said to be made in God's likeness, Judaeo-Christianity has accorded them dominion over nature. The basis for this interpretation, he says, is Genesis 1:26: "And God said, Let us make man in our image, after our likeness; and let them have dominion over the fish of the sea and over the fowl of the air, and over the cattle, and over all the earth, and over every creeping thing that creepeth on the earth." Donagan calls this "the fundamental principle governing man's relation to the earth," viz., the thesis that "the earth and all that is on it exist for the sake of the rational beings who live on it."[13]

Like any well-thought-out analysis in which one attempts to substantiate a cause-effect relationship, Donagan's and White's views have been the subject of much criticism. Who can, after all, conclusively *prove* that something (Judaeo-Christianity) has been a cause of something else (environmental deterioration), when causality itself cannot be established empirically? There is no conceivable experimental test of this hypothesis. Theologians, in particular, have taken issue with the White view of Genesis 1:26; several of them have claimed that this scripture passage supports stewardship of nature; they also have argued that other passages come close to pantheism, and that the Hebrew words for "dominion" and "subdue" do not preclude behaving in an environmentally responsible and benevolent way.[14] Moreover, points out D.E. Engel, Professor of Education at the University of Pittsburgh, biblical man was a dependent creature who was told by God to live as a farmer and preserver of his environment. Interpretations such as White's and Donagan's, says Engel, are wrong because they forget that the primitive relationship between man and nature was that of servant and master, not vice versa. The Hebrews moved to the "promised land," he says, so that they could dwell somewhere in harmony with nature.

Although all these reasons do not prove that Judaeo-Christian believers were obligated, by their faith, to care for nature, Engel believes that they do prevent drawing a negative conclusion. That is, one ought not to say that tenets of Judaeo-Christian belief are ultimately responsible for environmental problems or that they sanction man's absolute right to do whatever he wishes to nature.[15]

Another problem with the White-Donagan account, apart from the legitimacy of the biblical interpretations it presupposes, is that it leaves no room for the environmental stewardship suggested by Thomas Aquinas and realized in the thought of Francis of Assisi.[16] Benedictine monasteries, likewise, provide some excellent instances of explicit environmental ethics deep within the Judaeo-Christian tradition. According to one Abbot writing centuries ago, for example, humankind has "the most sacred obligation" to "replant and husband" pine forests.[17]

What these examples indicate is that Judaeo-Christian writers and believers have never been as single-mindedly anti-nature as White and Donagan suggest. Judaeo-Christian history, like all history, is complex and even contradictory, just as human beings are; hence it is probably not capable of being interpreted in the single way that White and Donagan wish.[18]

Rather than a religious tradition being responsible for environmental degradation and resource depletion, is it not true that greed, egoism, and short-sightedness have caused most environmental problems? Judaeo-Christian beliefs have never sanctioned such vices, even if it were the case that Christianity allowed man complete control over nature. Judaeo-Christians did not invent greed;[19] and so long as they did not, I suspect that the ultimate "blame" for environmental problems must be laid at the door of human behavior, in general, and not one religion, in particular. What may well be the case, however, is that Western Christianity exacerbated these human tendencies more than did Eastern and non-Judaeo-Christian religions. In the light of environmental problems in countries such as Japan, however, there is reason to wonder whether the West has aggravated nature more than the East has.

Another conceptual difficulty, involved in determining whether there is a causal link between Judaeo-Christian traditions and environmental destruction, is distinguishing chronologically between cause and effect. It may be the case, in this regard, that certain "anti-nature" themes in Judaeo-Christian writings and traditions have resulted from environmentally-destructive beliefs and practices rather than that they have caused those beliefs and practices. In other words, it may be truer to say that the existing cultural patterns determined religious belief and scriptural content more than this content determined cultural patterns of environmental degradation. Hence there are many reasons for believing that a "new" Judaeo-Christian ethic may not be needed in order to avoid environmental destruction.

Environmental Ethics and Utilitarianism

Like Judaeo-Christian beliefs, utilitarian ethical philosophies have been blamed for man's destruction of the environment.[20] I suspect, however, that just as the indictment against Western religion may have been too severe, so also has been the attack on this traditional ethical philosophy. To see why this is the case, let us be clear on what utilitarianism proposes by way of normative principles.

Utilitarianism was the creation of nineteenth-century English philosophers Jeremy Bentham and John Stuart Mill, who were eager to provide moral thought with a more "scientific" basis than it had previously enjoyed. According to the utilitarian account of moral obligation, the sole basic standard of right and wrong action is the "principle of utility." It posits, as the moral goal of all human actions, the greatest possible balance of "good" over "bad" for mankind as a whole.[21]

Since the moral end of utilitarianism is to maximize good for our species as a whole and to promote "social improvement," principles of individual rights and equal justice are recognized by utilitarians only to the extent that doing so will lead to the greatest amount of good for the greatest number of people.[22] This means that, by definition, members of the current generation are under moral obligation not to pollute the earth and deplete resources in such a way that the good of humanity as a whole will be harmed. If Ehrlich, Hardin, and the MIT authors of *The Limits to Growth* are correct in believing that current patterns of behavior toward nature will lead to planetary crisis within the next 100 years,[23] then obviously classical utilitarian theory prohibits such actions. Hence, to the extent that there is an environmental crisis threatening disaster to millions of members of future generations, the behavior responsible for that crisis is inimical to utilitarian norms. As R.M. Green puts it: "Since, from a utilitarian point of view, it is immaterial where or for whom happiness is produced, this requirement clearly extends to the future and helps explain our obligation to future persons."[24]

Given this brief account of utilitarianism, as defined by its proponents, it is evident that those who criticize it for failing to take account of long-term consequences of environmental actions do not really understand the philosophy.[25] The major flaw in utilitarianism is rendering the good of the individual subservient to the good of all mankind.[26] Hence, it is seriously wrong to suggest that utilitarians take inadequate account of future effects; they do so, perhaps even to a fault.

To the extent that environmental problems threaten future generations, they do so because persons have failed to follow the principle of utility, rather than because they have done so. As Passmore emphasizes, this in turn means that there is a need to change behavior, but not necessarily a need to change our ethical theories.[27] We might *not* need a new ethic. To suppose that we do is analogous to the error of claiming that because *de facto* integration has brought violence and other social problems, therefore we need a new *de jure* basis for integration. Both errors rest on the mistaken assumption that if an ethic or law exists, then it will be followed correctly, and if it is not followed correctly, then there must be a need for a new ethic or law. As the example shows, neither assumption is necessarily the case.

Environmental Ethics and Egalitarian Theories

Perhaps the clearest example of environmental damage resulting from failure to adhere to the principles of an established ethical system is given by disregard of egalitarian theories. According to noted environmentalist and attorney Joseph Sax, egalitarian ethical theories are the only ones able to explain theoretically what American nature writers such as Muir, Thoreau, Leopold, and Burroughs described.[28] In fact, Sax cites John Rawls, the most outstanding contemporary proponent of egalitarianism, to justify his environmental views.

According to this philosophy, the moral goal is not maximizing good for all mankind. Instead, egalitarians follow a principle based on "equality in the assignment of basic rights and duties," apart from the totality of good achieved by such a principle.[29] As a result of this tenet of equality, egalitarians hold, for example, that inequitable distributions of environmental costs and benefits are unethical.

As I argued in a previous work, because the US government permits emission of carcinogenic and mutagenic pollutants into the environment in amounts above their threshold for injury, it is violating principles of egalitarian ethics.[30] In the case of radiation emissions from nuclear power plants, allowed under Title 10, Part 20 of the US *Code of Federal Regulations,* reactor operators are permitted to cause members of the public to receive as much as 0.5 rem whole-body radiation dose in any calendar year.[31] Since radiation is long-lived (many isotopes, such as plutonium, retain their potential for carcinogenic and mutagenic injury for hundreds of thousands of years), cumulative in effect, and without a tolerance dose, such a policy is clearly questionable and dangerous. Public policy-makers admit that persons, living within several miles of a nuclear power plant, bear a cancer and genetic damage risk up to as much as 50 times greater than that borne by the general US public.[32]

According to government estimates, for example, every rad of radiation causes 0.002 genetic deaths among offspring of irradiated ancestors.[33] If, in the thirty-year lifetime of a particular reactor, 1000 persons in its environs annually received the maximum permissible yearly dose of low-level radiation (0.5 rem) from the plant, then at least 30 children born to these 1000 persons will die from genetic disorders induced by "allowable" radiation emissions.[34] Clearly, if one follows egalitarian ethics, then such emission standards are unethical, particularly because it is technically possible (but not cost-effective) to contain all low-level emissions from nuclear power plants.

As the ionizing radiation example shows, many environmental problems of resource depletion and pollution could be stopped, if government simply adhered to principles of egalitarian ethics in assessing the consequences of its environment-related decisions. Many philosophers have gone so far as to argue that, to the extent that government does not follow egalitarian principles in deciding public policy, to that degree its decisions are "incompatible with moral judgment and with the principles on which our Constitution rests."[35]

In arguing for egalitarian standards for assessing permissible levels of pollution, for example, one is really proposing acknowledgment of the constitutional rights of all persons. Equal protection is one such right, guaranteed by the Fifth and Fourteenth Amendments. Since its recognition is not contingent on whether it is economically beneficial to do so, this right is said to be inalienable. Hence if pollution endangers environmental values, it is at least in part because government is not adhering to a constitutionally guaranteed egalitarian ethic, and not necessarily because we need a new ethic.

Conclusion

Admittedly, however, even though adherence to egalitarian principles in environmental policy-making does not require a new ethic, it will certainly require new judicial interpretations of old doctrines, such as that of equal protection. In her classic work, *Silent Spring,* Rachel Carson argued this same point. She explained why the authors of the US *Constitution* would have meant to include protection from pesticidal injury, for example, under its due process and equal protection clauses, if they had foreseen the emergence of these chemicals in the environment.

What I have shown is that there is a strong rational foundation for using existing utilitarian and egalitarian ethical theories to safeguard the environment. Utilitarian doctrines clearly protect the interests of future generations and egalitarian schemes prohibit any environmental hazards against which persons cannot be assured equal protection. As I argued previously, moreover, it might also be the case that Judaeo-Christian traditions are not necessarily responsible for environmental degradation and resource depletion. If at least some of these considerations are correct, then it is not clear that a new ethic is needed to protect purely human interests in the environment. With Passmore, I believe that history contains the ethical seeds of a strong tradition of environmental protection.[36] The problem is, in part, that we have lost touch with many of these traditions.

As I argued, however, a "new ethic" for the environment may be needed if there are purely environmental interests separate from, or not capable of being included under, ecosystemic factors affecting human interests. One such environmental interest might be that of natural objects. In the course of developing a "new ethic," attorney Christopher Stone, for example, argues persuasively that natural objects ought to be accorded rights.[37] (This classic proposal will be discussed in detail in Chapter 4, which treats the rights of natural objects.) His argument is interesting because it presupposes an evolutionary view of ethics, and because it offers great promise for expanding the range of human sympathy and sensitivity. As Aldous Huxley once wrote, "Nothing short of everything will really do."

Notes

1. This point has been made by numerous writers. See, for example, J.S. Bowman, "The Environmental Movement: An Assessment of Ecological Politics," *Environmental Affairs,* V (Fall 1976): 663; hereafter cited as: Assessment. See also Alvin W. Weinberg, "Can Technology Replace Social Engineering?" in Noel de Nevers (ed.), *Technology and Society,* London, Addison-Wesley, 1972, p. 177; hereafter cited as: *Technology.*

2. Bowman, Assessment, p. 663.

3. Some of those who believe we need a "new" ethic to deal with the environment are, for example, D.H. Strong and E.S. Rosenfield, "Ethics or Expediency: An Environmental Question," *Environmental Affairs,* 5 (Spring 1976); 255-70; hereafter cited as Ethics; W.H. Ferry, "Must We Rewrite the Constitution To Control Technology?" in N. de Nevers (ed.), *Technology and Society,* London, Addison-Wesley, 1972, pp. 15, 19; hereafter cited as *Technology;* Stewart L. Udall, *The Quiet Crisis,* New York, Avon, 1963, pp. 200-202; hereafter cited as: *Crisis;* Frederick Elder, "The New Asceticism," in R.T. Roelofs, J.N. Crowley, D.L. Hardesty (eds.), *Environment and Society,* Englewood Cliffs, New Jersey, Prentice-Hall, 1974, pp. 319-24; hereafter cited as: Asceticism; C.D. Stone, *Should Trees Have Standing?* Los Altos, California, William Kaufmann, 1974, pp. 39, 42; hereafter cited as: *Trees;* W.T. Blackstone, "Ecology and Rights," in T.A. Mappes and J.S. Zembaty (eds.), *Social Ethics,* New York, McGraw Hill, 1977, pp. 342-3; hereafter cited as: Ecology; Holmes Rolston, "Is There an Ecological Ethic?" *Ethics,* 85 (January 1975): 104, 106; hereafter cited as: Ecological Ethic; N.O. Schedler, "Our Destruction of Tomorrow: A Philosophical Reflection on the Ecological Crisis," in W.R. Durland and W.H. Bruening (eds.), *Ethical Issues,* Palo Alto, Mayfield, 1975, pp. 245-47; hereafter cited as: *Crisis;* K.A. Manaster, "Law and the Dignity of Nature," *Environment Law Review—1978,* New York, Clark Boardman, 1978, p. 4; hereafter cited as: Nature; and S.C. Brown, "Challenges to Our Value Systems in the Scientific Age," *Zygon,* 6 (December 1971): 267-69; hereafter cited as: Challenges.

4. Stone, *Trees,* argues this point.

5. Several authors have made the distinction between an environmental ethic in a primary and a secondary sense. See, for example, Van Rensselaer Potter, "Evolving Ethical Concepts," *BioScience,* 27 (April 1977): 251-52; hereafter cited as: Concepts. See also L.H. Tribe, "Ways Not To Think about Plastic Trees," *Environment Law Review—1975,* New York, Clark Boardman, 1975, p. 133; hereafter cited as: Ways; Holmes Rolston, Ecological Ethic, pp. 104, 109.

6. John Passmore, *Man's Responsibility for Nature,* New York, Charles Scribner's Sons, 1974, p. 186; hereafter cited as: *Responsibility.*

7. See note 3.

8. See, for example, Richard Neuhaus, *In Defense of People,* New York, Macmillan, 1971, pp. 162-79, 192-201; hereafter cited as: *People;* and Frederick Elder, *Crisis in Eden,* New York, Abingdon, 1970, pp. 147-55; hereafter cited as: *Eden.*

9. J.A. Keller, "Types of Motives for Ecological Concern," *Zygon,* 6 (September 1971): 199, 207.

10. Rolston, Ecological Ethic, p. 109

11. Lynn White, "The Historical Roots of Our Ecologic Crisis," *Science,* 155 (1967): 1203-1207.

12. See Strong and Rosenfield, Ethics, pp. 257-59.

13. Alan Donagan, *The Theory of Morality,* Chicago, University Press, 1977, p. 96; hereafter cited as: *Morality.*

14. D.E. Engel, "Elements in a Theology of Environment," *Zygon,* 5 (September 1970): 219-221; hereafter cited as: Environment.

15. Engel, Environment, p. 222.

16. This point is made by Tribe, Ways, p. 736.

17. This example is given by Potter, Concepts, p. 251.

18. Passmore, *Responsibility,* p. 195, makes this same point.

19. Passmore, *Responsibility,* p. 195, makes this same point.

20. See, for example, Strong and Rosenfield, Ethics, pp. 259-260.

21. For discussion of the principle of utility and the utilitarian framework for ethics, see John Stuart Mill, *Utilitarianism, Liberty, and Representative Government,* New York, E.P. Dutton and Company, 1910, esp. pp. 6-24; hereafter cited as: *Utilitarianism.* See also Jeremy Bentham, *The Utilitarians: An Introduction to the Principles of Morals and Legislation,* Garden City, New York, Doubleday, 1961, esp. pp. 17-22; hereafter cited as: *Principles.* See P. Nowell-Smith, *Ethics,* Baltimore, Penguin, 1954, p. 34, and J.J.C. Smart, "An Outline of a System of Utilitarian Ethics," in J.J.C. Smart and B. Williams, *Utilitarianism: For and Against,* Cambridge, University Press, 1973, pp. 3-74; hereafter cited as: *Utilitarianism.*

22. Smart, *Utilitarianism,* pp. 69-71, and Mill, *Utilitarianism,* pp. 38-60.

23. See D.L. Meadows, *et al., The Limits to Growth,* New York, New American Library, 1974.

24. R.M. Green, "Intergenerational Distributive Justice and Environmental Responsibility," *BioScience,* 27 (April 1977): 260; hereafter cited as: Justice.

25. Strong and Rosenfield, Ethics, pp. 259-60.

26. Smart, *Utilitarianism,* pp. 69-72 and Mill, *Utilitarianism,* pp. 38-60, esp. 58-59.

27. Passmore, *Responsibility,* p. 188.

28. J.L. Sax, "Freedom: Voices from the Wilderness," *Environmental Law,* 7 (Spring 1977): 569; hereafter cited as: Wilderness.

29. John Rawls, *A Theory of Justice,* Cambridge, Harvard University Press, 1971, pp. 14-15; hereafter cited as: *Justice.* See Charles Fried, *An Anatomy of Values,* Cambridge, Harvard University Press, 1970, pp. 42-43; Charles Fried, *Right and Wrong,* Cambridge, Harvard University Press, 1978; pp. 116-17, 126-27, and Alan Donagan, *The Theory of Morality,* Chicago, University Press, 1977, pp. 221-39.

30. K.S. Shrader-Frechette, "Ecological Justice and Low-Level Radiation," *Journal of American Culture,* to appear in 1980. See also Shrader-Frechette, *Nuclear Power and Public Policy,* Dordrecht, D. Reidel, 1980, Chapter 2.

31. 10 CFR 20, p. 189. According to 10 CFR 20, p. 184, the rem is "a measure of the dose of any ionizing radiation to body tissues in terms of its estimated biological effect relative to a dose of one roentgen of X-rays." A rem is approximately equivalent to a rad.

32. US Atomic Energy Commission, *Comparative Risk-Cost-Benefit Study of Alternative Sources of Electrical Energy,* WASH-1224, Washington, D.C., US Government Printing Office, 1978, p. 4-16; hereafter cited as: WASH-1224.

33. WASH-1224, p. 4-14. See also note 31.

34. (0.002 genetic deaths/rad) $\times$ (30 years $\times$ 0.5 rad) = 3% risk per person or 30 deaths per 1000 exposures.

35. Marshall Cohen, "The Social Contract Explained and Defended," *The New York Times Book Review,* Section 7 (July 16, 1972), 1, 16-17.

36. Passmore, *Responsibility,* p. 186.

37. C.D. Stone, *Should Trees Have Standing?* Los Altos, California, William Kaufmann, 1974.

Further Reading

Abbey, E. *Desert Solitaire,* New York: Balantine.Barbour, I.G., ed. 1973. *Western Man and Environmental Ethics.* Reading, Massachusetts: Addison-Wesley.

Berry, T. 1988. *The Dream of the Earth.* San Francisco: *Sierra.*

Blackstone, W.T. 1977. Ecology and Rights. In Social Ethics, ed. T.A. Mappes and J.S. Zembaty, pp. 340-347. New York: McGraw-Hill.

Callahan, D. 1973. *The Tyranny of Survival.* New York: Macmillan.

Callicott, J. Baird. 1989. *In Defense of the Land Ethic.* Albany, New York: SUNY Press.

Callicott, J. Baird, and Roger T. Ames, eds. 1989. *Nature in Asian Traditions of Thought: Essays in Environmental Philosophy.* Albany, New York: SUNY Press.

Callicott, J. Baird. March 1990. Whither Conservation Ethics? *Conservation Biology,* 4, No. 2: 163-174.

Commoner, B. 1974a. The Dual Crisis in Science and Society, in *Environment and Society.* ed. R.T. Roelofs, J.N. Crowley, and D.L. Hardesty, pp. 57-62. Englewood Cliffs, New Jersey: Prentice-Hall.

______. 1974b. *The Closing Circle.* New York: Bantam.

Devall, B., and Sessions, G. eds. 1985. *Deep Ecology: Living As If Nature Mattered.* Salt Lake City: Gibbs Smith.

Disch, R., ed. 1970. *The Ecological Conscience: Values for Survival.* Englewood Cliffs, New Jersey: Prentice-Hall.

Ehrlich, P. 1988. The Loss of Diversity. In Wilson, *Biodiversity,* pp. 21-27.

Elder, F. 1970. *Crisis in Eden.* Nashville: Abingdon.

______. 1974. The New Asceticism. In *Environment and Society,* ed. R.T. Roelofs, J.N. Crowley, and D.L. Hardesty, pp. 319-324. Englewood Cliffs, New Jersey: Prentice-Hall.

Elliot, R., and Gare, A., eds., 1983. *Environmental Philosophy.* London: Open University Press.

Ferry, W.H. 1972. Must We Rewrite the Constitution To Control Technology? In *Technology and Society,* ed. N. deNevers, pp. 9-20. London: Addison-Wesley.

Fisher, J.L. 1970. Dimensions of the Environmental Crisis. *Zygon* 5 (December): 274-283.

Frankena, W. 1979. Ethics and Environment. In Goodpaster and Sayre, pp. 10-14.

Goodpaster, K.E. June 1978. On Being Morally Considerable. *The Journal of Philosophy* 75, No. 6: 308-234.

Goodpaster, K.E., and J.J. Sayre, eds. 1979. *Ethics and Problems of the 21st Century.* Notre Dame, Indiana: University of Notre Dame Press.

Green, R.M. 1977. Intergenerational Distributive Justice and Environmental Responsibility. *BioScience* 27 (April): 260-265.

Hargrove, E.C. 1989. *Foundations of Environmental Ethics.* Englewood Cliffs, New Jersey: Prentice Hall.

Heilbroner, R.L. 1974. *An Inquiry into the Human Prospect.* New York: Norton.

Helfrich, H.W., ed. 1970. *The Environmental Crisis.* New Haven, Connecticut: Yale University Press.

Huffman, J.L. 1977. Individual Liberty and Environmental Regulation. *Environmental Law* 7 (Spring): 431-447.

Huxley, A. 1974. The Politics of Ecology: The Question of Survival. In *Science, Technology, and Freedom,* ed. W.E. Truitt and T.G. Solomons, pp. 230-235. Boston: Houghton Mifflin.

Jonas, H. 1973. Technology and Responsibility: Reflections on the New Tasks of Ethics. *Social Research* 40 (Spring): 31-54.

Jonsen, A.R., and Butler, L.H. 1975. Public Ethics and Policy Making. *The Hastings Center Report* 5 (August): 19-31.

Keller, J.A. 1971. Types of Motives for Ecological Concern. *Zygon* 6 (September): 197-209.

Kennan, G.F. 1974. To Prevent a World Wasteland: A Proposal. In *Environment and Society,* R.T. Roelofs, J.N. Crowley, and D.L. Hardesty, pp. 263-274. Englewood Cliffs, New Jersey: Prentice-Hall.

Lane, L.W. 1974. An Environmental Ethic. In *Science, Technology, and Freedom,* ed. W.H. Truitt and T.W. Solomons, pp. 246-251, Palo Alto, California: Houghton Mifflin.

Leopold, A. 1949. *A Sand County Almanac.* New York: Oxford University Press.

Maddox, J. 1972. *The Doomsday Syndrome.* London: Macmillan.

Manaster, K.A. 1978. Law and the Dignity of Nature. In *Environment Law Review—1978,* New York: Clark Boardman.

Meadows, D.L., et al. 1974. *The Limits to Growth.* New York: New American Library.
Mesarovic, M., and Pestel, E. 1974. *Mankind at the Turning Point,* esp. pp. 10-17, 143-148. New York: New American Library.
Moncrief, L.W. 1970. The Cultural Basis for Our Environmental Crisis. *Science* 170: 508-512.
Mossman, A.S. 1970. Environmental Crisis and the Wildlife Ecologist. *BioScience* 20 (July 15): 813-14.
Murphy, E.F. 1970. A Law for Life. In *Environment Law Review—1970,* pp. 3-20. New York: Clark Boardman.
Nash, R. 1972. *Environment and Americans.* New York: Holt, Rinehart and Winston.
Nash, R. 1989. *The Rights of Nature.* Madison, Wisconsin: University of Wisconsin Press.
Neuhaus, R. 1971. *In Defense of People.* New York: Macmillan.
North, D.C. 1977. Political Economy and Environmental Policies. *Environmental Law* 7 (Spring): 449-462.
Passmore, J. 1974. *Man's Responsibility for Nature,* esp. pp. 186-195. New York: Scribners.
Pearson, J.Y. 1971. Toward a Constitutionally Protected Environment. In *Environment Law Review—1971,* pp. 53-83. New York: Clark Boardman.
Potter, V.R. 1977. Evolving Ethical Concepts. *BioScience* 27 (April): 251-253.
Rawls, J. 1971. *A Theory of Justice.* Cambridge, Massachusetts: Harvard University Press.
Regan, T., ed. 1984. *Earthbound.* New York: Random House.
Rolston, H. 1986. *Philosophy Gone Wild.* Buffalo, New York: Prometheus. 93-109.
Rolston, H. 1988. Environmental Ethics. Philadelphia: Temple University Press (unpublished report).
Sax, J.L. 1977. Freedom: Voices from the Wilderness. *Environmental Law* 7 (Spring): 565-574.
Schaeffer, R. 1990. Car Sick. *Greeenpeace* 15, No. 3 (May/June): 13-17.
Schedler, N.L. 1975. Our Destruction of Tomorrow: A Philosophical Reflection on the Ecological Crisis. In *Ethical Issues,* ed. W.R. Durland and W.H. Bruening, pp. 247-267. Palo Alto, California: Mayfield.
Shrader-Frechette, K.S. 1985a. *Science Policy, Ethics, and Economic Methodology.* Boston, Massachusetts: Reidel.
Shrader-Frechette, K.S. 1985b. Environmental Ethics and Global Imperatives. In *The Global Possible,* ed. R. Repetto, pp. 97-130. New Haven: Yale University Press.
Shrader-Frechette, K.S. 1989. Ecological Theories and Ethical Imperatives. In *Scientists and Their Responsibility,* ed. W. Shea and B. Sitter, pp. 73-104. Canton, Massachusetts: Watson.
Shrader-Frechette, K.S. and E, McCoy. May 1990. Theory Reduction and Explanation in Ecology. *Oikos* 58, No. 1: 13-114.
Stone, C.D. 1974. *Should Trees Have Standing?* Los Altos, California: William Kaufmann.
Stone, C.D. 1985. *Should Trees Have Standing?* Revisited: How Far Will Law and Morals Reach? A Pluralist Perspective. *Southern California Law Review* 59, No. 1: 6.
Strong, D.H., and Rosenfield, E.S. 1976. Ethics or Expediency: An Environmental Question, *Environmental Affairs* 5 (Spring): 255-270.
Taylor, P.W. 1986. *Respect for Nature.* Princeton, New Jersey: Princeton University Press.
Thompson, J. 1990. A Refutation of Environmental Ethics.*Environmental Ethics* 12: 2(Summer): 147-160.
Tribe, L.H. 1975. Ways Not To Think about Plastic Trees: New Foundations for Environmental Law. In *Environment Law Review—1975,* pp. 717-750. New York: Clark Boardman.
Udall, S.L. 1963. *The Quiet Crisis.* New York: Avon.
Whisenhunt, D.W. 1974. *The Environment and the American Experience.* Port Washington, New York: Kennikat.
White, L. 1967. The Historical Roots of Our Ecologic Crisis. *Science* 155: 1203-1207.
Wigner, E.P. The Limits of Science. In *Readings in the Philosophy of Science.* ed. H. Feigl and M. Brodbeck, pp. 757-765. New York: Appleton-Century-Crofts.

Wilson, E.O., ed. 1988. *Biodiversity.* Washington, D.C.: National Academy Press.

2

ALTERNATIVE ETHICS REGARDING THE ENVIRONMENT

Introduction

HOW TO VIEW man's relationship to the environment is one of the great moral problems of our time. As Buckminster Fuller put it, the most striking fact about planet earth is that "no instruction book came with it."

Knowing how to formulate this problem, however, is very likely the most trying part of solving it. In fact, metaphor is probably the only way of beginning such a difficult task. Perhaps this is why a number of environmental writers have chosen to describe the earth in metaphorical terms (e.g., as a spaceship or a lifeboat). Using their metaphors as a basis for building models of planetary relationships, they are then able to postulate ethical principles necessary for maintaining these relationships.

As was evident in the last chapter, a number of writers have argued that new models, and hence, new environmental ethics, are necessary. According to this view, Judaeo-Christian religions and pragmatic-utilitarian moral philosophies fail to provide a framework for addressing important ecological concerns. Other authors, however, have claimed that these traditional views *do* provide guidelines concerning the environment. For this reason they maintain that no new ethical system is needed to deal with contemporary ecological problems.

The purpose of this chapter is to describe several general environmental ethics, each based on a different model of the relationship between humans and nature. Some of these ethics are derived from traditional moral philosophies, while others are the result of relatively new beliefs. Apart from whether each of these ethical models may be said to be old or new, however, all of them raise classical philosophical issues as well as contemporary problems unique to the particular circumstances of today.

To some extent, the contrast between those desiring a new framework of environmental ethics and those defending traditional values is a conflict between "counterculture" and "establishment" ethics. T. Roszak, author of *The Making of the Counterculture,* characterized establishment ethics as emphasizing hard work, accumulation of material possessions, and consumption of the goods of the earth. He portrayed the counterculture morality as opposed to economic individualism, competition, and achievement, and as focusing on mutual dependency and concern for all things.

The controversy over which system of ethics ought to govern our relations with the environment is also in part a contest between East and West, or between a stereotype of East and West. If environmental ethics requires that one attempt to consume less, then it is close to the Buddhist teaching that one ought to purify himself by reducing personal wants. Conversely, if environmental ethics directs one to use the goods of the earth so as to yield the greatest profit, comfort, and efficiency for the current members of the human race, then this injunction is similar to that of Adam Smith and other western economists.

Also at the heart of the controversy, over which system of environmental ethics is superior, are a number of difficult moral issues, some of which will be treated specifically in later chapters. A few of these questions are whether one has environment-related obligations to future generations, whether nonhuman beings have rights, whether resource-rich nations have a duty to share their wealth with resource-poor countries, and whether all humans have a right to control their own reproductive activity.

Since the goal of this chapter is to present several general ethical models for man's relationship with the environment, the two readings in this chapter provide descriptions and evaluations of three major systems of environmental ethics. These are "Frontier or Cowboy Ethics," "Lifeboat Ethics," and "Spaceship Ethics."

In the first reading, one written especially for this book, I describe the dominant view of the relationship between humans and nature known as "Frontier or Cowboy Ethics." Arguing that this human-centered ethics has led the planet to environmental crisis, I discuss why a new model is needed for understanding the biosphere. One such alternative, "Lifeboat Ethics," has been proposed by Garrett Hardin, Professor Emeritus of Human Ecology at the University of California, Santa Barbara. After outlining the inadequacies of "Frontier Ethics," I present the "lifeboat" position.

According to Hardin, the planet may be thought of as a sea on which there are a number of lifeboats. The poor of the world are falling out of their crowded boats and trying to swim to the uncrowded, rich lifeboats of the developed nations. Hardin's "Lifeboat Ethics" is a summary of the moral actions which members of the rich nations ought to take in order to survive. These actions include mandatory population control, management of common resources, withdrawal of all contributions to world food banks, and strict immigration policies.

In response to "Lifeboat Ethics," I suggest that Hardin makes a number of questionable assumptions in defending his position, and I argue that the lifeboat morality is not defensible in terms of distributive justice. For me, survival (via "Lifeboat Ethics") is not worth the cost of the cruelty and callousness necessary to obtain it. Rather, survival is only one of a number of moral interests that ought to be served by an ethical system.

Following the insight of the first reading, that survival is not the *only* good, I argue in the second selection that "Spaceship Ethics" provides a workable model for environmental policy. Using a widely known metaphor also employed by Buckminster Fuller, Kenneth Boulding, Lynton Caldwell and others, I propose that earth be thought of as a finite, closed spacecraft. With malfunctioning life-support systems, low levels of vital supplies, and half of its passengers hungry, Spaceship Earth is in deep trouble. And, although the metaphor is limited, it can be used to suggest some solutions to these difficulties. I argue that we ought to follow "Spaceship Ethics," that is, to reduce consumption, reproduction, possessions, and expenditures on warfare, and to guarantee distributive justice. Unlike "Lifeboat Ethics," the spaceship view requires one neither to adopt involuntary limits on reproduction nor to allow inhabitants of Third-World nations to starve. Rather, its proponents call for a change in human behavior and for recognition that Spaceship Earth is a finite and closed system in which nothing short of brotherhood will solve the environmental crisis.

Whether one agrees with the principles of "Frontier or Cowboy Ethics," "Lifeboat Ethics," or "Spaceship Ethics," the important insight revealed by all three frameworks is that we cannot afford to ignore environmental values. We must have a system of principles, not only for helping us relate to our fellows, but also for helping us relate to the earth.

Readings

1. Shrader-Frechette, K.S. "Frontier Ethics" and "Lifeboat Ethics." Written especially for inclusion in this volume.
2. Shrader-Frechette, K.S. "Spaceship Ethics." Written especially for inclusion in this volume.

"FRONTIER OR COWBOY ETHICS" AND "LIFEBOAT ETHICS"

K.S. Shrader-Frechette

"WHY IS IT," asked Barry Commoner recently, "that we can nourish a man in outer space but cannot feed the children of Calcutta or Harlem?" Why are we able to analyze life on other planets but unable to understand our own neighbors? The usual explanation of this paradox is that, while humankind is competent in the areas of science and technology, we are tragically incompetent in ethical thinking and decision-making. We do not know how to deal with problems requiring value judgments about personal and social goods.

Part of the reason for our difficulty with ethical questions is that moral principles do not have the same degree of acceptance and factual support as scientific or technical truths. This is because no amount of empirical data, about what people *in fact* do, can ever constitute sufficient grounds for justifying a claim about what people *ought* to do. Since moral principles cannot be confirmed by empirical data, ethical statements are often relegated to the realm of what is "subjective," while scientific statements are said to be "objective." This oversimplification and misrepresentation of the relationship between science and ethics encourages a concern for what is said to be "objective" and a disavowal of what is thought to be "subjective." As a result, the ethical consequences of certain scientific and technological activities are often ignored.

Environmental policy-making is one of the most important areas in which both moral considerations, as well as those of science and technology, must be employed to produce sound societal directives. On the one hand, decisions involving nature cannot be made in the absence of hard technological data about what is scientifically feasible. On the other hand, however, such decisions cannot be made solely on the basis of what is technically most efficient. The former procedure is unrealistic and unworkable, while the latter might result in trading the rights of one class of individuals for technological or economic benefits to another group.

Formulating a sound system of environmental ethics demands that both errors be avoided. The purpose here is to give accounts of two different ethics regarding the environment and to show whether, and to what degree, they provide good frameworks for understanding the way in which humans ought to relate to nature. Both of these ethical models offer a general account of the relationship between ourselves and nature and of the ethical norms following from their respective theories. Neither of them, however, provides detailed analyses of specific issues (e.g., whether members of future generations have the same rights as we). Such particular environment-related questions will be dealt with later in the book. For the present, we need to gain several broad, but different, contexts for understanding both the relationships among the many issues to be discussed later and the assumptions in terms of which they are debated. Hence, we need to examine alternative models for environmental ethics.

Frontier or Cowboy Ethics

Perhaps the dominant model, in terms of which environmental policy has been formulated, is that of "Frontier or Cowboy Ethics." Most traditional philosophical theories and much of Judaeo-Christian thought are built on the premise that man is master of nature and that nonhuman entities cannot be said to have any rights that humans are obligated to respect.[1]

In part because of this justification of human supremacy, and in part as a consequence of human greed and egoism, many individuals have followed "Frontier Ethics." Like some American cowboys who nearly destroyed the bison, and countless Spanish colonialists who raped the lands they discovered, throughout history numerous persons have plundered resources for immediate gain. Those who pushed back geographical frontiers and exploited the wealth of new territories and continents usually were looked upon as noble and courageous. They were often believed to be fighting a "holy battle" to control what was "wild" and "savage," and many of their actions served the human goals of rapid economic expansion and increased production of goods.

In the presence of abundant resources and boundless frontiers, the "Cowboy Ethics" (of exploiting the earth in order to produce human prosperity and economic development) appears to succeed. In fact, it is in large part responsible for giving us such a high standard of living that we are currently able to question its value.[2]

Now that resources are scarcer and we do not need to fight for survival in a wilderness, it is appropriate to ask whether we have been following a legitimate set of moral principles. Ought we to continue, like the frontiersmen of earlier times, to promote geographical and economic expansion and their attendant rapid consumption of resources?

As Stewart L. Udall says, acceptance of the "Frontier Ethic" is dependant upon the acceptance of the "Myth of Superabundance." Whenever this myth is followed, wise management of the land and provident husbandry are superfluous.[3] Obviously, however, the myth is wrong, and ethics on a finite planet cannot be based on the factual assumption that there will always be more resources to be utilized. Instead of the "Myth of Superabundance," proponents of "Frontier Ethics" have lately turned to an alternative rationale. This is the "Myth of Scientific Supremacy"—the belief that, even though resources are finite, we can still pursue "Frontier Ethics" because science and technology will provide ways of solving the problems of pollution and resource depletion.[4]

The Transition to Lifeboat Ethics

"The Myth of Scientific Supremacy," however, does not resolve the difficulties of "Frontier or Cowboy Ethics." If science, alone, were as successful as the myth suggests, then the human race would not be in the throes of a crisis of pollution, population, industrialization, and resource depletion. As many writers have argued, environmental problems have become so extreme that the very survival of the human race, as we know it, is now at stake.[5] The population of some nations is doubling every 20 years and, says Garrett Hardin, even if the technology of the "Green Revolution" were successful enough to feed all these people, the problem would not be solved. This is because, "If we satisfy the need for food in a growing population, we necessarily decrease the supply of other goods and thereby increase the difficulty of equitably allocating scarce goods."[6]

Another reason why scientific and technological advances, alone, cannot solve the problems of environmental deterioration and resource depletion is that there is no way to avoid "the tragedy of the commons." The tragedy consists in the fact that most persons constantly attempt to maximize their own well-being, even at the expense of all other inhabitants of the planet; and, as a result, all persons suffer. For example, most people in the US drive highly polluting individual vehicles for their transportation, in order to maximize their own convenience and efficiency. As a consequence, however, air quality is worsened for everyone. The tragedy is that very few people will act to preserve the common property of everyone (e.g., air quality) so long as they believe they can use this commons for their own self interest, although in a destructive way. For this reason, notes Hardin, a "tragedy of the commons" is unavoidable unless man is forced to put planetary interests above his own. "No technological invention can prevent this fate."[7]

Hardin's solution to the catastrophe induced by "Frontier Ethics" is to reverse its system of values; instead of following environmental policy based on the premise that nature is subordinate to man and has no rights, Hardin proposes that environmental well-being be served before that of individual humans. He calls this new system of environmental policies, "Lifeboat Ethics."

"Lifeboat Ethics" is based on the view that the rich nations of the world may be thought of as lifeboats having moderate numbers of persons on board, while the poor countries are severely crowded lifeboats. Continuously, says Hardin, the poor fall out of their lifeboats, swim for a while, and hope to be admitted to one of the rich lifeboats. The ethical dilemma is whether the passengers on the affluent, less crowded boats should help the swimmers or allow them to come aboard. Like the land of every nation, each lifeboat has a limited carrying capacity.

Hardin proposes that this ethical dilemma be resolved by looking at only one lifeboat, that of the US. According to his calculations, there are twice as many poor people outside our boat as there are rich inside it, and it is loaded to capacity if one considers "the safety factor" necessary to insure the well-being of its passengers. If one follows Christian or Marxist ethics, says Hardin, then—since the needs of those inside and outside the boat are the same—one is bound to admit everyone to the boat. This, however, swamps the boat. "Complete justice, complete catastrophe."[8]

A second alternative might be to admit only a few people, and thereby get rid of "the safety factor" by filling the boat to capacity. This option, says Hardin, has two problems. First, which persons ought to be let on board? Secondly, eventually all passengers on the boat will be put in jeopardy because the safety factor has been ignored.

Finally, there is a third possibility. We could protect the survival of those on board, preserve the safety factor, and admit no more people to the rich boats. This is Hardin's solution. He justifies it by a number of considerations. First, he maintains, if rich nations gave resources to poor ones or admitted many of their people through generous immigration policies, then disaster would eventually strike the wealthy countries. This is because the population doubling time of poor nations is roughly one-quarter that of rich ones. If these same trends continued, then—eventually—the prolific passengers would swamp the life boat.[9]

A second rationale Hardin uses to justify his position is the thesis that, if the goods of the earth are assumed to be part of a commons open to all, then (as was mentioned earlier) "the tragedy of the commons" will occur. People will attempt to safeguard their own interests, and in doing so, will bring ruin to all. Pollution, for example, is increasing drastically because air and water are treated as commons. Fish populations are being depleted because the oceans are exploited as commons. Even the noble idea of a "world food bank," says Hardin, to which nations contribute according to their abilities and from which they draw according to their needs, will lead to the tragedy of the commons. Although special interest groups (e.g., farmers, pesticide manufacturers) gain through such policy, all US taxpayers must pay for it. Thus, he maintains, this allegedly humanitarian scheme really hides a powerful special interest "lobby for extracting money from taxpayers."[10] Moreover, he says, if poor countries are "bailed out" whenever there is an emergency, they will not learn from their experience. Rather, explains Hardin, if their emergencies are not met by outside help, their populations will drop back to the carrying capacity of the land. Relief from the outside ultimately hurts poor nations because it escalates population. The input of food thus acts, he says, as the pawl of a ratchet, "preventing the population from retracting."[11]

In spite of this reasoning regarding "the tragedy of the commons," says Hardin, people are still following an ethic of economic growth. This is what was described earlier as "Cowboy or Frontier Ethics." Such a policy will not help either population or other environmental problems, he notes, because a rising gross national product is usually accompanied by increasing population growth, and together they lead to "the tragedy of the commons."[12]

The environmental load of increasing population, of world food banks, of unrestricted immigration, and of fostering economic development, is already greater than it should be, concludes Hardin. We must follow "Lifeboat Ethics" and restrict all these trends; if we do not, then we will be violating the rights of posterity to a clean and abundant environment.

Although he admits that drawing such lines may be unfair, from the point of view of distributive justice, Hardin believes that "any other action is practically worse." It is impossible, he maintains, "without violent disorder and suffering," to distribute the world's resources equitably. Hence, argues Hardin, we must "begin the journey to tomorrow from the point where we are today." In order to guarantee that our grandchildren do not inhabit a ruined world, we must follow "Lifeboat Ethics."[13]

Hardin's views regarding economic and population growth, world food banks, unrestricted immigration, and allocating resources by distributive justice, are similar to those of Paul Ehrlich in *The Population Bomb.* Like Hardin, Ehrlich criticizes "Frontier or Cowboy Ethics" and argues that foreign aid to underdeveloped countries is wrong, both because it exacerbates the environmental crisis and because it threatens ultimate planetary well-being. Also like Hardin, Ehrlich takes as his starting point the belief that, unless one follows "Lifeboat Ethics," human survival is questionable. Ehrlich illustrates his points by means of the concept of "triage."

Borrowed from military medicine, this concept is used to describe a means of allocating any type of scarce resources. On the basis of a method developed by the French in World War I, for example, battlefield casualties are divided into three categories: "those who will die regardless of treatment," those "who will survive regardless of treatment," and those who can be saved only if they are given prompt treatment.[14] Because medical resources on the battlefield are scarce, they are allocated according to a principle of triage, of helping only the members of the third group and leaving those in the other categories untreated. When triage is employed as a solution to population and environmental problems, it is taken to mean the policy of denying foreign aid and food assistance to poor countries such as India, who are not making enough progress in the environmental battle to control population, develop natural resources, and expand food production. Although "triage" runs counter to traditional ethical notions, argues Ehrlich, it is necessary if we are to provide for our own survival, to discourage unlimited population growth, and to insure environmental well-being.

An important component of Ehrlich's triage solution is to sanction abortion and involuntary limitation of procreation in both developed and underdeveloped nations. Arguing that fetuses represent only *potential* humans, Ehrlich maintains that termination of pregnancies is a lesser evil than widespread starvation and eventual threats to global survival. Likewise, involuntary limits on reproduction represent "coercion in a good cause," namely, the well-being of the planet.[15]

The key questions raised, by both Ehrlich's and Hardin's views of "Lifeboat Ethics," are whether they are correct in placing the interests of current passengers on the rich lifeboats ahead of those of persons on the poor boats, and whether they are right in affirming that ultimate ecosystemic well-being is a higher value than equitable distribution of resources in the present.

Determining the answers to these ethical questions requires, in part, an understanding of several factual or biological issues. Scientifically and ecologically speaking, is the health and well-being of the entire planet essential to human survival and spiritual growth? If so, then clearly the welfare of nature cannot unequivocally be placed second to that of humanity. If not, then it is possible that human good might be served simply by placing it above that of nature. Ecologically speaking, we must try to discover all the consequences of putting our own interests first, last, or on the same level as those of nature. Once these factual issues are clearer, the ethical questions, such as which of the three stances one ought to follow, and which represents the greatest good, can be answered more easily.

Those who argue that human interests ought to be served before or above those of nature usually do so because they believe that humankind is superior to everything else on the planet and that, as a result of this, justice demands that our well-being be safeguarded before that of any other being(s). Ecologists who claim, on the other hand, that human interests ought to be considered after those of nature, often maintain that preserving the entire "web of creation" is a greater good than safeguarding only one part of it. They argue, consequently, that environmental ethics must focus on the welfare of the entire ecosphere (the living and nonliving environment and its various populations), and not just on our own good. In other words, they claim that environmental ethics ought to be *ecocentric* and not *anthropocentric*. As Hugh Iltis, Professor of Botany at the University of Wisconsin, put it: "The earth and the web of life come first, man comes second."[16]

A third category of writers believes in neither our ethical primacy nor that of nature. They maintain that one should have a reverence for everything in the biosphere, and that a recognition of the community of all members precludes giving precedence either to humans or to nature. They reason that any attempt, either to endow one group with a moral superiority or to allow some members of the planet to exercise inordinate power over others, will result only in planetary disaster, since all beings are interdependent. Likewise, they maintain that the only way to serve the interests, either of man or of nature, is to seek the well-being of both and to give neither a position of primacy.

As a result of the environmental crisis, the basic question of what ought to be "the place of man" has taken at least two distinct forms. One variant of the issue deals with how we ought to treat other humans, while the other addresses how we ought to behave toward nature.

Largely in response to the food shortage, the scarcity of natural resources, and the population problem, a number of people have asked whether there ought to be a new ethic to govern our responsibility to each other. On the one hand, writers such as Hardin have charged that "a sharing ethic" will eventually destroy those who succumb to humanitarian impulses.[17] On the other hand, Daniel Callahan and others have argued that the present set of environmental circumstances does not justify an ethic of cruelty or callousness to fellow human beings.[18]

The second question regarding "the place of man" has arisen, in great part, out of the problems of species extinction. This issue has forced us to ask whether there ought to be a new ethic to govern our responsibility to nonhuman members of the ecoshpere. Writers such as James Keller have argued that a single human being has more value than any member of another species, and hence that we still ought to follow a "man-first" ethic. On the contrary, Hugh Iltis and other environmentalists have claimed that humans have responsibility for animals as well as for people; they maintain that the collective well-being of nature must be placed ahead of that of humanity, both because all things are interdependent and because nature has no advocate to defend its interests.

Difficulties with the Lifeboat Ethic

Who is right in the humans-versus-nature argument? How does one determine whether the proponents of "Lifeboat Ethics" are correct in giving primacy to ecosystemic well-being over that of individual humans? Part of the answer to these questions can be obtained by examining the arguments of Ehrlich and Hardin more closely and by analyzing the ethical assumptions implicit in them.

Neuhaus' Charge: Lifeboat Ethics Is Anti-People

One of the most severe critics of "Lifeboat Ethics" is Richard Neuhaus, a New York clergyman and pacifist. He argues that Ehrlich and Hardin are "anti-people." According to Neuhaus, who specifically attacks the work of Ehrlich, no "Lifeboat" revolution is desirable as a result of the environmental crisis. Not only has the seriousness of the problem been overstated, he says, but also the "lifeboat" position is beset by a fundamental contradiction. Neuhaus is puzzled about how Ehrlich and others can maintain what he calls an "awesome veneration for apparently marginal forms of plant and animal life and an almost fanatical determination to belittle any sense of mystery about human gestation." Ehrlich's promotion of abortion and triage as solutions to environmental overload, says Neuhaus, is inconsistent with his apparent respect for nonhuman forms of life. As Neuhaus puts it, "It is intolerable that brothers and sisters should be dying from hunger while we fret about being overweight and organize 'radical' coalitions to save the Pine Martins."[19]

While Neuhaus' point seems insightful in that, often, people who revere nature also despise their brothers and sisters, it misses the logic of a deeper point made by Ehrlich and Hardin. This point is that a situation of triage is a "no-win" case and that, regardless of what action is taken, some human lives are bound to be lost in such an instance. The most ethical course of action, therefore, is to recognize the equal rights of all persons and then to do whatever will cause the least amount of harm and violate the fewest rights. (This claim will be discussed in more detail later.)

Neuhaus' argument, that Ehrlich's triage proposal is inconsistent with his apparent respect for nonhuman life, misses the point. This is that, regardless of whether one is for or against nonhuman life, failure to practice triage in a crisis situation will surely lead to greater losses of human life than will its practice. Hence, a proponent of triage is not necessarily anti-life, but perhaps merely in favor of saving the greatest possible number of lives.

The same sort of point can be made with respect to Neuhaus' arguments regarding abortion. Even if one assumes that abortion is unethical, if one also presupposes that the situation at hand requires triage, then abortion might be allowable if it prevented greater losses of human life than would failure to practice it. The key question, therefore, is not whether triage is pro-life or anti-life. Obviously, if it is used correctly, it is pro-life. The real question is whether Ehrlich and Hardin have accurately described the world situation as one of crisis and one which therefore requires the practice of triage, not whether triage denigrates human life.

The issue of whether the current environmental state is indeed one of crisis is the lynchpin of "Lifeboat Ethics." It is also a question which is difficult to resolve. There have been endless debates over whether the ozone layer has been depleted, for example, and over whether technology can deliver us from resource depletion and pollution. The one clear thing, however, is that a finite planet cannot tolerate ever-increasing population, pollution, and resource depletion, simply because it is finite. Hence, even if the crisis is not yet upon us, it will be someday, if present trends of growth continue.[20]

The difficulty with knowing that current rates of growth will surely lead to catastrophe, however, is that we thereby know neither precisely when that crisis will occur nor how close we are to it. In the absence of these certainties, the logic of "Lifeboat Ethics" can be quite problematic. Politically speaking, it is dangerous to be manipulated by a crisis mentality. What "crisis" usually suggests is that an imminent hazard somehow justifies circumventing the time-consuming and sometimes-frustrating processes of democracy and the inconvenient dictates of justice. If we are in the midst of Ehrlich's "crisis," then democracy can be forgetten in order to obtain the benefits of coercive limits on population. Likewise, ordinarily one would not argue that it is morally desirable to allow thousands of persons in underdeveloped countries to starve to death. If a worse situation could be averted by doing so, however, then this act might be said to be tolerable. This mentality is dangerous precisely because one might see a crisis where there is none, or if there is one, might overestimate it. The political and ethical consequences of subverting the normal dictates of democracy and distributive justice could be quite extreme, as I will argue in the following paragraphs.

In discussing the "Survival Ethic" of the "Lifeboat" position, Neuhaus tries to draw some parallels between National Socialism (Nazism) and the policies advocated by Ehrlich and Hardin. He alleges that both philosophies are inimical to liberal democracy, both have an anti-technological prejudice, both adhere to extremely "negative" views of life, and both justify great horrors on the grounds of obeying "nature's bidding."[21] Hardin and Ehrlich are willing to let persons die in the name of following nature's interests. These interests include saving the world from environmental ruin and limiting the number of people that the land must "carry." Neuhaus believes that there is more than accidental similarity between the views of "Lifeboat"-ethicians and Hitler, who called for "unconditional submission to the divine law of existence" and for reverence for the "fundamental necessity of the rule of nature."[22]

Neuhaus' fundamental insight appears legitimate (viz., that "following nature's bidding" is dangerous), both because it is difficult to know precisely what is in nature's interest, and because following it could lead to great harm to human beings. Where his analysis goes wrong, I think, is in ignoring the fact that Hardin defends "Lifeboat Ethics," not merely by asserting the primacy of nature's well-being over that of humanity, but also by defending the rights of members of future generations against encroachments by those of us who live in the present. Hardin speaks very clearly of the necessity to protect "subsequent generations" from threats to their survival.[23] Hence the Hardin discussion is placed in the framework of contests among rights of persons, not merely in the context of conflicts between man and nature.

Lifeboat Ethics and Violations of Distributive Justice

The critical point at issue between Neuhaus and Hardin is, I think, whether one has an ethical obligation to maximize ecosystemic excellence even in order to benefit future persons, if present individuals will suffer because of it. More will be said later, in Chapter 3, about the rights of posterity. What is clear, however, is that "no amount of research can verify that the right [i.e., what is right] is the optimum biotic community."[24] Throughout the history of ethical thought, there have been great debates over what was "right" or "good." It has been defined as pleasure, happiness, utility, and duty, among other things. Hence, if what is right is said to be "optimizing the biotic community," there might be disagreement about why some other thing was not said to be of higher priority. There are reasonable grounds for aruging, for example, that realizing distributive justice is a higher priority than maximizing the well-being of the biotic community. In other words, one might choose an ethic based on *equity* rather than on optimizing overall well-being, apart from individual welfare.

There is a long tradition, arising out of both social-contract theories and Judaeo-Christian views, that equity ought to be maximized before general well-being. Perhaps the greatest contemporary exponent of the contractarian, egalitarian view is John Rawls, Professor of Philosophy at Harvard University.[25] Rawls maintains that it is irrational for any objective person to choose an ethic other than one based on absolute equality and distributive justice, because it is always possible that another ethic could result in his social disenfranchisement. Hence, says Rawls, the only reason one might choose a nonegalitarian ethic is that, as a result of personal advantages (e.g., education, money), he thinks it is possible to obtain more than his fair share of wealth. A purely disinterested (and therefore objective) person, who did not base his moral principles on his own hope for unfair advantage, however, would choose an egalitarian ethic, says Rawls.

Rawl's views appear reasonable, particularly in the light of the American history of recognizing the rights to equal opportunity, equal justice, and equal protection, especially as guaranteed in the Fifth and Fourteenth Amendments to the US *Constitution.* Rawls maintains that no rational and purely objective person would sanction inequalities in wealth and authority, unless they would "result in compensating benefits for everyone, and in particular for the least advantaged members of society."[26] If he is correct in this belief, then Hardin and Ehrlich might not have defined accurately what is "right." Instead, one might have to replace "the myth of survival" and "the myth of social Darwinism" that they advocate with what Neuhaus calls "the myth of the covenant with the poor."

Lifeboat Ethics and the Distinction between Duty and Charity

Neuhaus believes that the authentic touchstone of morality is not survival but concern for the poor. In addition to the reasons given by Rawls for such concern, he maintains that this new myth, which rejects social Darwinism and a survival ethic, is one of "healing."[27] Like Frederick Elder in *Crisis in Eden,* Neuhaus wants "to turn the environmental ethic toward reverence for ... human life."[28]

One defense of the position of Elder and Neuhaus is given by Peter Singer, in his classic article, "Famine, Affluence, and Morality." Here Singer argues against the "triage" solution of "Lifeboat Ethics" by undercutting the traditional distinction between duty and charity. His major thesis is that we have a moral obligation to prevent suffering from happening, if we can do so without sacrificing "anything of comparable moral importance."[29] Since this precept does not require one to promote good, Singer accepts it as "uncontroversial."[30] Moreover, he continues, if we "accept any principle of impartiality, universalizability, equality, or whatever, we cannot discriminate against someone merely because he is far away from us (e.g., in India).[31] Therefore, he says, we have a moral obligation to prevent famine and poverty, if we can help to do so without sacrificing anything of comparable moral importance.

The main difficulty with Singer's response, to the dictates of "Lifeboat Ethics," is that his discussion does not consider Hardin's and Ehrlich's point that, in the long run, more lives will be saved by the practice of "Lifeboat Ethics" than by the failure to follow it. This same weakness can be levelled at Neuhaus and Elder. Perhaps "Lifeboat Ethics" does save more lives because it staves off a worse crisis for members of future generations.

Daniel Callahan, however, is one ethician who disagrees. He believes that, even if one admits that our obligations to future generations are greater than those to present individuals, Lifeboat Ethics is based on some erroneous ethical assumptions. For one thing, says Callahan, Hardin assumes that wealthy nations have "self-sufficient lifeboats."[32] This is clearly a false presupposition as the oil crisis shows. With 50% of US oil now being imported, some of it from underdeveloped countries, it is wrong to picture the US as a "have" who is not in any way dependent upon the "have nots." Hardin's lifeboat scenario, moreover, does not take into account the fact that many of the wealthy countries were helped to prosperity because they were able to buy resources cheaply from poor nations and then to sell finished products back to them at a high cost.[33]

Related to these assumptions, that the rich lifeboats are self-sufficient, and that they have not exploited the poor, is the additional presupposition of Hardin's that the poor countries deserve to be in their state. He suggests that their people are not as wise or as competent as those in rich nations, and completely ignores the factors which are beyond the control of the poor (e.g., employment opportunities). Instead, Hardin's view is that, if developed countries "get tough," poor nations will mend their ways and decrease their population. It is hard to believe that this assumption is correct. If bread becomes scarcer, then, if anything, this might be a greater incentive to have more children. After all, in a poverty-stricken nation, what better assurance does one have of being cared for in old age than by having many children and attempting to help them survive? It seems that Hardin's own analysis of "the tragedy of the commons" provides ample evidence that persons will not voluntarily put limits on their own self-interest. For this reason, it is difficult to believe that a tough foreign aid policy will cause the poor to be less prolific.

Perhaps the most disturbing assumption in the "Lifeboat Ethics" analysis is Hardin's belief that, since perfect justice is impossible (e.g., we can't give America back to the Indians), we have no obligations to justice at all. In other words, since it is impossible to redistribute the wealth of the world equitably, we don't have to worry about any claims made on us by the poor nations.[34] Contrary to Hardin's view, it seems reasonable to me that, although perfect justice is impossible, one does have an obligation, so far as possible, to cease exploitation and repression of disadvantaged countries. Otherwise, any duty could be dispensed with on the grounds that attaining perfect justice is impossible. This point, however, is not addressed by Hardin.

From the political and ethical point of view, another questionable presupposition made by Hardin is that the rich nations have the right to decide who will live and who will die.[35] Even if proponents of "Lifeboat Ethics" are correct in assuming that present levels of population, pollution, and resource depletion ought not to continue, and that growth in all three ought to be stopped, what gives the "rich lifeboats" the right to allow passengers on poor lifeboats to die? Why is it that the wealthy are allowed to "play God"? This ethic sounds suspiciously like the view criticized long ago by Plato, who said that, contrary to Gorgias, might did not make right.

Another interesting omission in the Hardin discussion of "Lifeboat Ethics" is the fact that wealthy nations are using a disproportionate share of the planet's resources, and that their depletion of nonrenewable materials might also cause harm to future generations. If the "Lifeboat" position is argued in the name of posterity, then one wonders why other alternatives (e.g., decreasing US consumption), to allowing persons to die, were not discussed. Besides, if Wayne Davis is correct, then it would take several hundred poverty-stricken Indians to have the same adverse effect on the environment as one average American.[36] In view of the differing levels of consumption practiced in the rich and the poor lifeboats, one wonders why Hardin believes that the poor constitute a threat to future generations, but that, so long as they do not reproduce to excess or promote economic expansion, the rich nations can go on with "business as usual."

Pragmatically speaking, another deficiency with the "Lifeboat" proposal is that it ignores the fact that failure to help poor nations might result in political crisis. This point has been mentioned both by the authors of the famous MIT-Club of Rome report, *The Limits to Growth,* and by experts in development.[37] If they are correct, then it is possible that the political instability created by "Lifeboat Ethics" might be as great a threat to posterity as the environmental crisis allegedly caused by failure to practice it.

The big question, of course, underlying "Lifeboat Ethics," is whether Hardin and Ehrlich are correct in assuming that the problems in poor nations are so extreme that they must be "written off." As Callahan suggests, perhaps we should act *as if* they can be saved; otherwise, he says, we will simply be following a self-fulfilling prophecy. How could we know whether they can be saved, except by trying to do so?[38] Moreover, even if poor countries do represent a great drain on the planet's ability to survive in the future, simply cutting them off from aid presupposes that survival is the only, or at least the most important, moral principle to be served. As was pointed out in the preceding subsection, it might be just as necessary to serve the interests of equity or distributive justice. Likewise, it might be just as important, if not more so, to serve one's more immediate obligations to people now alive, as to serve one's more remote duties to posterity.

Conclusion

To the extent that assumptions such as those just discussed are questionable, then to that degree, "Lifeboat Ethics" is problematic. It leaves many issues, regarding equity, distributive justice, democratic decision-making, and the value of survival insufficiently resolved.

Apart from these difficulties, however, what is clear is that ethical decision-making cannot ignore environmental values, as it has so often in the past when persons followed "Frontier or Cowboy Ethics." Although "we yet lack a clear account of the logic by which we get our values, it seems undeniable that we shape them in significant measure in accord with our notion of the kind of universe that we live in."[39] The emergence of "Lifeboat Ethics," for all its problems, shows that our notion of the universe is undergoing an ecological revolution. Our standard of morality is changing, rising higher, to include environmental values.

Whether one believes, with Ehrlich, that our ability to survive the environmental crisis demands an ethics which some have called "anti-people," or with Neuhaus, that such a "survival ethic" is immoral and inconsistent, their views illustrate an important point: One ought to be *concerned with the morality* of personal response to the environmental crisis. Whether for our own survival or for our moral growth, we can no longer afford a purely anthropocentric ethics.

Notes

1. See, for example, Lynn White, "The Historical Roots of Our Ecologic Crisis," *Science*, 155 (March 10, 1967): 1203-7 and Alan Donagan, *The Theory of Morality*, Chicago, University Press, 1977, both of whom discuss this point. Kenneth Boulding was the first to use the terms "Cowboy Ethics" and "Cowboy Economy." See his works cited at the end of this chapter.

2. Duane S. Elgin and Arnold Mitchell, "Voluntary Simplicity," *The Futurist*, 11 (August 1977): 256. See also Stewart L. Udall, *The Quiet Crisis*, New York, Avon, 1963, pp. 66-80, who also discusses Frontier or Cowboy Ethics.

3. Udall, *The Quiet Crisis*, pp. 66-67.

4. See Udall, *The Quiet Crisis*, p. 190, for discussion of "the Myth of Scientific Supremacy."

5. See, for example, Garrett Hardin, *Exploring New Ethics for Survival*, New York, Viking, 1972, and Hardin, "Living on a Lifeboat," *BioScience*, 24 (October 1974): 561-68; hereafter cited as: Lifeboat. See also Paul R. Ehrlich, *The Population Bomb*, New York, Ballantine, 1968 and D. Meadows, *et al.*, *The Limits to Growth*, New York, New American Library, 1974; hereafter cited as: *Growth*.

6. Hardin, Lifeboat, p. 565.

7. Hardin, Lifeboat, p. 563.

8. Hardin, Lifeboat, p. 562.

9. Hardin, Lifeboat, p. 562.

10. Hardin, Lifeboat, pp. 562-63.

11. Hardin, Lifeboat, p. 564.

12. Hardin, Lifeboat, p. 565.

13. Hardin, Lifeboat, pp. 565-68.

14. Paul Ehrlich, *The Population Bomb*, New York, Ballantine, 1968, p. 159; hereafter cited as: *Bomb*.

15. Ehrlich, *Bomb*, pp. 147-73.

16. Hugh H. Iltis, "Man First? Man Last? The Paradox of Human Ecology," from *BioScience*, 20 (July 1970): 820.

17. Hardin, Lifeboat, pp. 561-68.

18. See Daniel Callahan, "Doing Well by Doing Good: Garrett Hardin's 'Lifeboat Ethic'," *The Hastings Center Report*, 4 (December 1974): 1-4; hereafter cited as: Doing Well.

19. Richard Neuhaus, *In Defense of People*, New York, Macmillan, 1971, p. 265. See also pp. 162-79, 192-201; hereafter cited as: *People*.

20. Meadows, *Growth*.

21. Neuhaus, *People*, pp. 152-62.

22. Neuhaus, *People*, p. 155.

23. Hardin, Lifeboat, pp. 566, 67, 68.

24. Holmes Rolston, "Is There an Ecological Ethic?" *Ethics*, 85 (1975): 101; hereafter cited as: Ethic.

25. See John Rawls, *A Theory of Justice*, Cambridge, Harvard University Press, 1971. See also Charles Fried, *An Anatomy of Values*, Cambridge, Harvard University Press, 1970; Charles Fried, *Right and Wrong*, Cambridge, Harvard University Press, 1978; and Alan Donagan, *The Theory of Morality*, Chicago, University Press, 1977.

26. Rawls, *A Theory of Justice*, pp. 14-15.

27. Neuhaus, *People,* pp. 301-303.

28. Neuhaus, *People,* p. 197. See also Frederick Elder, *Crisis in Eden,* New York, Abingdon, 1970, pp. 147-55.

29. Peter Singer, "Famine, Affluence, and Morality," in T.A. Mappes and J.S. Zembaty, *Social Ethics,* New York, McGraw-Hill, 1977, p. 319; hereafter cited as: Famine.

30. Singer, Famine, p. 317.

31. Singer, Famine, p. 318.

32. Callahan, Doing Well, p. 3.

33. Callahan, Doing Well, p. 3.

34. This same point is made by Callahan, Doing Well, p. 3.

35. This point is also made by Callahan, Doing Well, p. 3.

36. See Wayne H. Davis, "The Land Must Live," *Environment Action Bulletin,* 3 (July 15, 1972): 4-5, 7.

37. See D. Meadows, *et al., Growth,* and Denis Goulet, *The Cruel Choice: A New Concept in the Theory of Development,* New York, Atheneum, 1971.

38. Callahan, Doing Well, pp. 3-4.

39. Rolston, Ethic, p. 108.

SPACESHIP ETHICS

K.S. Shrader-Frechette

CONCERN FOR NATURE is viewed with suspicion by many people who fear it is misplaced and often a substitute for interest in our fellow humans. Critics cite the millions of dollars spent on pet cemeteries, and point out that individuals in the US spend more per year to feed and house their own animals, than the government allocates for care of this country's needy children.

Perhaps one of the paradigm cases of putting the interests of nature above those of humanity was exhibited by Adolf Hitler. Psychologists and historians have described repeatedly how the Nazi leader manifested an inordinate form of sentimentalism and emotionalism regarding animals. Apparently, he was extremely cautious to protect wildlife, but flagrantly insensitive to basic human rights. And, although we can readily see that his system of ethics is unacceptable, it is not easy to determine the right balance between respect for nature and respect for humans. What is our place within the ecological framework? How ought we to behave both toward other humans and toward the living and nonliving beings on our planet?

As the last essay revealed, the dominant environmental policy we seem to have followed, "Frontier or Cowboy Ethics," has accorded humans a place of superiority and control over the rest of nature. Observant of the planetary destruction and resource depletion encouraged by this view, Garrett Hardin has proposed that we adopt a new system, "Lifeboat Ethics." Arguing that the very survival of the earth and its ecosystems is at stake, proponents of "Lifeboat Ethics" subscribe to the belief that the well-being of nature as a whole has primacy over the interests of individual men. This stance, however, leads one into direct conflict with all moral codes based on the principles of distributive justice.

Spaceship Ethics

Assuming that there are grave difficulties with both "Frontier or Cowboy Ethics" and "Lifeboat Ethics," it would be well to consider a third framework for environmental behavior, that of "Spaceship Ethics." Unlike the two previous theories, this account accords primacy neither to man over nature, nor to passengers of rich lifeboats over those on poor ones. Instead, proponents of "Spaceship Ethics" believe that the authentic welfare of both humanity and nature is intimately bound together, and that neither can be given superiority without ultimately hurting the other.

The spaceship has become a symbol for man's relationship with the environment ever since 1965 when Adlai Stevenson invoked the image in an address to the United Nations.[1] Since that time, environmental writers Kenneth Boulding, Buckminster Fuller, William Pollard, Barbara Ward, and others have developed the model as a basis for understanding the biosphere.[2]

The Meaning of the Spaceship Metaphor

As Fuller revealed in his seminal work, earth may be thought of as a spaceship only 8,000 miles in diameter. Its energy-supplying mothership, the Sun, is 92,000,000 miles away, and our spacecraft is flying at 60,000 miles per hour around it. Our ship has been so well designed that, during the 2,000,000 years humans have been passengers, they have not realized they were on a spacecraft. Life has been regenerating on board despite the phenomenon of entropy, according to which all physical systems lose energy. Employing photosynthesis, vegetation on the land and algae in the sea "impound" life-regenerating energy for us. Animals thrive on this plant life; we, in turn, take such items as the milk and meat from animals, and the energy is relayed to us.[3]

Despite the fact that spaceship earth is endowed with unique internal support systems, up to now humans have been misusing and polluting the on-board energy-exchange structures to such an extent that it is becoming more difficult to regenerate life aboard the ship. We know what to do to keep our automobiles in working order, but as the environmental crisis shows, very few of us know how to keep spaceship earth running smoothly. One reason, why we have not learned this, is that the total wealth of our ship has provided an enormous safety factor which has allowed most of us to remain ignorant of the environmental problems resulting from abuse of the spacecraft.[4] In fact, this safety factor of abundant resources is the main why, for so many years, we have followed "Frontier or Cowboy Ethics." We have assumed that we could always find a new natural frontier, replete with more planetary riches. As we have discovered that resources are finite, we have come to realize that we need a new system of ethics and that our "cowboy" policies must be abandoned.

As an alternative to "Frontier or Cowboy Ethics," one new system of values is built upon the consequences of employing the spaceship metaphor to describe the earth. First, both the spacecraft and our planet are relatively closed systems, even though the earth is dependent upon inputs of solar radiation. Secondly, both are finite; this precludes any indefinite expansion of activity on either the spaceship or the earth, and it means that the carrying capacity of both is limited. If resources are not renewed or recycled, then, as a result of the finite and closed nature of the systems of spaceship earth, twin disasters of pollution and resource depletion will occur. Thirdly, both our planet and a spacecraft support life through balanced, interrelating, self-sustaining methods. If the critical systems fail, either on earth or on the spaceship, then the passengers will not survive.[5]

Perhaps the most basic difference between the earth and the spaceship is that the spacecraft has a home base from which it came and to which it can return, while the planet does not. Earth, on the other hand, "is its own base."[6] Another deficiency in the spaceship metaphor is that it is quite simplified when one considers the variety and complexity of all the systems of the planet. A third difference is that, whereas we are the creators of the spaceship, which we use for our purposes, we have not created the earth; we cannot control it completely. If we are to survive, we ought not to use it for our own purposes, apart from the good of nature. The spaceship obviously cannot go on without us; it needs human agency for its control. The earth, however, does not need the human species in order to survive, at least in some minimal sense.

In spite of these points of disanalogy, the spaceship metaphor provides a useful vehicle for thinking about the relationship between man and the biosphere. The points of similarity between the earth and the spacecraft (e.g., both are closed and finite systems) enable us to think more realistically about ecological conditions. The spaceship model also makes it possible for persons to recognize the long-term effects of habits (e.g., rapid consumption of resources) whose significance might not be recognized, if we had to use a purely abstract model of the planet.

An Operating Manual for Spaceship Earth

Even though the spaceship model allows mankind to come to new realizations regarding the nature of the earth, because of the points of disanalogy between the two, it is not possible to formulate extremely precise ethical principles which follow from acceptance of the model. As was noted earlier, "There is one outstandingly important fact regarding Spaceship Earth, and that is that no instruction book came with it."[7] A number of proponents of the spaceship metaphor have attempted, however, to spell out some of the general ethical norms which ought to be followed if one recognizes the appropriateness of the model.

Lynton Caldwell, in a classic essay on spaceship ethics and politics, pointed out that our new morality must include the principle that we ought to limit population, industrialization, and economic expansion; further, he says, we ought to allow resources to renew themselves or, if this is not possible, we ought to recycle nonrenewable materials. All of these injunctions follow from the recognition that our spaceship earth is a finite and closed system. Moreover, maintains Caldwell, "The tighter the fit between man and his necessities, the greater the need for regimentation, rationing, regulation, and control."[8] This means that limiting population, economic expansion, and use of resources serves two goals. It not only prevents man from exceeding the carrying capacity of the planet, but also provides for a maximum amount of personal freedom.

Obviously, persons are freer to make their own choices when they are not living under conditions such as scarcity of food or overcrowding of housing. Hence, if they accept the ethical responsibility to limit their reproduction, economic growth, and use of planetary resources, they will ultimately help prevent dangerous expansion which, on a finite spaceship, can lead to crisis.

In addition to these general principles of limiting various forms of growth, Fuller's interpretation of "Spaceship Ethics" includes a number of particular precepts. On the whole, these are quite unlike the norms defended as part of "Lifeboat Ethics." Instead of maintaining that the poor nations must be accorded no help, so as to force them to live within the carrying capacity of the planet, Fuller argues that, with appropriate moderate patterns of consumption and growth throughout the world, all the inhabitants of the earth can be supported. He believes that it is only our unwillingness to spend the money and make the effort that causes us to believe, falsely, that some underdeveloped nations need to be "written off." The chief culprit in such self-serving tendencies, says Fuller, is the inability to think on a worldwide or planetary scale. We are all on spaceship earth together, and it is unrealistic to think that the entire craft can make a successful voyage when some passengers on it are suffering extreme forms of deprivation. If nothing else, this will lead to political crisis when the poor revolt. (For further discussion of this point, see the last article in Chapter 7.) Moreover, says Fuller, it is only because of national "piracy" toward poor countries that their inhabitants have become so impoverished, "underfed, and physically afflicted."[9] Because wealthy nations have, in part, attained their superior positions by exploiting Third-World countries, Fuller believes that the rich have a duty to help the poor. Unlike the proponents of "Lifeboat Ethics," Fuller believes in large-scale distributive justice; he says "that wealth is as much everybody's as is the air and sunlight."[10]

Whereas some of the ethical precepts of Fuller (e.g., the practice of distributive justice) are central to classical interpretations of traditional moral philosophy, some of his principles of "Spaceship Ethics" are quite new, and even radical. In order for us to survive on the finite spacecraft which is a closed system, Fuller believes that everyone must practice conservation, and not be "so foolish as to continue to exhaust in a split second of astronomical history" the resource savings of "billions of years of conservation." In particular, he maintains that humankind on the spaceship should use no fossil fuels, but should "operate exclusively" on the vast daily energy income from the powers of wind, tide, water, and the sun. If we use energy sources other than these, says Fuller, it is analogous to running our automobiles on batteries instead of liquid fuel.[11]

Despite the automobile analogy, Fuller also believes that the practitioners of "Spaceship Ethics" ought not to own automobiles, but instead should rent them. Some social planners have argued that such a rental system would serve the interests of environmental quality and resource conservation, since the number of vehicles would be reduced by 40 percent and people would tend to use less fuel.[12] In fact, says Fuller, "Spaceship Ethics" requires that one divest himself of all his outmoded habits regarding most forms of ownership. Simply using material things would enable them to be recycled, and it would free persons from the obsolete, "burdensome and wasteful" practice of possession of goods.[13]

All these new ethical precepts of Fuller's come down to what Caldwell has called the need for a "fundamental" change in our values and habits, rather than merely for an incremental "adjustment." Unfortunately, says Caldwell, when something like the Santa Barbara oil spill or the pesticide fish kills in the Mississippi River occur, the response of the public is simply to use *ad hoc* remedies to prevent their recurrence and to clean up the specific damage at hand. What is needed instead is a new "general conclusion" or "comprehensive policy." Rather than these "incremental adjustments," says Caldwell, "Spaceship Ethics" requires a fundamental change in the system itself. "Superficial palliatives" will not solve the problem.[14]

Because "Spaceship Ethics" requires a new world view, one based on the finite and closed nature of our planet, it also demands a completely new type of political leader. There is an environmental crisis, in part, because there is a "crisis of leadership." Leaders have ignored recycling, promoted economic expansion, and, worst of all, they have spent billions of dollars on war and military proliferation. At the same time, says Fuller, they have argued that there is neither enough money to clean up the environment nor enough capital and technology to insure a decent standard of living for everyone. As Fuller points out, governments have continually put off environmental reforms and foreign aid; then, later, they have claimed that the price tags for both were too high. As a consequence, says Fuller, we have never attained the goals of "Spaceship Ethics."[15]

Part of the reason why our leaders have ignored implementation of "Spaceship Ethics" is that developed peoples have become too enamored of the individual freedoms guaranteed by their courts and constitutions. If we want to begin to think on a worldwide or planetary scale, then we must be willing to relinquish certain patterns of inordinate consumption and economic growth. Although abandonment of these old modes of behavior may appear to represent a limitation on freedom which many people are loathe to accept, the real truth is that foregoing some of these lesser individual freedoms (e.g., the right to consume as much as one wishes) will pave the way for a more authentic experience of liberty. (See the article on "Voluntary Simplicity" in Chapter 7 for arguments to this effect.) As Caldwell points out, freedom of *everyone* is threatened by conditions of overcrowding and scarce resources. This is precisely the political point not recognized by those who practice "Lifeboat Ethics," rather than "Spaceship Ethics." As Fuller explains, if one follows the spaceship view, then all the human passengers will be able to enjoy the whole craft (planet) "without any individual being advanced at the expense of another."[16] What has happened instead is that most people employ too shortsighted a view; they have forgotten about collective planetary good and that short-term restraint in minor affairs (e.g., energy consumption) can lead to greater freedom for everyone in the long run.

What these general and specific precepts of "Spaceship Ethics" reveal is that humankind must undergo a change of habit and a change of heart. As Caldwell puts it: "Modern man must somehow become better than he is; this is not a counsel of utopian idealism; it is a counsel for survival."[17]

Problems with Spaceship Ethics

Despite the plausibility of, and the need for, the revolution in values that is defended by proponents of "Spaceship Ethics," these environmental precepts have come under significant attack. In fact their greatest liability is the very thing that separates them from the principles of "Lifeboat Ethics." As was indicated in the previous essay, lifeboat behavior calls for coercive means of population control and deliberate refusal to aid underdeveloped countries so as to allow the number of their inhabitants to fall back to a level capable of being carried by the environment. Proponents of "Spaceship Ethics," on the other hand, do not sanction involuntary limits on reproduction or consumption. They believe the land can carry all its people, provided that resources are distributed justly, provided that humankind refrains from expensive tooling for war, and provided that we learn to consume only what we need.[18] In other words, lifeboat ethicians call primarily for a change in policy toward the needs of the poor, as a means of preserving the environment, while spaceship ethicians call for a conversion in the hearts of all people, especially the wealthy, in order to secure environmental stability.

Perhaps the most serious indictment of the latter position is that its proponents overestimate the voluntary "human capacity for restraint and for collective self-control." As Caldwell expresses it, "there is a dangerous innocence in Fuller's [spaceship] optimism."[19] Or, as Hardin might put it, there is insufficient reason to believe, either that humans will not fall victim to the demands of their own self-interest, or that they will avoid "the tragedy of the commons,"[20] and, hence, that they will live according to the dictates of "Spaceship Ethics." Without coercive enforcement of limits on population, immigration, and foreign aid to underdeveloped countries, he believes that humans will not willingly make the right choices in these matters. This is where "Spaceship Ethics" fails, says Hardin. It will lead to environmental suicide, because the earth cannot support everyone on it, especially the poor. Moreover, he argues, the spaceship is under no unitary sovereign control; it has no captain and no executive committee. "Spacemen" are therefore likely to claim all the rights of spacecraft passengers, without also accepting the responsibilities of their state.[21]

Hardin's charges of utopianism and unworkability are difficult to answer. Essentially he believes that the environmental crisis can only be brought under control by a strong central authority who makes hard decisions allowing Third-World peoples to die. To some extent, his beliefs may be countered by the discussion in the second half of the previous essay; the argument there reaffirms the desirability of adhering to principles of distributive justice, despite the attacks made on it by proponents of "Lifeboat Ethics."

Is the "Spaceship Ethics" position, however, unworkable as an ethical framework? There are several reasons why I think this is not the case. First, I believe spaceship ethicians are correct in refusing to sanction coercive means of limiting population and Third-World consumption of resources, since it is virtually impossible to determine what is "needed," either by an individual or by a nation.[22] Granted, there is a difference between needs and wants, but as the arguments in Chapter 7 indicate, it is difficult in any given instance to distinguish the two. It is also questionable whether, except in dire cases, one person ever has the right to determine what patterns of reproduction and consumption should be followed by other persons or nations. If this is true, then it is extremely dangerous to enforce coercively certain limits on population and resource consumption.

If Hardin and other lifeboat ethicians are correct in affirming that Third-World peoples are beyond the point of being able to be helped by rich nations, then to demand a mandatory halt to aid for these lands ignores the fact that their poverty and crisis may have been brought about, not only by their own lack of environmental sensitivity, but also because of their being "raped" by wealthy nations. Fuller points out, for example, that the plight of certain countries is the consequence, not of their environmental naivete, but of "the Great Pirates robbing Indo-China for centuries and cashing in their booty in Europe."[23] If he is at least partially correct, then one ought to be doubly careful in imposing involuntary restrictions on people who do not bear full responsibility for being in their desperate situation.

A third difficulty, with the position that one ought to follow "lifeboat" norms because "spaceship" principles are too idealistic, is that there are often other ways of implementing the spaceship position, short of the direct coercion sanctioned by proponents of the lifeboat view. One might help to control population, for example, by using a system of tax incentives for small families or by providing governmental means to care adequately for all persons in their old age.[24] Likewise one might reform the current system of property laws, and recognize what Ezra Mishan calls "amenity rights," the rights to a liveable environment.[25] In this way, one might open up an ethical framework, short of direct coercion, for avoiding "the tragedy of the commons." Another alternative might be to recognize a "Land Ethic," as Aldo Leopold urged decades ago, or to implement a system of rights for natural objects, as was proposed by Christopher Stone in his classic work, *Should Trees Have Standing?* [26]

What all these suggestions come down to is that there are many legal and political means of halting pollution and resource depletion, short of allowing inhabitants of overpopulated countries to die. Clearly, however, the implementation of some or all of these "spaceship" alternatives requires more effort and imagination than would the practice of "Lifeboat Ethics."

There are also some basic factual disagreements between proponents of the spaceship, as opposed to the lifeboat, position. Hardin believes that improving the standard of living of the poor will accelerate population growth, for example, while Fuller maintains that it will decrease the rate of population increase . Conflicts such as these have to be resolved before one can evaluate the effects of practicing either "Spaceship" or "Lifeboat Ethics." What is clear at the outset, however, is that social systems (such as "Spaceship Ethics") whose goal is maximization of equity through democratic controls, are much more difficult to implement than are schemes (such as "Lifeboat Ethics") which attempt neither to maximize equity nor to achieve environmental reform in democratic ways.

Insofar as "Spaceship Ethics" relies on democratic political means of implementing its principles, rather than on whatever coercion which rich nations are able to inflict on the poor ones, it is not likely to provide as efficient a method of controlling population and resource depletion as is "Lifeboat Ethics." Nonetheless, it appears desirable in this case to sacrifice efficiency in the interests of democratic control, free choice, and equitable distribution of goods. To see why this is so, consider another notoriously inefficient system of social protection, that of trial by jury.

Long a part of traditional legal means for promoting justice and ethical decision-making, trial by jury is recognized as being a cumbersome, expensive, and often untrustworthy vehicle for social protection. Despite these liabilities, however, trial by jury is in no danger of being replaced in the western world by a more "efficient" system. Just as its shortcomings are outweighed by its benefits, so also the difficulties with the democratic, largely noncoercive, and "incentive" schemes of "Spaceship Ethics" are offset by its benefits (viz., providing for distributive justice and equity).

Obviously a philosopher king, in Plato's sense, or even a benevolent dictator, if either were possible, could operate a criminal justice system in a more efficient manner than is now accomplished by trial by jury. Direct, undemocratic, governmental coercion would likewise probably be more effective in reducing population and resource depletion than would programs based on environmental education, positive incentives, democratic regulation, and free choice. Neither of these facts, however, constitutes sufficient cause for abandoning either trial by jury or the principles of "Spaceship Ethics." Most reasonable people would probably agree that, only under the most severe constraints, ought one to trade equity for efficiency.

If "Spaceship Ethics" is charged with failure to guarantee environmental quality, then its chief alternative, "Lifeboat Ethics," falls victim to a worse indictment, viz., it justifies purchasing environmental benefits at the price of individual liberty. This accusation, in turn, raises a whole host of classic questions concerning the desirability of seeking relief from social ills by undemocratic means.[27] Although governmental regulation is essential to insure environmental well-being, it becomes questionable in cases like that of "Lifeboat Ethics," in which the coercion is neither mutually agreed upon nor designed to maximize equity.

If citizens were coerced into obeying the norms of "Lifeboat Ethics," rather than provided with incentives for complying with principles of "Spaceship Ethics," several undesirable consequences might result. For one thing, it would be difficult to police any system, especially one based on coercion, given our human penchant for finding financial loopholes and for obtaining goods on the black market. This suggests that, even if "Spaceship Ethics" is too idealistic in requiring voluntary compliance with its precepts, it recognizes the wisdom of the belief that any ethical system is only as effective as the behavior of those who follow it. Moreover, in light of the earlier discussion regarding incentives and amenity rights, there may be several legal and political ways to protect the environment, short of following "Lifeboat Ethics" and therefore ignoring distributive justice and certain types of individual freedom.

In calling for incentives to limit consumption, growth, possessions, and reproduction, proponents of "Spaceship Ethics" are promoting acceptance of a system somewhat similar to that of the US income tax. Like this form of taxation, principles of "Spaceship Ethics" depend for their success upon the voluntary compliance of the public. Neither the tax nor this framework of environmental ethics specifies the same behavior or identical obligations for any two individuals. Both recognize the necessity of a system which is voluntary and yet designed to meet the individual needs of each person.

In the lifeboat model, however, considerations of individual needs and equity are ignored. Decisions are made across-the-board to deny resources to Third-World nations and to prevent immigration of their peoples to developed countries. The income-tax analog for such environmental policies would be to tax the poor more heavily than the rich, to prevent their upward mobility into more desirable social classes, and to ignore the need for a graduated scale of taxation. Apart from the problems of equity that such policies present, they are all the more onerous if they are imposed involuntarily, as in a system of "Lifeboat Ethics." This suggests that, to the extent that a particular policy is suspected of being unjust or inequitable, to that same degree ought one to be wary of implementing it coercively. This is one reason why, within the framework of "Spaceship Ethics," decisions involving resource allocation, population control, and immigration are not imposed in undemocratic ways.

What all this suggests is that there is a partially valid analogy between practicing "Spaceship Ethics" and raising a child. Both require patient adherence to long-term, general principles in order to insure the future survival, either of the land or the child. Just as one would not necessarily criticize a parent for employing noncoercive, allegedly utopian or idealistic ways of dealing with a particular teenager, so also one ought not necessarily to fault an environmentalist for subscribing to "Spaceship Ethics" as a noncoercive means of protecting the environment. With the land as with the child, the only real test, of one's success in teaching the ways of stewardship or adulthood, is whether the desirable values have been internalized. Acceptance of externally imposed coercion, either from environmentalist or parental "dictators," does not guarantee any change of heart. Yet, ultimately, only one's heart can sustain one in the endeavor, either to be humane or to protect the earth. Perhaps this is the reason why proponents of traditional ethical theories and representatives of international bodies, such as the United Nations, have affirmed the noncoercive, democratic principles of "Spaceship Ethics" and criticized the coercive, undemocratic tenets of "Lifeboat Ethics."[28]

Conclusion

The heart of the controversy, over which ethics are best able to deal with the environmental crisis, is not merely the question of how to resolve the *physical* problems of resource depletion and pollution. An equally pressing issue is how to resolve the *spiritual* crisis posed by undemocratic means of safeguarding environmental stability.

If legal and political institutions do not provide incentives for voluntary limits on production, reproduction, consumption and growth, in accord with "Spaceship Ethics," then humanity may ultimately be forced to move toward more coercive forms of environmental protection. With effort and with education, however, we can avoid the problems posed by adoption of "Lifeboat Ethics." If we recognize the spiritual and personal dimensions of this crisis, we may be able to use "Spaceship Ethics" to change both our hearts and our environmental habits.

Notes

1. Cited by Lynton K. Caldwell, "The Coming Polity of Spaceship Earth," in R.T. Roelofs, J.N. Crowley, D.L. Hardesty (eds.), *Environment and Society,* Englewood Cliffs, New Jersey, Prentice-Hall, 1974, p. 250; hereafter cited as: Polity.

2. See Kenneth Boulding, "The Economics of the Coming Spaceship Earth," in Henry Jarrett (ed.), *Environmental Quality in a Growing Economy,* Baltimore, Johns Hopkins University Press, 1966, pp. 3-14; William G. Pollard, *Man on a Spaceship,* Claremont, California, The Claremont Colleges, 1967; and Barbara Ward, *Spaceship Earth,* New York, Columbia University Press, 1966. See also R. Buckminster Fuller, *Operating Manual for Spaceship Earth,* New York, Pocket Books, 1969; hereafter cited as: *Manual.*

3. Fuller, *Manual,* pp. 44-46.

4. Fuller, *Manual,* pp. 46-48 makes this same point.

5. Caldwell, Polity, p. 251, gives an account of the similarities between the earth and a spaceship.

6. Caldwell, Polity, p. 251.

7. Fuller, *Manual,* p. 47.

8. Caldwell, Polity, p. 252.

9. Fuller, *Manual,* pp. 97-99, 109-12; see also p. 12.

10. Fuller, *Manual,* p. 119; see also p. 117.

11. Fuller, *Manual,* pp. 111-13.

12. Fuller, *Manual,* p. 117. This same point is also made by Goran Backstrand and Lars Ingelstam, "Should We Put Limits on Consumption?" *The Futurist,* 8 (June 1977): 160; hereafter cited as: Consumption.

13. Fuller, *Manual,* p. 117.

14. Caldwell, Polity, pp. 258-59, and Fuller, *Manual,* p. 12.

15. Caldwell, Polity, p. 258, and Fuller, *Manual,* p. 98.

16. Caldwell, Polity, p. 252, and Fuller, *Manual,* p. 112.

17. Caldwell, Polity, p. 262.

18. See Fuller, *Manual,* pp. 96-98, 111-112, 117.

19. Caldwell, Polity, p. 261.

20. See the previous essay in this chapter.

21. Garrett Hardin, "Living on a Lifeboat," *BioScience,* 24 (October 1974): 561.

22. See Chapter 7 as well as E.C. Pasour, "Austerity, Waste, and Need," *The Intercollegiate Review,* 13 (Winter-Spring, 1978): 79-85.

23. Fuller, *Manual,* p. 97.

24. In this regard, see Backstrand and Ingelstam, Consumption, pp. 157-62.

25. See the readings in Chapters 6, 7, and 8 for discussion of this point.

26. For analysis of this point, see Chapter 4 and C.D. Stone, *Should Trees Have Standing?* Los Altos, California, William Kaufmann, 1974. See also Aldo Leopold, *A Sand County Almanac,* New York, Oxford University Press, 1949, esp. pp. 201-26.

27. For discussion of the conflict between environment-related coercion and individual liberty, see J.L. Huffman, "Individual Liberty and Environmental Regulation: Can We Protect People While Preserving the Environment?" *Environmental Law,* 7 (Spring 1977): 431-47.

28. See "Declaration on Social Progress and Development," from *Human Rights: A Compilation of International Instruments of the United Nations,* New York, United Nations, 1973, pp. 97-102.

Further Reading

Adler, C.A. 1973. *Ecological Fantasies,* esp. pp. 313-317. New York: Delta.

Allen, D.L. 1975. Man and the Environmental Ethic. In *Environmental Quality and Society,* ed. R.A. Tybout, pp. 26-43. Columbus, Ohio: Ohio State University Press.

Baier, K., and Rescher, N., eds. 1969. *Values and the Future.* New York: The Free Press.

Barbour, I.G. ed. 1972. *Earth Might Be Fair: Reflections on Ethics, Religion, and Ecology.* Englewood Cliffs, New Jersey: Prentice-Hall.

______. 1973. *Western Man and Environmental Ethics.* Reading, Massachusetts: Addison-Wesley.

Blackstone, W.T. ed. 1974. *Philosophy and Environmental Crisis.* Athens, Georgia: University of Georgia Press.

Boulding, K. 1969. The Economics of Coming Spaceship Earth. In *Environmental Quality in a Growing Economy,* ed. H. Jarrett, pp. 3-14. Baltimore: Johns Hopkins University Press.

______. 1966. *Human Values on the Spaceship Earth.* New York: National Council of Churches.

Bowman, J.S. 1976. The Environmental Movement. *Environmental Affairs* 5 (Fall): 649-667.

Buchsbaum, R., and Buchsbaum, M. 1957. *Basic Ecology.* Pittsburgh: The Boxwood Press.

Caldwell, L.K. 1974. The Coming Polity of Spaceship Earth. In *Environment and Society,* ed. R.T. Roelofs, J.N. Crowley, and D.L. Hardesty, pp. 57-62. Englewood Cliffs, New Jersey: Prentice-Hall.

______. 1970. *Environment.* Garden City, New York: Natural History Press.

Callahan, D. 1973. *The Tyranny of Survival.* New York: Macmillan.

______. 1974. Doing Well by Doing Good: Garrett Hardin's "Lifeboat Ethic." *The Hastings Center Report* 4 (December): 1-4.

Donagan, A. 1977. *The Theory of Morality.* Chicago: University of Chicago Press.

Ehrlich, P.R. 1968. *The Population Bomb.* New York: Ballantine.

______., and Harriman, R.L. 1971. Spaceship Ethics. Selection from *How To Be a Survivor: A Plan To Save Spaceship Earth,* pp. 1-18, 124-139. New York: Ballantine.

Einstein, A. 1953. The Laws of Science and the Laws of Ethics. In *Readings in the Philosophy of Science,* ed. H. Feigl and M. Brodbeck, pp. 779-780. New York: Appleton-Century-Crofts.

Elder, F. 1970. *Crisis in Eden.* New York: Abingdon.

Ellul, J. 1964. *The Technological Society.* New York: Random House.

Engel, D.E. 1970. Elements in a Theology of Environment. *Zygon* 5 (September): 216-227.

Environmental Action Foundation. 1970. *Earth Day—The Beginning.* New York: Arno.

Fackre, G. 1971. Ecology and Theology. *Religion in Life* 40: 210-224.

Ferkiss, V. 1969. *Technological Man.* New York: George Braziller.

Fuller, R.B. 1969. *Operating Manual for Spaceship Earth.* Carbondale, Illinois: Southern Illinois University Press.

Gabel, M. 1975. *Energy, Earth, and Everyone: A Global Energy Strategy for Spaceship Earth.* San Francisco: Straight Arrow Books.

Galbraith, J.K. 1967. *The New Industrial State.* Boston: Houghton Mifflin.

Glacken, C.J. 1974. Man against Nature: An Outmoded Concept. In *Environment and Society*, ed. R.T. Roelofs, J.N. Crowley, and D.L. Hardesty, pp. 17-28. Englewood Cliffs, New Jersey: Prentice-Hall.

Goodman, P. 1969. Can Technology be Humane? *New York Review of Books*, Section 13 (November 20): 27-34.

Gordon, M. 1975. Ecology as Ideology.In *Environmental Quality and Society*, ed. R.A. Tybout, pp. 59-69. Columbus: Ohio State University Press.

Grayson, M.J., and Shepard, T.R. 1973. *The Disaster Lobby: Prophets of Ecological Doom and Other Absurdities*, esp. pp. 1-20. Chicago: Follett.

Hardin, G. 1972. *Exploring New Ethics for Survival*. New York: Viking.

______. 1974. Living on a Lifeboat. *BioScience* 24 (October): 561-568.

Heilbroner, R. 1974. *An Inquiry into the Human Prospect*. New York: Norton.

Hollander, R. 1976. Ecologists, Ethical Codes, and the Struggles of a New Profession. *Hastings Center Report* 6 (February): 45-46.

Iltis, H.H. 1970. Man First? Man Last? The Paradox of Human Ecology. *BioScience* 20 (July): 820.

Kozlovsky, D.G. 1974. *An Ecological and Evolutionary Ethic*. Englewood Cliffs, New Jersey: Prentice-Hall.

Lane, L.W. 1974. An Environmental Ethic. In *Science, Technology, and Freedom*, ed. W.H. Truitt and T.W. Solomons, pp. 246-251. Boston: Houghton Mifflin.

Leonard, J.W. 1975. Ethics and Realities. In *Environmental Quality and Society*, ed. R.A. Tybout, pp. 44-58. Columbus: Ohio State University Press.

Leopold, A. 1949. The Land Ethic. From *A Sand County Almanac*, pp. 201-226. New York: Oxford University Press.

Lewis, C.S. 1947. *The Abolition of Man*. New York: Macmillan.

Meyers, C.J. 1976. An Introduction to Environmental Thought: Some Sources and Some Criticisms. *Environment Law Review—1976*, pp. 3-48. New York: Clark Boardman.

Miller, G.T. 1975. Environmental Ethics and Hope. In *Living in the Environment*, Belmont: Wadsworth, 1975, pp. 365-380.

Mitcham, C., and Mackey, R. eds. 1972. *Philosophy and Technology*. New York: Macmillan.

Nelkin, D. 1976. Ecologists and the Public Interest. *Hastings Center Report* 6 (February): 38-44.

Neuhaus, R. 1971. *In Defense of People*. New York: Macmillan.

Passmore, J. 1974. *Man's Responsibility for Nature*. New York: Scribners.

Pollard, G. 1967. *Man on a Spaceship*. Claremont, California: Claremont Colleges.

Potter, V.R. 1977. Evolving Ethical Concepts. *BioScience* 27 (April): 251-253.

Rawls, J. 1971. *A Theory of Justice*. Cambridge, Massachusetts: Harvard University Press.

Roelofs, R.T., Crowley, J.N., and Hardesty, D.L., eds. 1974. Toward an Environmental Ethic. *Environment and Society*, pp. 315-573. Englewood Cliffs, New Jersey: Prentice-Hall.

Rolston, H. 1975. Is There an Ecological Ethic? *Ethics* 85: 93-109.

Russell, B. 1951. *The Social Impact of Science*. New York: Columbia University Press.

Schedler, N.O. 1975. Our Destruction of Tomorrow: A Philosophical Reflection on the Ecological Crisis. In *Ethical Issues*, ed. W.R. Durland and W.H. Bruening, pp. 247-269. Palo Alto, California: Mayfield.

Schumacher, J.F. 1973. *Small Is Beautiful: Economics as if People Mattered*. New York: Harper and Row.

Shepard, P. 1970. Ecology and Man. In *The Ecological Conscience*, ed. Robert Disch, pp. 56-66. Englewood Cliffs, New Jersey: Prentice-Hall.

Smith, H. 1972. Tao Now: An Ecological Testament. In *Earth Might Be Fair*, ed. I.G. Barbour, pp. 62-81. Englewood Cliffs, New Jersey: Prentice-Hall.

Steffenson, D., Herrscher, W., and Cook, R. eds. 1973. *Ethics for Environment*. Green Bay, Wisconsin: University of Wisconsin.

Stone, G.C., ed. 1971. *A New Ethic for a New Earth*. New York: Friendship Press.

Ward, B. 1966. *Spaceship Earth*. New York: Columbia University Press.

SECTION TWO:

Rights, Duties, and the Environment

3

THE RIGHTS OF FUTURE GENERATIONS

Introduction

ECONOMISTS traditionally discount costs or risks that occur in the future, especially when they are evaluating long-term environmental impacts. For example, if a toxic waste dump kills one person this year, and that person's life is economically counted as $40,000 in the benefit-cost analysis, then at a 6 percent discount rate, one death 13 years from now is economically counted as approximately $20,000 today. Likewise, at a 6 percent discount rate, if one death now is counted at $60,000, then one death 20 years later is counted now as $20,000.

In other words, future costs—like death or environmental pollution—are counted by economists, in the present, as the amount of money (e.g., $20,000) needed to be invested now at 6 percent would yield approximately $40,000 in 13 years and approximately $60,000 in 20 years. In this typical example, economists assume that the value of human life, at a 6% discount rate, is worth half as much in 13 years, and one-third as much in 20 years. After several hundred years, a human life would be counted as worth less than one-millionth of what it is worth now.

Because classical economists use social discounting techniques for future risks and costs, they do not assume that the lives of future persons are as valuable as our own. Philosophers like Partridge would claim that such economists violate principles of intergenerational equity. Moreover, for exceedingly long-term environmental hazards, like storing high-level radioactive wastes, economists effectively count the risks as zero. This is because, invested at some traditional rate over thousands of years, only a tiny sum—less than a dollar—would be necessary to yield an amount valued later at billions of dollars.

Are the lives of future persons worth the same as ours? Do they have rights equal to ours? Those who claim that future persons do not have rights equal to our own use one or more variants of two common arguments. (1) They say that future persons are contingent, not actual, beings and that we do not have duties to contingent persons. Or (2) They maintain that we are ignorant of the needs and desires of members of future generations and hence have no duties to them in the present.

Defenders of our duties to posterity typically use one or more variants of three common arguments. (1) They claim that, because we often have duties not to jeopardize the welfare of contingent or unidentifiable persons in the present, likewise we have similar duties to such persons in the future. Or (2) They argue that, despite our ignorance of the future, we know enough to avid certain actions, such as storing long-lived toxins. (3) Still other thinkers argue that future persons need our care, and that our ability to meet their needs entails our duty to do so.

The two authors represented in this chapter each come to somewhat different conclusions regarding the rights of future generations and they arrive at their results by quite diverse philosophical arguments. Walter C. Wagner, a social scientist from the University of the Pacific, views possible obligations to future generations from a perspective of egoistic ethics; he denies a contractarian foundation for obligations to future generations, while I strongly affirm a social contract obliging us to posterity.

Arguing that the social contract among humans is "time bound," Wagner concludes that is no rational or logical basis for obligations to future generations. There is, however, a basis for "concern" about our descendants, he says, but only because such regard is in the self-interest of present generations. If current inhabitants of the earth engage in "empathetic" attitudes toward posterity, he maintains we will become "more dynamic, self-actualizing, mentally healthy, goal directed, organized, and integrated people."

Contrary to Wagner, I believe that there is a social contract among all humans, and that membership in the moral community need not be limited by considerations of time. This position, however, is contingent upon a number of assumptions (e.g., that past, present, and future generations share relatively the same conception of "the good life"). I maintain that, although we do have limited knowledge regarding the social ideal of posterity, our partial ignorance does not free us completely from an ethical obligation to our remote descendants. Rather, our ignorance in this respect necessitates our acting on the assumption that future social ideals are not radically different from our own. I conclude, therefore, that we must ascribe the same basic rights to future generations as those we claim for ourselves.

The interesting ethical controversy raised by my own discussion is whether one ought to adopt the morally "liberal" position of allowing our incomplete knowledge to justify our failure to consider obligations to future generations. Or, ought one to adopt the morally "conservative" position of disallowing partial ignorance as a justification for failure to consider these obligations? In other words, when in doubt, does one affirm or deny the duties in question?

One reason why it is difficult to answer this question is that affirming obligations to future generations might necessitate denying certain rights to present persons. As we shall see, I consider the ethical basis for making a tradeoff between present and furture individuals, and conclude by providing several ethical principles which have implications for the future.

Whether one agrees with Wagner, with me, or with one of the authors whose work is cited in the references following these selections, the question of rights of future generations must be considered by anyone attempting to develop a system of environmental ethics. The more basic issue, at the heart of the questions, is how society is to compute the costs and benefits of its energy/environment policies. If one maintains that a desirable ethical policy is one which maximizes the good and minimizes the evil to society, then one must decide how to define "society" and membership in it. Does one limit, in time, the class of persons whose interests are considered and the span over which costs and benefits are calculated? How we answer this question depends in part on whether our environmental ethics is based on considerations of equal justice for all generations or on time-bound calculations of the greatest good for the greatest number of those presently alive. This means that the question of the rights of future individuals inevitably draws us into a classic conflict in ethical philosophy. Shall we serve equity, and call all generations equal, or serve utility, and limit the rights of our descendants? If we listen to Edmund Burke, in *Reflections on the Revolution in France,* then "Society" is indeed a contract.

What binds present to future persons, one to another, he says, is neither temporary nor utilitarian. It is instead a primeval partnership, incapable of dissolution, which ought to be lived out in reference for all who are or will be. Let us see if Burke is correct.

Readings

1. Wagner, W.C. 1971. Futurity Morality. *The Futurist* 5 (October): 197-199. Reprinted by permission of *The Futurist,* the World Future Society, 4916 St. Elmo Avenue, Washington, D.C. 20014.

2. Shrader-Frechette, K.S. 1979. Technology, the Environment, and Intergenerational Equity. A much shorter version of this essay, "Technology Assessment, Future Generations, and the Social Contract," appeared in *Journal for the Humanities and Technology.* 1 (1979): 3 ff.

FUTURITY MORALITY

Walter C. Wagner

SHOULD WE HAVE a moral sense of obligation to deny ourselves in the present in order that men will live better at a time when we will not be here to participate? This question poses, in what may be its most puzzling form, the old issue of why we should feel moral obligation.

Illustrations of this dilemma are legion: Should disposal of war gases be so careful as to protect the future in perpetuity? Should radioactive polluting be kept at a level that will *never* become cumulatively dangerous? Why conserve our natural resources so they will outlast us? Why forgo having children to forestall over-population a hundred years hence?

If we are to have an obligation to posterity what is the basis of that moral imperative? My answer is exploratory, highly speculative and, given the complexity of the issue, frustratingly inade quate.

I am specifically focusing on the situation where the future's gains are our losses. I am excluding from this analysis problems of weighing degrees of present sacrificing for degrees of future advantages in a cost benefit conceptualization. And as a final delineation, the intriguing topic of values and morality of the future is set outside the present definition of the problem.

There are many irrational dispositions of the problem:

1. The belief in a Divine plan that will care for all of God's children who may populate the globe threatens us with disaster.

2. An ideological ghost of beneficent Natural laws purports to automatically bring men to their maximum well-being. The rise of concern for planning is a monument on the grave of this once-convincing notion.

3. The belief that we will be rewarded in an afterlife for our worldly virtues is weak, on the wane, or dead.

4. The emotionally persuasive stand that the future will be populated by our great grandchildren on whom we cannot turn our backs begs the question and is fraught with animism. Why can't we turn our backs? *That* is the question at issue. Because they are our childrens' children? If we accept such animism, why not build the case on reincarnation?

5. Possibly the most pervasive and tenacious disposition of the problem is the one that I suspect supports the sense of obligation to the future that many of us hold. The commitment is held as an imponderable; a conviction, transmitted in the cultural heritage, that is so covertly and extensively accepted (as marrying and having children) that it never occurs, within the conventional ethos, to submit the issue to analysis. It is a deeply held conviction that is a non-issue.

The anti- or non-rationalists may contend that we should feel obligation to the future because of an irrational love of man. To which a rationalist responds: "Why love irrationally? Why love men of the future? And why sacrifice for irrational reasons?" The non-rationalist rebuttal is something like this. "Of course, you rationalists can't understand. Your question, 'Why sacrifice for irrational reasons?' is stuck in the rational realm. You ask for a cold logical defense for that which has none and needs none. Here again is the limitation of rationality and science. Both are laudable in their proper place. But the realm of morality is not such a place."

It is exactly at this point that the need for fresh speculations gave birth to this paper. In light of the non-rationalist argument one could ask: Why not reject an obligation to the future irrationally and defend the rejection by a plea for irrationality? **For want of a strongly felt, explicit and defensible moral stance, sacrifices for the future may be rejected much as we seem to neglect a moral obligation toward suffering in distant parts of the globe today.** This presents us rationalists, who feel irrationally sentimental about the future, with a crucial question. Is there any relevance for the present in the threats to man's existence in the future? In a *direct dependency* sense I think not! To repeat, the present is not *directly* dependent upon the future. This is not to say, however, that we in the present derive nothing of importance to our self-love by altruism toward the future.

Rationale for Obligation to Future Lies in Nature of Man

We do have rational reasons for obligating ourselves toward a distant future. The reasons lie in the very nature of man. Though some of the details of human nature which I will develop may be open to question, I think the fundamental structure is defensible and adequate to the task.

Man has evolved bio-culturally. He is culturally dependent as well as genetically dependent. Through evolution he has become the animal genetically programmed as a tool user and tool dependent. He is most sharply distinguished from most other animals by the highly developed genetic and cultural capacity for sophisticated symbolic behavior. He is also "homo ludens." As with other animals he finds satisfaction in being active, in playing. He is possessed of "idle curiosity." Man's playing with tools and symbols has had different consequences from other animal play. He evolved as a problem solver and this has become integral to his evolution, adaptation and survival as a species. But as a symbolizing, problem-solving and play animal he has also emerged as a highly abstract puzzle poser and puzzle solving animal.

Man is also a social animal. The verdict is not in as to the influence of genes versus the influence of culture on his being socially interdependent. I do not think, however, that the central stance of this paper is dependent upon that verdict. Man, to a considerable degree, is socially organized by culturally prescribed *roles.* Societal living is role-*performing* and role-*taking.* That each is socialized by learning to *perform* many roles is now quite common knowledge. Not nearly so clearly grasped is our social and emotional dependence on socializing individuals in role-*taking.* Role-taking is the putting of oneself in the role of another in the process of social intercourse. Learning role-taking is learning to empathize, to feel for others. It has a powerful emotional as well as rational ingredient. The capacity of members of a society to relate well is a function of the sensitivity of its members in role-taking. Failure to rear individuals empathetic in role-*taking* is a failure to rear compassionate and mature individuals. They are callous and cold. Learning appropriate role-performance and role-taking requires reinforcement of good responses and discouragement of poor responses. The achievement of a more humane society will come about only by recognizing and acting on these characteristics of role.

"Social Contract" Has Rational Basis

Man, in addition to being a socially dependent role-performing, role-taking animal is also a self-loving animal with self-will, self-interest, egoistic drive and a quest for self-actualization. These self-love needs are dependent, for their realization, upon all individuals sacrificing some immediate desires in order that all may live in a society where each is sacrificing to serve others' self interests more effectively. Each has a self-love stake in his society, which transmits a sense of self-restraint and supports the imposition of constraint. The necessity that individuals be socialized to feel and act on obligations and respond to social sanctions is essential to a social order and self-actualization. The "social contract" is the pervasiveness in a social ethos of attitudes supporting socialization of society's members in obligation and sanction acceptance. The imperative for accepting moral obligation or sanctions is a derivative of the very nature of man as a culturally and socially dependent animal with strong ego-needs which can only be realized within a cultural pattern which socializes individuals to feel the importance of those conditions which indeed are important to individual and collective well-being. The imperative of moral obligation to cooperate, even through self-sacrifice, need not be founded in irrational leaps of faith. It can be well founded in self-love and the practical necessity for well cultivated empathetic or compassionate capacities of role-taking. "Man is his brother's keeper" and "no man is an island unto himself" express in edifying phraseology of fairly popular currency this rational basis for social contract.

The argument for rational moral obligation is finally reinforced by what I have termed the "morality of mutual adaptation." It is a concept of morality founded in a post-Darwinian perspective of evolution, adaptation and a kind of ecological mutuality of dependency.

The analysis thus far is inadequate to establish a *futurity* morality. Its deficiency, for this purpose, is that it is time bound, current; it establishes no dependency of the present on the future and no rational self-love grounds for altruism toward the future. The analysis to this point, if applied to the future, has the ring of the answer to the question: "Given your agnosticism, why did you attend your friend's funeral?" "Oh, if I didn't, he wouldn't attend mine."

Man Is Future-Oriented by Nature

This time-bound deficiency must now be surmounted. It is in the very nature of man to have a future focus. Decisions are normally conclusions for future actions in terms of future consequences. Even in traditional societies there was a preoccupation with futures as far forward as heaven, hell, happy hunting ground or reincarnation. Homo ludens, the culture animal, is a symbol-user, problem-poser, puzzle-solver. He probes the future in work and play, through symbolic extrapolation. We call the process "conjecture," "speculation" or "forecasting." It is inevitable in man's bio-cultural nature. The invention, spread and refinement of the symbolic processes of logic, mathematics, statistics, alphabet, printing, literacy, methods of science have led modern man to increasingly sophisticated means of probing the future (simulation games, systems analysis, Delphi technique). The decline of the heavy hand of traditionalism accompanying the emergence of secularism has given rise to a shift in man's future focus from an animistic life after death to the continuities and discontinuities of life on earth. Modern societies justify themselves in terms of their future, whether capitalists, communists, or open society advocates. We are now very much alive to a living future whose fate is out of the hands of the gods and in our own. We have met the gods and they are us. We have visited the future in our forecasts. We dwell in future communities through symbolic conjecturing. We encounter their probable problems and possibilities. We live in an ethos of increasing identification with future roles.

We Are Becoming Intimate with the Future

Forecasting, by bringing a sense of life in the future into the present, affects us in many ways. We have an enhanced or closer role intimacy with the inhabitants. We take their roles. We experience intensified awareness and identification; social role distance is narrowed. Forecasting heightens our perception of their problems. Problems are made explicit. Puzzles are posed that challenge our problem-solving intellects and tantalize our play proclivities. Our "instinct of workmanship" is activated. We press for long-range planning with future problems in mind. Homo ludens seeks high adventure in the future. In place of historical novels we have science fiction. Vicariously we have box seats from which to view the drama for which we sacrifice now to stage in the future. Emotionally we feel vicarious involvement; we enjoy and suffer for future people through role-taking empathy. Inevitably we feel an inner need of the role-performing, role-taking human animal to respond to the roles of others. It is natural to human nature.

To this point I have not established a rational reason why a self-loving animal should hold a sense of the imperative that we obligate ourselves to the future. I have indicated how we become sensitized to life in the future and why we tend to be responsive. I could suggest how we become sensitized to female pulchritude and why we may tend to act on the excitation. Devoutly as it may be wished, this does not establish a moral imperative to act!

As previously noted, the "social contract," "man is his brother's keeper," "no man is an island" and "mutual adaptation" arguments for obligation are time-bound and fail to rationally stretch current self-love altruism into the distant future. The resolution of the dilemma turns on the issue of whether there are current self-love needs of individuals and societies which make a futurity morality desirable. The case for altruism toward the distant future is based not on enhancing future well-being or avoiding future suffering per se. Instead, the dependence of the self-loving present on the future is a function of how our reacting to the future enhances self-actualization of ourselves in the present. For self-love reasons we should act toward the future so as to most fully actualize ourselves in the present. This linkage is the foundation for futurity obligation.

Concern for Future Helps Us in the Present

How then do we benefit in the present from concern for the future? Symbolic conjecturing increases our role intimacy with the future. This involves us in the intimacy of empathetic role-taking toward future peoples. We can reject responses of empathy and compassion only at the expense of present callousness. Callousness endangers present social intercourse. A compassionate humanity is imperative to the safe and satisfying self-actualization of each of us. It really does no direct harm to man to pull off one leg and wing of a fly at a time. It even helps rid us of flies. But not to correct such wanton lack of compassion in a culture and its individuals is a threat to each of us. Compassion and its sentiments are highly adaptive for the human species. Such sensitivity is instrumental to the good life now. Compassion in role-taking in the present is reinforced by understanding and empathetic role-taking responses to future roles. Goodness begets goodness.

This concludes the major argument upon which is based the contention that a futurity morality is imperative to our self-love needs.

There are supplementary supports to this futurity morality imperative. They too are based upon the fulfillments to be experienced in the present by a humanity concerned for the future.

1. Man in the quest of purpose can find fulfillments for his work and play impulses by posing and solving puzzling problems for the future.

2. Forecasting, speculating, and conjecturing about the future have a very direct aesthetic gratification, as each of us knows from pondering alternative futures.

3. Alvin Toffler notes a very utilitarian reason to express concern for the future by systematically studying it: "We are experimenting with potentially powerful new tools for probing the future; even when grossly in error, the very effort helps us identify key variables in change, it helps clarify goals, and it forces more careful evaluation of policy alternatives."

4. Abraham Maslow's observations about the focus on one's personal future can readily be seen in role theory terms to have longer-range social application. **Futurity concern makes for more dynamic, self-actualizing, mentally healthy, goal-directed, organized and integrated people.**

These many practical and inspiring influences on the present from a futurity commitment provide grounds for cultivating in each individual a deeply held sense of moral obligation to those who will be victims of our folly or beneficiaries of our well-reasoned concern. I have, I believe, done what I set out to do. I have developed a rational stand that **modern man must be moral man, toward the future as well as the present, because the practical necessity and self-interest of each make it desirable that we cultivate compassion and morality in us all.**

(Walter C. Wagner is Professor of Social Science at Raymond College, University of the Pacific, Stockton, California 95204. His paper was presented at the session on "Man's Nature and Spirit" at the First General Assembly of the World Future Society, Washington, D.C. May 14, 1971.)

TECHNOLOGY, THE ENVIRONMENT, AND INTERGENERATIONAL EQUITY

K.S. Shrader-Frechette

Introduction

IN 1942, Paul Muller received patents for his discovery of DDT; since that time more than 4.4 billion pounds of the toxic chemical have been spread around the globe. It has saved many thousands of persons from death by malaria or typhus, and has protected the food supplies of millions from the ravages of insects. Also in 1942, Enrico Fermi and his team produced the world's first nuclear chain reaction. As a consequence of this technological advance, the US was able to obtain needed plutonium for the two atomic bombs used in World War II and to encourage commercial uses of nuclear energy; currently, 65 fission reactors produce approximately 8 percent of this country's electricity.[1]

Now almost forty years later, in the afterglow of the Nobel Prize awarded for Muller's discovery and the acclaim given to the weapon that allegedly ended the war, doubts have arisen regarding both these technologies. DDT has been discovered in the polar ice caps, and radiation has been detected in mothers' milk. Perhaps the most important questions, about both the fission and the DDT technology, are a consequence of the longevity of the hazards they have introduced. Both DDT and the plutonium byproduct of nuclear fission have essentially permanent toxicity; they can, and possibly will, cause cancer, genetic damage, and even death among humans inhabiting the earth a hundred thousand years fom now. For this reason they provide archetypal cases for discussing an unanswered ethical question. Do we have obligations to future generations?

Answering this question is not difficult, if one thinks only of his own children or grandchildren. In fact, many of us have experienced that magnificent moment when the cry of our just-born, unwashed child pierces our fragile and sometimes uncommitted presentness and exchanges it suddenly for a stake in the future. Thinking of generations thousands of years hence, however, is difficult. Do they have any claims on our moral sensibilities?

On the one hand, it seems puzzling to affirm that *future* generations have *present* rights to such things as equal protection guaranteed by the Fifth and Fourteenth Amendments. Moreover, although it is certainly desirable that future individuals have the same protection from carcinogens that we now enjoy, their safety might be bought at the expense of technological developments whose benefits to future people could outweigh the costs. The security of remote descendants might also be bought at the expense of the lives of members of present generations who depend, for example, on DDT for disease control and for achieving the green revolution in a world in which two-thirds of the population is malnourished.

On the other hand, perhaps Albert Schweitzer was right when he urged humankind to adopt a "reverence for all life," and not to limit ethical considerations merely to present persons in specified situations. Even a utilitarian ethic based on "the greatest good for the greatest number" seems to require one to recognize the interests of future generations. Besides, is not individual and societal progress measured in part by a heightened moral sensitivity to the interests of a growing number of persons? If it is, and if future generations are said to have certain rights, then determining ethical answers to pressing problems of technology assessment will be far more complex an enterprise then it would have been otherwise. Conversely, if future generations are said not to have certain rights, then public policy regarding technology may be formulated under considerably fewer constraints than it might if all effects on future persons had to be evaluated.

According to several policymakers from the Environmental Protection Agency,

> ... the degree of risk that the ... generation passes on to the future represents an important legacy .... An ethical basis for decisions regarding risk transference is needed not only for philosophical reasons, but also for the practical purpose of implementing evaluation techniques such as cost-effectiveness and risk-cost analyses. Unfortunately, society has not established clear approaches for dealing with the imposition of such risks far into the future.[2]

This is a significant omission in current policy regarding technology, say government officials, since "a long term view of risk may result in different choices [of technologies] than would result from a shorter term consideration."[3]

If the rights of both present and future generations are not clarified relative to each other, then there is great danger that US public policy will continue on its present incoherent path. This course is unsatisfactory, in part, because its ends are often not consistent with the means taken to achieve them. Current US public policy, for example, affirms that technology should not impose risks on "a future generation ... greater than those acceptable to the current generation."[4] Yet, at the same time, government regulations allow emissions of pollutants whose carcinogenic and mutagenic effects are essentially permanent (i.e., lasting a million years) and cumulative; hence it is impossible that such emissions will not put a greater risk on posterity than would be acceptable to current persons.[5] For all these reasons, it is vitally important to determine whether future generations may be said to have rights.

In the ensuing considerations, we will assume that 'future persons' refers to all really existing individuals who will eventually be born as viable human beings. Thus those people whose existence is either debatable or merely possible are not the subject of our inquiry; obviously it would be difficult to speak of the rights of "possible" persons. (This definition will also allow the discussion to sidestep a lengthy treatment of a difficult but only partially related topic, viz., abortion.[6])

Further, when we ask whether future individuals may be said to have rights, we mean by 'rights' those which are fundamental to human existence.[7] These include, primarily, the rights to life, to freedom of speech, press, ownership, and assembly, as well as those to due process and to equal protection under the law. Although justification for terming these "fundamental rights" is too lengthy to be given here, we have defined them as such, both because they are contained in the US Bill of Rights and because the governments of many peoples have affirmed that all persons possess them.

Future Generations and the Social Contract

When one raises the question of the rights of future generations, at least three important issues come to mind. First, is there an ethical framework within which the existence of such rights can be rationally substantiated? Secondly, assuming these rights exist, is it possible either to calculate or to meet the legitimate interests of all persons, present and future? Thirdly, would recognition of the rights of future generations diminish the extent to which the needs of current individuals (especially the poor or socially disenfranchised) were served?

In the following pages I will discuss all three of these questions, although my emphasis will be on the first one, since it is both logically and chronologically prior to the other two. Moreover, although there are numerous ethical frameworks (e.g., altruism, natural law theory, rule- or act-utilitarianism) within which the question of the rights of future persons may be investigated, I will analyze primarily the social-contract basis for affirming such rights. The main reasons for discussing the contractarian view are that it is easily understood and common-sensical, it has a long tradition in the history of ethics, and most importantly, it is the camp from which many of the major attacks on the existence of the rights of future persons have issued. Hence, an important parameter for determining whether present individuals have certain obligations, because of the rights of future persons, is the extent to which current and forthcoming generations share a social contract. Are both members of the same "moral community"?

According to M.P. Golding, who wrote one of the first and major analyses of the rights of future generations, a moral community may be constituted in one of two ways. It may be initiated either by "an explicit contract between its members" or by "a social arrangement in which each member derives benefits from the efforts of the other members."[8] Since it is impossible for present persons to enter into an explicit, legal bargain with future individuals, several authors have maintained that social contracts between current and forthcoming persons are a chronological impossibility; moreover, they argue, the second type of moral community is also impossible, since present individuals can benefit future ones, but not vice versa. Because remote generations can share neither type of moral community with us, they reason, members of distant posterity cannot be said to have rights based on a social contract or arrangement.[9]

Some authors maintain that, even though future persons do not have rights based either on an explicit contract or on a reciprocal agreement, they might have rights equal to ours, provided that they share the same interests or social ideal as we.[10] In other words, if our remote descendants share the same conception as we for the good life of humans, then altruism and fellow-feeling urge us to affirm that they are members of the same moral community, and therefore have the same rights as we. In his classic article, Golding argued that, as a result of the complexity and rapidity of technological change, it is impossible to know the conditions of life of our remote descendants. Because it is impossible to foresee whether they will maintain the same conception of the good life as we, Golding concluded that they could not be said to share our social ideal and hence the same rights as we.[11]

Interesting as is the case against the existence of rights for future persons, it suggests investigation of at least three questions whose answers could lead one to affirm that members of future generations do have rights. First, is intergenerational reciprocity impossible, as Golding and others say it is? Secondly, is explicit reciprocity a necessary condition for all social contracts based on self-interest? And thirdly, is it plausible to reason that we do not know whether future persons will have the same conception of the good life, and therefore that they cannot be said to have rights bearing on our present obligations? Let us examine these questions in sequence.

There are several considerations suggesting that intergenerational reciprocity might be possible and future generations might be said, indirectly, to benefit us in exchange for our recognition of their alleged rights. The first type of reciprocity has been described by social scientist Walter Wagner, who argues that if we recognize the rights of future persons, we obtain in return a greater degree of happiness and self-actualization. Active concern for the interests of our remote descendants, says Wagner, increases our empathy and our compassion and thus benefits us as individuals.[12] Hence, indirectly, future persons might be said to have a reciprocal relationship with some current individuals.

A more frequently discussed, and perhaps more important, case of intergenerational reciprocity is based on the notion that since our forebears have benefitted us in numerous ways, we have obligations to aid remote posterity. In other words, the presupposition of this social contract is not that generation A benefits generation B and vice versa, but that A benefits B, B benefits C, C benefits D, and so on. One writer has cited the rationale behind this ancient notion of intergenerational dependence by quoting an old manuscript of Benedictine monks, "On the Conservation of Pine Forests":

> ... no one who plants a fir-tree can hope to fell it when it is fully grown, no matter how youthful the person is. In spite of this the most *sacred obligation* is to replant and husband these pine forests. If we sweat for the benefit of posterity, we should not complain as we reap the results of the efforts of our forefathers.[13]

Or, as Peter Faulkner put it: "This generation owes to posterity concern of the same quality and degree that [our ancestors] ... devoted to all generations following theirs and that made our present happiness possible."[14]

This particular concept of intergenerational reciprocity appears to have been even more explicitly formulated in terms of the Japanese concept of *on,* whose meaning is close to the Western notion of "obligation"; one author, discussing *on,* wrote: "One makes past payment on *on"* to one's ancestors "by giving equally good or better" things to posterity. Thus the obligations one has to past generations "are merely subsumed under *on*" to future generations.[15] In other words, future persons may be thought of as proxies for past persons to whom we are indebted; we repay these debts by doing for our descendants what our ancestors have done for us.

If one accepts either Wagner's, Faulkner's, or the Japanese notion of intergenerational reciprocity, then perhaps there are grounds for affirming that present and future persons share a social contract and hence some bases for affirming that future generations may be said to have rights. Even if one does not accept their reasoning, however, one might still have reasons for believing that social contracts can exist between present and future generations. This brings us to a second question raised by the case against the rights of remote descendants. Is explicit reciprocity a necessary condition, as Golding and others say it is, for all social contracts based on self-interest? There are a number of considerations suggesting that reciprocity is not necessary for all moral communities. John Rawls, for example, author of perhaps the most widely discussed current ethical theory, proposes a "thought experiment" which shows that the social contract is not based on reciprocity but on moral reasoning. Although Rawls' account is somewhat complex and hypothetical, the ethical reasoning he advocates is based on a familiar insight (viz., the duty to base one's contracts on the golden rule, on walking in the other person's shoes).

According to his view, one may think of a purely hypothetical situation (1) in which each of us seeks to protect our possible interests but (2) in which none of us knows his particular place in society, natural ability, assets, or liabilities. Rawls calls a situation in which (1) and (2) are the case, "the original position." If one were in "the original position," he says, one would formulate a social contract based both on reason and on self-interest. Rawls says this is because:

> ... principles of justice are chosen behind a veil of ignorance. This ensures that no one is advantaged or disadvantaged in the choice of principles by the outcome of natural chance or the contingency of social circumstances. Since all are similarly situated and no one is able to design principles to favor his particular condition, the principles of justice are the result of a fair agreement or bargain.[16]

Thus, if one thinks of all humans as being in "the original position," it is clear, provided that they are capable of rational deliberation, that they all would agree on the same set of moral principles based on equity. For example, if all members of the human race were in the hypothetical "original position," then no one would know of which generation he were a member.[17] Because of this ignorance, the only reasonable moral principle for any person to follow would be that members of all generations ought to have equal rights. As Rawls puts it, no rational person would agree to a contract that would allow "an enduring loss for himself in order to bring about a greater net balance of satisfaction"; hence, all persons would choose to follow principles of "equality in the assignment of basic rights and duties."[18] Thus if one conceives of an alternative account of social contracts, not based on reciprocity but instead on rationality, self-interest, and justice, then one has a contractarian foundation for rights of future generations.

Less abstract and hypothetical than Rawls' account, Daniel Callahan's view of the social contract also does not involve notions of reciprocity. His example is of the parent-child relationship; it deserves special treatment here, not only because it might provide a counter-example to Golding's claim, but also because it might serve as a model for understanding the rights of present and future persons.

According to Callahan, certain types of contracts come into being, not because of any agreement based on prearranged reciprocal benefits, but simply because one party to the contract chooses to accept an obligation. Children are not asked whether they wish to be born, says Callahan, but the fact that parents take upon themselves an obligation toward them means that a contract between the two has been initiated. Anyone who values his life, he says, has thereby accepted an implicit obligation to his mother and father. This partnership or contract exists, says Callahan, because children are in debt to their parents.[19]

In describing the parent-child relationship, Callahan has provided an example of a social contract with at least two distinctive features. First, the children are not asked if they wish to be a part of the agreement. Secondly, the parents' obligation is not contingent on the fact that the child reciprocates what has been done for him, since presumably they would still have the duty to care for their offspring even if he or she were unable or unwilling to reciprocate in any way. Hence, if one accepts Callahan's apparent assumption, that an implicit contract may be initiated when one person is knowingly or unknowingly put in debt to another,[20] then he has an example of a contract not necessarily involving notions of reciprocity. As Callahan suggests, this model may provide a prototype for relationships between present and future generations, since the latter (like offspring) are not asked whether they wish to enter an agreement and either cannot or may not return the benefits given them. Moreover present generations (like parents) may not always know what future generations (like children) need. Just as lack of this complete knowledge does not remove parental obligations, so also it could be argued that it does not eliminate duties to the future.[21]

What of the contention that, because we do not know whether future persons will have our same conception of the good life, therefore they cannot be said to have rights bearing on our present obligations? On the one hand, it appears plausible to affirm that we do not know what to desire for the future. For example, we do not even know whether we have an obligation to future generations to maintain current genetic quality, because a particular "chromosome is 'deleterious' or 'advantageous' only relative to given circumstances," and we do not know the conditions of the future.[22] On the other hand, however, it seems that members of future generations do have a number of "proxies" to speak now about their conception of the good life. Although we proxies are uncertain about whether the good life in the future will include use of great quantities of fossil fuel, for example, we do know, says Joel Feinberg, that future persons "will have an interest in living space, fertile soil, fresh air, and the like."[23] Hence, he concludes, "Whoever these human beings may turn out to be, and whatever they might reasonably be expected to be like, they will have interests that we can affect, for better or worse right now ... . The fact of their interest-ownership is crystal clear, and that is all that is necessary to certify the coherence of present talk about their rights."[24] In other words, Feinberg claims that the specific nature of the interests of future generations is not significant, but only that they have interests. Hence he believes that knowing the exact nature of our descendants' notion of the good life is not necessary for awarding them rights; rather, all that is required is our knowledge that they will have a conception of the social ideal which we can affect.

Feinberg's answer, however, does not resolve the difficulty raised by those who deny that future persons have rights. What good does it do to affirm that future generations have interests and rights which we can affect, and therefore that we have a duty to safeguard them, if we do not know, specifically, what we are obligated to protect? Also, if we assume, with Feinberg, that future persons will have an interest in "living space, fertile soil, fresh air, and the like," then we will not have answered the critical question but only begged it. In other words, we will not have said how or why we know what the interests of distant generations are, but we will have taken it for granted that some of their needs will be the same as ours. Given rapid technological change and our ability to compensate for certain environmental mistakes, this might be a false assumption.

The real error of the argument Feinberg attempts to refute, however, is the presupposition that, because we are in ignorance about the specific welfare of future generations, we are therefore justified in assuming that their interests will not be the same as ours. This presupposition comes down to commission of the fallacy of *argumentum ad ignorantiam,* the argument from ignorance. From the premise that we are ignorant about *x,* it never follows that *x* is or is not of a particular nature or that we are justified in assuming that *x* is or is not of a particular nature. From such a premise, nothing follows. On the same basis, our ignorance about the specific welfare of future generations would not logically justify our concluding that those interests are the same as ours. Where, then, does all this leave us? Is it accurate to assume that, because we do not know whether future persons will have our same conception of the good life, therefore they do not belong to the same moral community and consequently they cannot be said to have rights?

There are several reasons for believing that our ignorance in this regard does not enable us to affirm that future generations have no rights. Although it is true that we do not know what, positively, is in the interests of our descendants, we *do* have some idea of what might be very dangerous to them.[25] We know, for example, of the long-lived toxicity of both plutonium and DDT. Since both hazards have harmful mutagenic and carcinogenic consequences, it is implausible to assume that increasing the permanent contamination of the planet by these substances will not adversely affect the interests of future generations. Thus, in at least some circumstances, present knowledge of harmful, long-term consequences might be sufficient to enable us to claim that future generations have an interest in not being subjected to higher incidences of cancer and genetic damage. And if they have such an interest in common with us, then it is conceivable that, at least in this sense, we are members of the same moral community and have equal rights to protection against similar damages.

In the same vein, although the particulars of some future conception of the good life might not be known to us now, on the basis of Rawls' thought experiment, we do know that any person (past, present, or future) is likely to desire an ethical code based on equity. This means, for example, one which does not permit treating humans as means, leaving one's debts to others to pay, distributing resources inequitably, ignoring due process, or failing to protect the utterly helpless members of society. In other words, we still have a great deal of information about what we ought not to do, even though we do not (specifically) know what to do on behalf of later generations. If present persons pollute the atmosphere with growing amounts of permanent carcinogenic and mutagenic materials, then it is difficult to see how this behavior would not constitute both using future generations as means for the ends of present generations, as well as leaving our debts for our descendants to pay.[26] Obviously, people in Rawls' "original position" would condemn such an action. Likewise, if present individuals were to use the lion's share of nonrenewable resources for which there were no known substitutes, then this would be a clear violation of principles of distributive justice which everyone would condemn, were they in "the original position." Similarly, since it will be impossible for future persons to collect damages for what current individuals have inflicted on them (e.g., permanent carcinogens), then it is likely that a great many actions of this generation constitute imposing a degraded environment on posterity without due process, a constitutional right. Because of the impossibility of providing complete due process to all our descendants, they are, in at least one sense, in a position of utter helplessness. And, as Hans Jonas points out, "Utter helplessness demands utter protection"; this is an "inflexible principle."[27] Thus if we are in ignorance regarding the social ideal of future generations, we nevertheless have an obligation to provide them with "utter protection." This is another way of saying that, since future generations are helpless to change what we force on them, it is incumbent on us to impose lesser risks on them than those we would voluntarily take on ourselves, since they must accept such hazards involuntarily. As William

Lowrance put it, "Risks that citizens have to bear involuntarily are the ones the government has the greatest responsibility to regulate."[28]

There is also another reason for arguing that, even if we are in ignorance about the social ideal of our remote descendants, this is not sufficient grounds for concluding that their possible interests ought to be ignored. If we admit that since we do not know "the good life" for future generations, therefore we cannot "plan for them" by giving them certain rights,[29] then by virtue of consistency, it is also incumbent on us not to plan for them in a different way. In other words, if one follows the logic of those who deny that future persons have rights, then one is also bound not to destroy the possibility of distant generations' planning for themselves. Obviously certain of our acts in the present could preempt the choices of future generations.[30] But such preemption would be morally unjustified if we did not know whether we were destroying the possibility for future persons' attaining their social ideal, whatever it might be. Thus even if we admit our ignorance about the interests of later generations, they have rights not to have damages inflicted upon them, and rights not to have their choice of "the good life" preempted.

A final consideration is that, since we cannot know the social ideal of future generations, the morally safe course to follow is to assume it will not be different from our own. If we do something in the present which would be extremely hazardous to persons in the future, if they shared our conception of the good life, then it would be morally irresponsible to assume that our behavior is totally acceptable. It would not be unlike using a gun to shoot at moving underbrush, before deciding whether the source of the motion were our hunting partner or a game animal. In other words, when a situation of ignorance exists, the morally responsible course is to follow the position least likely to violate possible rights. As Callahan says, since we cannot know what the social ideal will be for future generations, "we should act on the assumption that it will not be all that dissimilar from our own."[31]

The Problem of Priorities: Present Persons vs. Future Ones

There are at least two problems, however, with affirming the morally conservative position of behaving as if the rights of later generations were equal to our own. These are that, first, it is impossible to calculate, much less meet, the interests of all future individuals;[32] secondly, the rights of existing persons, especially the poor or socially disenfranchised, might be thereby diminished.[33] Suppose, for example, that present persons were said to have an obligation, on behalf of the interests of distant generations, to limit the number of their offspring. Although zero population growth might seem to provide for intergenerational equity and for more equal opportunity to use the goods of the earth, such a policy might also violate the rights to life of current members of Third-World or agrarian societies. Their only hope for survival in old age might be to produce many children who could care for their long-term interests.[34] As Golding points out, "We shall have to be highly scrupulous in regard to anything we do for any future generation that could adversely affect the rights of an intervening generation. Anything else would be 'gambling in futures'."[35]

Although there is no simple solution to the problem of conflicts between the interests of present and future generations, several considerations provide a helpful perspective on this situation. First, if Rawls' reasoning (discussed earlier) is correct, then the only rational and disinterested way to resolve such a controversy would be to agree to principles of equal rights governing all generations. This does not mitigate the suffering of present persons, but it does force one to realize that failure to assent to principles of equity might result in even greater suffering.

A second consideration is that the moral choices available are not either to recognize the interests of present individuals, exclusively, or to acknowledge those of future persons. The situation is far more complex than this. For example, suppose one argued that since amounts of fossil fuel are finite, therefore we ought to recognize the interests of later generations and limit use of such fuel by increasing its price. Suppose, on the other hand, one argued that, since accelerating energy costs hurt primarily the poor and the middle class,[36] therefore we ought to acknowledge the well-being of present persons and not use price increases as a means of lowering fuel consumption. From this example, it is clear that many other alternatives are possible besides the two mentioned. Some of these options include reducing the use of fossil fuel by means other than increasing its price (e.g., by conservation, rationing, energy stamps, or utility rate reform).[37] Hence, a second insight about intergenerational conflicts is that they become more difficult the less we consider all alternative ways of resolving them. Moreover, perhaps the best of such options provides for the interests of all generations.

If Barry Commoner and others are correct, protecting the rights of future generations to the goods of the earth is the only authentic way to safeguard the well-being of present persons. When the productive system as a whole "borrows" from the ecosystem and incurs a debt to the future represented by pollution, says Commoner, it is true that this debt is "an immediate saving for the producer." He points out, however, "Pollution often adds to the living costs of the population as a whole, most of whom are wage earners rather than entrepreneurs."[38] Thus, when the workers in the vicinity of a power plant find their laundry bills or medical expenses increased because of soot emitted by its stacks, their wages are reduced, even if only by a small degree. Especially in the case of air pollution, it is the poor who bear the health risks associated with it.[39] If this is true, then protecting the rights of the poor is also a means of safeguarding those of future generations and vice versa.

Commoner's analysis suggests a third insight (viz., that distribution of the goods of the earth is not basically a question of whether to serve present or future generations, but a problem of whether to allocate goods in terms of profit or need). If profits determine who receives the costs of pollution and the benefits of resources, then difficulties with distribution are inevitable, regardless of whether one is concerned with intergenerational or intragenerational conflicts. In our own generation, for example, we have a moral dilemma posed by finite energy resources. On the one hand, the US, with 5.8 percent of the world's population, consumes 33 percent of its fossil fuels. On the other hand, 80 of the 100 developing nations of the world are nonexporters of petroleum.[40] Which countries have the right to the remaining energy reserves, those who can pay for them or those who (so far) have had the least share of them? Likewise, should reserves be given to those countries who need them the most, who have access to them, or who would waste them the least? When considering questions of intergenerational distribution, the same problems arise. One key to reconciling the interests of present and future generations, it follows, is to realize that, like intragenerational balancing, it is both an old problem and a difficult one. Violating principles of justice in intragenerational conflicts, moreover, can exacerbate similar intergenerational conflicts. If urban poor, for example, are treated unjustly with respect to their equal rights to protection from polluted air, then it is difficult to see how they can take seriously the rights of remote descendants to clean air. For this reason, injustice to present persons has to be attacked along with problems of injustice to future individuals.

While the preceding considerations do not answer the question of how to resolve intergenerational conflicts, they do given some perspective on the problem. Moreover, persons involved in both intra- and intergenerational dilemmas would probably argue that although equity ought to be pursued as a goal, those individuals with the most immediate needs ought to be recognized before those with more remote interests. That is, in problems of intragenerational ethics, we would probably sanction the necessity to meet actual requirements (e.g., for food today) before alleviating more remote needs (e.g., for food next year). This situation, however, would probably cause us to say, neither that no requirements existed for next year, nor that they did exist but could be completely ignored. Likewise, in problems of intergenerational ethics, one should rightly sanction the necessity of meeting actual exigencies of present individuals before attempting to alleviate the potential needs of future persons. This situation, however, ought not to cause us to say, either that there are no requirements for future persons or that they do exist but can be completely ignored. In other words, just as there is no irreconcilable opposition between current and remote requirements of present generations, so also there is no irreconcilable opposition between the actual needs of present individuals and the potential interests of future persons.

Conclusion

If the arguments provided here are correct, then there is both a contractarian basis for affirming that members of future generations may be said to have rights, and some precedent in *intragenerational* conflicts for gaining insight into *intergenerational* disputes. Moreover there is a philosophical basis for recent government policy which speaks of the "responsibility each generation has to succeeding ones."[41]

If my remarks are correct, then government policymakers, attorneys, political scientists, and humanists can no longer assume that there is no specific social contract binding members of present and future generations. But if there is an intergenerational social contract, then public policy concerning both the environment and technology must change, and change drastically. The Nuclear Regulatory Commission can no longer refuse to calculate carcinogenic effects of radioactive pollution for periods longer than 100 years, on the grounds that its refusal constitutes no violations of constitutional rights. The Department of Agriculture and the Food and Drug Administration can no longer ignore long-term mutagenic effects of pesticide use on future generations, on the grounds that such considerations are not part of Fifth- and Fourteenth-Amendment provisions for equal protection.

If there is a rational basis for extending the notion of "social contract" to include posterity, then widespread changes must be made, not only in the legal and constitutional theories which define "equal protection," but also in the technological and environmental practices whose morality and legality are specified by such theories. This means that long-term technological "debts" (e.g., from pesticides or radioactive wastes) can no longer be "charged" to future generations. If they cannot, then society must abandon all technologies (e.g., nuclear energy) which involuntarily impose their own debts on future persons. Rather, as John Buchan, former Governor General of Canada put it, we ought instead to bestow gifts on those who come after us, in exchange for what we have received from those who have come before us.

Notes

1. Statistics regarding use of DDT were obtained from William W. Lowrance, *Of Acceptable Risk: Science and the Determination of Safety* (Los Altos, William Kaufmann, 1976), p. 173; hereafter cited as: *Safety*. Information about nuclear installations and capacity were found in the testimony of L.V. Gossick, Executive Director for Operations, Nuclear Regulatory Commission, to the Committee on Interstate and Foreign Commerce, *Nuclear Waste Management and Disposal,* Hearings Before the Subcommittee on Oversight and Investigations of the Committee on Interstate and Foreign Commerce, House of Representatives, 95th Congress, First Session, July 29 and August 1, 1977, Serial 95-67 (Washington, Government Printing Office, 1977), p. 103.

2. J.E. Martin, US EPA, *Considerations of Environmental Protection Criteria for Radioactive Waste* (Washington, Government Printing Office, February, 1978), p. 17.

3. Stanley Lichtman, US EPA, "Risk in Radioactive Waste Management," in US EPA, *Proceedings of a Public Forum on Environmental Protection Criteria for Radioactive Wastes,* 30 March-1 April 1978, ORP/CSD-78-2 (Washington, Government Printing Office, May 1978), p. 33

4. Environmental Protection Agency, "Criteria for Radioactive Wastes," *Federal Register,* 43, No. 221, Part 9 (November 15, 1978): 53267; hereafter cited as: Criteria. The National Environmental Policy Act (NEPA), enacted in 1969 (42 U.S.C. § 4321 *et seq.* 1970), also points out that it is an important national goal to "fulfill the responsibilities of each generation as trustee of the environment for succeeding generations." (EPA, Criteria, p. 53262.)

5. One example of legally-allowed emissions of pollutants (placing a greater burden on future generations than on present ones) is provided by radioactive isotopes dispersed throughout the nuclear fuel cycle. Although small amounts of tritium, ^{85}kr, ^{129}I, and ^{239}Pu, for example, are released annually and said to cause "no immediate health hazard," in the next 100 years, the cumulative and permanent toxicity of allowable emissions of these four isotopes from all nuclear fuel cycles will have resulted in a total of 30,530 deaths worldwide. (ERDA, *Final Environmental Impact Statement: Waste Management Operations, Idaho National Engineering Laboratory,* ERDA-1536 (Springfield, Virginia, National Technical Information Service, September, 1977), pp. III-103 and III-104.)

6. This means that we will sidestep the question of whether there is a "right to be born" or a "right to come into being," and that instead we will ask whether real future persons have rights. For further discussion of the relation of the abortion issue to that of the rights of future generations, see R.M. Green, "Intergenerational Distributive Justice and Environmental Responsibility," *Bioscience,* 27, No. 4 (April 1977): 261; hereafter cited as: Justice. See also Joel Feinberg, "The Rights of Animals and Unborn Generations," in T.A. Mappes and J.S. Zembaty (eds.) *Social Ethics* (New York, McGraw-Hill, 1977), pp. 358-59; hereafter cited as: Rights; and Daniel Callahan, "What Obligations Do We Have to Future Generations?" *The American Ecclesiastical Review,* 164, No. 4 (April 1971): 265-68; hereafter cited as: Obligations.

7. Callahan, Obligations, pp. 265-80, also defines the rights of future generations in terms of fundamental rights possessed by present persons. He argues that "while the rights of the living should take precedence over the rights of unborn generations, the living have an obligation to refrain from actions that would endanger future generations' enjoyment of the same rights that the living now enjoy."

8. M.P. Golding, "Obligations to Future Generations," *The Monist,* 56, No. 1 (January 1972): 91; hereafter cited as: Obligations.

9. Golding, Obligations, p. 91, and J.B. Stearns, "Ecology and the Indefinite Unborn," *The Monist,* 56, No. 4 (October 1972): 612-25.

10. Golding, Obligations, pp. 91-95, and Feinberg, Rights, pp. 358-59.

11. Obligations, p. 98.

12. W.C. Wagner, "Futurity Morality," *The Futurist,* 5, No. 5 (October 1971): 199; hereafter cited as: Morality.

13. Van Rensselaer Potter, "Evolving Ethical Concepts," *BioScience,* 27, No. 4 (April 1977): 251.

14. P. Faulkner, "Protection for Future Generations," *Frontiers,* 42, No. 4 (Summer 1978): 35.

15. Cited in Callahan, Obligations, p. 271.

16. John Rawls, *A Theory of Justice* (Cambridge, Harvard University Press, 1971), p. 12; hereafter cited as: *Justice.*

17. Rawls, *Justice,* pp. 284-93.

18. Rawls, *Justice,* pp. 14-15. An additional consideration, suggested by Green, Justice, p. 261, is that persons in the "original position" would choose to be governed by principles of equity across generations, since such a plan would reduce societal conflict. Were such principles not chosen, says Green, than all persons might be subject to the chaotic, Hobbesian "war of all with all."

19. Callahan, Obligations, p. 270.

20. Vincent Ruggiero and W.D. Ross are two ethicians who share Callahan's apparent belief that an implicit contract is initiated whenever one person is knowingly or unknowingly put in debt to another person. Ruggiero and Ross speak of the "obligation of gratitude" that follows from such a relationship, even when both parties to the "contract" do not explicitly agree, beforehand, on its terms or even on its existence. See V.R. Ruggiero, *The Moral Imperative* (Port Washington, N.Y., Alfred, 1973), pp. 75-80, and W.D. Ross, *The Right and the Good* (Oxford, Clarendon Press, 1930), Chapter 2.

21. Callahan, Obligations, p. 280, discusses this point.

22. Golding, Obligations, p. 97.

23. "The Rights of Animals and Unborn Generations," in T.A. Mappes and J.S. Zembaty (eds.), *Social Ethics: Morality and Social Policy* (New York, McGraw-Hill, 1977), p. 358; hereafter cited as: Rights.

24. Joel Feinberg, Rights, p. 359.

25. This same point is also made in Callahan, What Obligations, p. 272.

26. J.L. Huffman, "Individual Liberty and Environmental Regulation," *Environmental Law,* 7, No. 3 (Spring 1977): 442, for example, says that nuclear waste is a debt charged by this generation against future generations.

27. Hans Jonas, "Philosophical Reflections on Experimenting with Human Subjects," in K.J. and P.R. Struhl (eds.), *Ethics in Perspective* (New York, Random House, 1975), p. 348.

28. Lowrance, *Safety,* p. 88.

29. Golding, Obligations, pp. 97-98.

30. A similar point is made by Callahan, What Obligations, p. 273.

31. Callahan, What Obligations, p. 274. Some persons have argued, however, that since technology will be more highly developed in the future, therefore our descendants will not necessarily be harmed for our failure to look after their interests. While this argument is correct, it still misses the point, viz., "we cannot be at all certain when and whether those advances will take place, or what form they will assume ... . The uncertainties ... remain." (John Passmore, "Conservation and Posterity," in T.A. Mappes and J.S. Zembaty (eds.), *Social Ethics,* p. 349.

32. Although they do not justify the morality of doing so, several authors point out why such calculations are difficult. See, for example, Kenneth Goodpaster and Kenneth Sayre, "An Ethical Analysis of Power Company Decision-Making," in K.M. Sayre (ed.), *Values in the Electric Power Industry* (Notre Dame, University Press, 1977), p. 268.

33. A number of authors have gone so far as to term environmentalists' concern for future generations "elitist." See John Maddox, *The Doomsday Syndrome* (London, Macmillan, 1972), esp. pp. 234-37; and Richard Neuhaus, *In Defense of People* (New York, Macmillan, 1971), esp. pp. 106-139; 152-81.

34. Green, Distributive Justice, p. 261, makes a similar point.

35. Golding, Obligations, p. 78.

36. In the recent oil crisis, for example, "price increases resulted in a 4.8 percent reduction in the disposable income of poor families, while the affluent had only a 0.5 percent reduction." (K. Seelman, "The Ethical Implications of Energy Production and Use," *Solar Age,* 3, No. 10 (October 1978): 18; hereafter cited as: Implications.)

37. Some far-reaching alternatives are mentioned by K. Seelman, Implications, pp. 16-21.

38. Barry Commoner, *The Closing Circle* (New York, Bantam, 1974), pp. 270-71.

39. Julian McCaull, "Discriminatory Air Pollution," *Environment,* 18, No. 2 (March 1976): 26-31; Sam Love, "Ecology and Social Justice: Is There a Conflict?" *Environmental Action,* 4, No. 6 (August 5, 1972): 3-6; Virginia Brodine, "A Special Burden," *Environment,* 13, No. 2 (March 1971): 22-33; P.P. Craig and E. Berlin, "The Air of Poverty," *Environment,* 13, No. 3 (June 1971): 2-9.

40. Cited by Seelman, Implications, p. 19.

41. Environmental Protection Agency, "Criteria for Radioactive Wastes," *Federal Register,* 43, No. 221, Part 9 (November 15, 1978): 53266.

Further Reading

Blackstone, W. 1977. "Ecology and Rights, In *Social Ethics,* ed. T.A. Mappes and J.S. Zembaty, pp. 340-347. New York: McGraw-Hill.

Cooley, R.A., and Wandesforde-Smith, G. 1970. Congress and the Environment of the Future. In *Congress and the Environment,* ed. R.A. Cooley, and G. Wandesforde-Smith. Seattle: University of Washington Press.

Delattre, E. 1972. Rights, Responsibilities, and Future Persons. *Ethics* 82 (April): 254-258.

Doeleman, J. 1980. On the Social Rate of Discount. *Environmental Ethics* 2, No. 1 (Spring): 45-58.

Faulkner, P. 1978. Protection for Future Generations. *Frontiers* 42 (Summer): 35-37.

Feinberg, J. 1974. The Rights of Animals and Unborn Generations. In *Philosophy and Environmental Crisis,* ed. William T. Blackstone, pp. 43-68. Athens, Georgia: University of Georgia Press.

Golding, M.P. 1972. Obligations to Future Generations. *The Monist* 56 (January): 85-99.

Goodpaster, K., and Sayre, K. 1977. An Ethical Analysis of Power Company Decision-Making. In *Values in the Electric Power Industry,* ed. K. Sayre, pp. 238-288. Notre Dame, Indiana: University of Notre Dame Press.

Green, R.M. 1977. Intergenerational Distributive Justice and Environmental Responsibility. *BioScience,* 27 (April): 260-265.

Heilbroner, R.L. 1974. *An Inquiry into the Human Prospect.* New York: Norton.

MacIntyre, A. 1979. Seven Traits for the Future. *The Hastings Center Report* 9 (February): 5-7.

Manaster, K.A. 1978. Law and the Dignity of Nature: Foundations of Environmental Law. *Land Use and Environmental Law Review—1978,* pp. 3-26. New York: Clark Boardman.

Murphy, E.F. 1970. A Law for Life. *Environment Law Review—1970,* pp. 3-20. New York: Clark Boardman.

Nozick, R. 1973. Distributive Justice. *Philosophy and Public Affairs* 3 (Fall): 45-126.

Partridge, E., ed. 1981. *Responsibilities to Future Generations.* Buffalo, New York: Prometheus.

Passmore, J. 1974. *Man's Responsibility for Nature,* esp. pp. 74-92. New York: Scribners.

Rawls, J. 1971. *A Theory of Justice.* Cambridge, Massachusetts: Harvard University Press.

Rescher, N. 1966. *Distributive Justice.* Indianapolis: Bobbs-Merrill.

Schedler, N.O. 1975. Our Destruction of Tommorrow: A Philosophical Reflection on the Ecological Crisis. In *Ethical Issues,* ed. W.R. Durland and W.H. Bruening. Palo Alto, California: Mayfield.

Seelman, K. 1978. The Ethical Implications of Energy Production and Use. *Solar Age* 3 (October): 16-21.

Sikkora, R., and Barry, B., eds. 1978. *Obligations to Future Generations.* Philadelphia, Pennsylvania: Temple Univcersity Press.

Singer, M.G. 1972. The Basis of Rights and Duties. *Philosophical Studies* 23 (February): 48-57.

Sterba, J. 1980. Abortion and Future Generations. *Journal of Philosophy* 77 (July): 424 ff.

Stewart, R.B. 1977. Paradoxes of Liberty, Integrity, and Fraternity: The Collective Nature of Environmental Quality and Judicial Review of Administrative Action. *Environmental Law* 7 (Spring): 463-484.

Wandesforde-Smith, G. 1977. National Policy for the Environment: Politics and the Concept of Stewardship. In *Congress and the Environment,* ed. R. Cooley and G. Wandesforde-Smith, pp. 205-226. Seattle: University of Washington Press.

Wenz, P. 1983. Ethics, Energy Policy, and Future Generations. *Environmental Ethics* 5, No. 3 (Fall): 195-210.

Westra, L. 1985. Let It Be: Heidegger and Future Generations. *Environmental Ethics* 7, No. 4 (Winter): 341-350.

Williams, M. 1978. Discounting Versus Maximum Sustainable Yield. In Sikora and Barry, *Oblications to Future Generations.*

Young, G.L. 1976. Environmental Law: Perspectives from Human Ecology. *Environmental Law* 6 (Winter): 289-308.

4

THE RIGHTS OF NATURAL OBJECTS

Introduction

BEFORE it was covered in 1982 with hundreds of feet of water behind the New Melones Dam, California's Stanislaus River was the most-rafted stretch of USA whitewater west of the Mississippi. Environmental activist Mark Dubois vowed to "save" the river, and in 1979 he brought the flooding of the canyon to a temporary halt by chaining himself to a rock while the reservoir waters rose.

Dubois' action raises important questions for anyone concerned with environmental welfare. Do rivers and deserts, rocks and mountains have a continued right to existence and to noninterference by humans? Do persons have duties to protect natural environments, even those that are nonliving? Are natural objects intrinsically valuable? Does wilderness deserve special protection because of its pristine uniqueness?

Biocentric thinkers, like Taylor (see Chapter 1), argue that because living things have a good of their own—continued existence—they have inherent worth and we humans have duties to protect them. Following G.E. Moore, philosophers like Hill have argued that some natural phenomena are intrinsically valuable, not just valued, and not merely because of their worth to living, feeling beings. Historians like Rod Nash have gone so far as to say that even "rocks have rights." Deep ecologists and radical environmentalists such as Foreman and Naess have argued that massive portions of the earth should be completely cordoned off from humans and that population should be reduced by as much as 90 percent to preserve wilderness.

Those who argue that we humans have no duties to natural objects claim that only living, sentient beings have "interests" and can be harmed, and that it is impossible to harm a nonliving being, like a river or a desert. Other opponents of duties to natural objects use arguments similar to those invoked against biocentric ethics. Still other thinkers say that the founding of the *Wilderness Society* in 1935 and the passage of the 1964 US Wilderness Act are rich-nation phenomena. Because so many human needs are not met in the third world and among the poor of every country, Guha and others claim that questions of survival, equity, and redistribution come before "elitist" concerns about duties to the natural environment.

Controversy over duties regarding natural objects is unlikely to be settled soon because it strikes at the heart of important, legal, philosophical, and political issues. These include human rights to private property in land and resources; philosophical distinctions between subjects and objects, and the ethical precept that objects (things) may be treated as mere instruments by humans. Moreover, arguments about whether natural objects ought to be given *legal rights* touch largely on pragmatic issues and interpretations of "standing" in court. The more difficult issue, whether natural objects have *moral rights,* involves questions of alleged moral criteria for rights holders, such as sentience and the ability to have interests or to be harmed.

In the two selections in this chapter, both authors argue for taking greater account of natural objects. Biologist Davis argues that, because of our ecological interdependence, all living beings are equal, and presumably equal as holders of moral and legal rights. Nonliving entities like water or earth deserve our respect, he says; even though they are not equals with us, they are part of our life-support system. Far from seeing humans as superior because of their self-consciousness, Davis sees them, in one sense, as inferior. Only plants can make food; people cannot. Hence, for Davis, environmental ethics must begin, not with the illusion that the planet is human-centered, but with the recognition that it is a dependent ecosystem of producers and consumers who must work together.

My own commentary follows many of the lines of argument found in Christopher Stone's classic, *Should Trees have Standing?* Side-stgepping the issue of moral rights, I neither relgate natural objects to a position of intrinsic inferiority to humans, nor raise them to a level of absolute equality to humans. The argument, instead, is that natural objects ought to be accorded three legal rights: (1) "standing," such that legal actions can be instituted on their behalf, rather than merely at the behest of humans; (2) recognition of injuries done to them; and (3) the ability to act as beneficiaries when damages are proved and relief provided for them

Whether one agrees with classical philosophers, with Davis, or with me, societal policies regarding natural objects will surely affect the environment, whether for good or for ill. Perhaps 400 years from now, posterity will wonder why we spoke of the rights of natural objects, when most of the humans on this planet do not live in societies where their basic rights are recognized. Or, on the contrary, perhaps our descendants will think us insensitive and short-sighted for taking so long to recognize the rights of the earth.

Readings

1. Davis, W.H. 1972. The Land Must Live. *Environment Action Bulletin* 3 (July 15): 4-5, 7. Reprinted by permission of *Rodale's Environment Action Bulletin.*

2. Shrader-Frechette, K.S. Ethis and the Rights of Natural Objects. Written especially for inclusion in this volume.

THE LAND MUST LIVE

Wayne H. Davis — University of Kentucky

I WAS born in 1930 and thus had the wonderful opportunities of a child growing up during the Great Depression. Ours was the magnificent world of work, fun, games, good times, and happy children. I never saw a whining fussy child until the Great Economic Boom brought us Future Shock along with television, swimming pools, built in obsolescence, and finally, the use-it-once-and-throw-away economy.

The Depression gave the nation its lowest fertility rates. Children grew up with a sense of stability of the land and their environment, a sense which contributes strongly to the feeling of security and the development of sound mental health. We saw firsthand the fundamental relationship between man and the land, a land that was not being destroyed to build new shopping centers and superhighways. We also learned the basic laws of ecology by simply observing the fundamental interrelationships among living things. We knew what type of place the catbird would choose for its nest and where it would forage for food. Without television and swimming pools to distract us, we were able to spend all our summer spare time playing outdoors.

Perhaps of greatest value to us kids growing up during the Depression was the land we had. Around my home there were many lots whose titles had reverted to the city in default of taxes. This land grew up in wild cherry and locust trees. It was used by no one but the kids and it belonged to us as completely as if we had owned title to the land. Here we built our tree houses, kept trails cleared, planted apple cores and peach seeds, cut a tree now and then, made slingshots, and had corn roasts. The land was ours to manage as we chose so as to keep it wooded and reasonably wild. It was large and diverse enough that we always had a wood thrush, a towhee and a field sparrow nesting there in the city.

Urban Sprawl

As I watched the drastic reshaping of our Earth as urban sprawl engulfed the nation following World War II, I often wondered what the effect would have been on us kids had the bulldozers moved in and converted our woods and fields into a parking lot for a new shopping center. At the very least such a traumatic experience would have alienated us against the Establishment and caused us to question the values and the concept of progress held by the older generation.

Yet today such an emotional shock has come to be a routine part of the growing up of most children. I suspect that it is also a major cause of such problems as the generation gap, rising rates of delinquency, crime, drug addiction, mental health problems, and the general rebellion against society and its lack of values.

Consider the changes in our ability to destroy the land that have developed within one generation. There were no bulldozers when I was a boy. Roads were built by men using a pick and shovel. Coal was mined with the same tools. Today we have giant earthmoving machinery that can pick up 200 cubic yards of earth in a single scoop to get to the coal seam beneath it. In building highways we do not hestitate to move mountains and fill in valleys.

Progress?

I can remember when it was a major job to fell a tree. In cutting timber, two men with broad axes would chop into one side of the tree and then use a hand-drawn cross cut saw from the other side until the tree fell. With today's power saws two men could clear a substantial area in the same time. To destroy a forest and replace it with a paved parking lot is not nearly the job today that it was 30 years ago. This is a major reason why we are now paving the land surface of this country at a rate of over a million acres a year. In California alone land is going under at a rate of 375 acres a day.

Ten years ago I bought a home in Lexington, Ky. Behind the house is a stream and small open field. Although I realized that in the nation's 14th fastest growing city, the last of such green open spaces would soon be destroyed, I decided that it would be nice for the children while it lasted.

The little stream died last year. Even though it had been barely limping by from one year to the next, the end came as a shock for me and the children. We had known it well and had come to be close friends.

When we arrived in Lexington I was surprised to find that this little stream was alive. Although it had lived happily for several thousand years before I first saw it, the impact of man had hit it pretty hard by then. It arises from a culvert leading from beneath a parking lot of a shopping center, flows about 50 yards, and then enters another culvert which carries it underground so that the beauty of suburban Lexington need not be marred by a free-flowing stream.

In this short section of stream dwelled an abundance of crayfish, snails, and leaches. All three repeatedly found their way into my home in glass jars as the children successively went through stages of their development. The stream was a laboratory for the study of life, much to the delight of the neighborhood children, and the unfortunate consternation of most of the mothers.

Few care about the loss of this tiny stream and others of its kind. But I think its loss is an important symptom of the disease that affects this dying Earth and represents a cause of our ever growing social degeneration and loss of humanity among mankind. The next generation of children will never see a living thing in this stream. As they grow up with stinking, dead and trash-filled streams, would you expect them to develop an appreciation of life and a humane attitude toward animals and their fellow man? I don't see how they could unless they rebel against the ugly, meaningless, and self-destructive life style that my generation has left for them.

Two Messages

There are two other messages that come from my little stream to you. First is that if you want your new baby ever to be able to fish in the streams, you had better join the fight for clean water now. Life is disappearing from our watercourses at a rapidly accelerating pace.

Second is a broader message from Earth to all its human inhabitants. Earth is giving us a nonnegotiable demand. Either we change our ways or she will roll over and die. If we continue to destroy all life on successive pieces of land and water on this finite Earth, the end result is known: It will be only a matter of time until we have destroyed ourselves.

Next to go will be the open field behind my house. I would love to buy it as an investment in the mental health and stability of my children and future generations. I would plant a vegetable garden out there, but most of the land would be undisturbed, to grow grasshoppers and lightning bugs.

But I cannot buy the land, for it would cost about $100,000, which is many times my annual income. The irony is that in the depth of the Depression my father was able to buy a piece of land across from our house to prevent someone's building on it. He paid $150 and $200 per lot, or about one-tenth of his annual income, for it.

Thus we we see that, although modern mass production and a growing economy has given us more material things and has allowed us to turn our natural resources into junk at an ever accelerating rate, we are now less able to afford those things needed to maintain the quality of life than we were during the Depression.

The reason we have been gulled into thinking growth is progress is that we have a basic misunderstanding of the difference between cost and value. When speaking of land and homes, we use these words as interchangeable synonyms whereas they are actually antonyms. This was best brought out by naturalist writer Joseph Wood Krutch in the last paper he wrote before he died.

Twenty years ago Krutch built a home on five acres outside Tucson, Ariz. As the city crept out to surround his little oasis in the desert wilds, real-estate developers tried to get him to sell his land, telling him that it had risen in value to many times what he had paid for it.

"Value for what?" he asked. "Certainly not as a place to live." Now that the traffic congestion, noise, smog, high crime rate, and drug addiction had come out to join him, he claimed the value of the land had gone down. The developers found it hard to believe that a man would rather live on a piece of land than to sell it at a profit.

What goes up is not value, but cost—cost to everybody. The owner does not gain by selling, for to get a comparable home he must pay the same high cost plus the agent's fees.

Cost goes up because there are more and more people each year entering the housing market. Since they are not making any more land these days, but are destroying it at an ever-increasing rate, this means that each year more people are snapping at an ever-smaller piece of the pie. Naturally the price goes up.

Inverse Relationship

So what has developed as a result of urban growth in America is an inverse relationship between the cost and value of land and homes. Cost goes up, especially for young people trying to acquire their first home, and the value goes down for everyone, as the quality of life deteriorates in a cloud of smog, noise, traffic, and the other unpleasant by-products of growth.

An earlier writer who recognized the value of land as different from the cost was Aldo Leopold. "There is much confusion" he wrote, "between land and country. Land is the place where corn, gullies, and mortgages grow. Country is the personality of land, the collective harmony of its soil, life, and weather. Country knows no mortgages, no alphabetical agencies, no tobacco road; it is calmly aloof to these petty exigencies of its alleged owners.

"Poor land may be rich country and vice versa. Only economists mistake physical opulence for riches. Country may be rich despite a conspicuous poverty of physical endowment, and its quality may not be apparent at first glance, nor at all times." *(A Sand County Almanac,* Sierra Club/Ballantine, New York, 1970.)

The land behind my home is zoned for agricultural use. Land bought for that use sells for about $300 an acre. With a zoning change for residential housing use, the cost jumps to $15,000 an acre. Changed to professional or business use, the cost of the land would take another quantum jump.

Open Space is Highest Value

That the highest value of the land, on the other hand, is as open space, is evident from the unanimous desires of the people of the neighborhood to have it remain as such. The people of this city do not need more office buildings, more service stations, and drive-in restaurants, or more traffic problems It is ironic that, with their demand for housing, new people moving to the cities are paying for the destruction of those values that we most wish to keep. And the cost of our housing is so high that we of the affluent society cannot afford to buy additional land in order to save it.

In 1949 Aldo Leopold wrote that we must develop a land ethic. The truth of his message comes across more urgently with every passing day. The destruction of the land for personal financial profit is a behavior that is simply not ethical and cannot long be tolerated.

Ethics in human behavior have probably been with us since our earliest civilization, although what we consider ethical has changed somewhat with time. When Odysseus returned from the wars of Troy as hero of the wooden-horse episode, he hanged on one rope a dozen of his slave girls whom he thought to have been unfaithful during his absence.

There was no question of right or wrong, for no ethic is involved in the manner in which a man disposes of his property. The act would be considered wrong today only because people are no longer slaves and women are finally being recognized as individual people entitled to the same personal consideration as everyone else. But the lack of relationship between morality and that which the law considers private property remains unchanged since the time of Homer, and is a major factor which prevents any real progress in dealing with some of the most obscene aspects of the environmental crisis.

Today's equivalent of the Greek slave girl is the land. This land is your land and my land, and like air and water it is a basic element upon which our survival depends. On it stand our cities and our homes, and from it we derive nearly all our food. And the land, more than any other factor, is the measure of the quality of life now available to you and me. It is the landscape, and the trees, and the rabbits thereon. It is also the city park, or the one that should have been there. For the farmer and his wife it is peace of mind as they gaze out over the fields on a quiet evening.

Yet, as we enter the last part of the 20th century we still have no land ethic. Although we sing our praises to the amber waves of grain, we consider the land upon which it grows as no more than dirt, to be subdivided and covered with asphalt as soon as the profit margin becomes satisfactory. Land is treated as a commodity, to be exchanged for money, with the sole object being profit for buyer and seller. The fate of the land at the end of the hangman's rope is supposed to be of no concern to anyone in spite of the fact that it is of value to all mankind.

Farmers 'Kicked Around'

No one on earth is more derserving to stand tall and proud than the American farmer, who has fed you and me and most of the rest of the people on Earth. Yet farmers are kicked around economically and forced to flee to the cities by the thousands each year to earn a living while the giant land companies get an ever-tightening stranglehold on the nation. It is a sad commentary that the poorest investment one can make is to buy a farm and farm it, while his best investment is to buy a farm and subdivide it into a real estate development.

To the ecologist perhaps the most insane of all meanderings in the business world is that which equates land to paper money and makes the two freely interchangeable. Once sold, the land can be flooded behind a dam, blasted away to open a mine or road or sent helter skelter down into the streams via erosion. We consider no moral question to be involved in this. The only value judgment for society is made on the other side of the transaction: The money received for the land is viewed in a positive light as a stimulus to the local economy.

Perhaps the time is ripe for a new religion based upon a reverence for the land and all the life which springs therefrom. Is it not every bit as logical to worship the solid earth beneath your feet from which the mystery of life springs eternal with each vernal equinox as to have to imagine some unseen being in the sky? If the Great Spirit still stalks this Dark and Bloody Ground so long after the last Indian fled to the West from his Happy Hunting Ground ahead of Boone and Kenton, surely He must reside in the land and trees as He always did. If we make the land our temple, perhaps at last our fellowman will respect it as he does the man-made structures with their spires rising helplessly in the wrong direction in their search for hope and eternal life.

The time has come when we must all develop respect for the air, water, land, and the living things thereon. We are all dependent upon the same life support system to Earth and must protect if if we are to survive.

I hold these truths to be self evident: All living things are created equal and are interdependent upon one another. All flesh is grass. Only plants can make food. Man and all other animals are totally dependent upon the plants which we so casually push aside in pursuit of the ever greater megalopolis. Animals need their oxygen and the plants our carbon dioxide. Both are dependent upon numerous species of microbes which make amino acids and vitamins, digest food, fix nitrogen for our use, and return it to the air when we die. And all are dependent upon the exceedingly complex ecosystem of producers and consumers, predators and prey, herbivores and carnivores, and parasites and disease, to provide for their needs for survival and to control their numbers. Man cannot survive alone. Nor can he continue to increase his numbers at the expense of other living things.

But man is arrogant. He refuses to face reality. Four centuries after Copernicus he still really believes that the earth is the center of the universe and that God's only concern is with his welfare. A century after Darwin man still thinks of himself as apart from nature, with a divine destiny to conquer nature and exploit the other creatures for his own use.

No wonder our young people rebel. They are concerned not only about the deterioration of the quality of life, but also about the prospects of surviving on the dying planet they will inherit. Young activist Penfield Jensen recognized the real issues when he wrote, "We will stop the destruction of this planet even at the cost of our own futures, careers, and blood. The situation is simply like that. If you are not going to live for the earth, what are you going to live for?" *(No Deposit, No Return,* Addison-Wesley, Reading, Mass. 1970)

ETHICS AND THE RIGHTS OF NATURAL OBJECTS

K.S. Shrader-Frechette

SEVERAL DECADES AGO the American naturalist, Aldo Leopold, pointed out the dangerous implications of the word "property." He told how, when Odysseus returned home from the wars in Troy, he hanged a dozen slave girls whom he suspected of misbehavior during his absence. "This hanging involved no question of propriety. The girls were property. The disposal of property was then, as now, a matter of expedience, not of right and wrong." Although the Greeks of Odysseus' time had a well developed ethical framework, it had not been extended to include "human chattels."[1]

In the three thousand or more years which have elapsed since the time of the myths of Homer, ethics has been extended to include many areas of human behavior.[2] Unfortunately, however, our ethical framework has not been extended far enough to differentiate all forms of social from antisocial conduct. Land and other natural objects are still viewed as property, just as were Odysseus' slave girls. For persons like Leopold, however, extending ethics to include the rights of the environment is "an evolutionary possibility and an ecological necessity."[3]

My point of view is that there are ethical and legal grounds for recognizing the rights of the environment. Although this thesis cannot be proved conclusively, my reflections on it are designed to achieve three goals. First, they will remove some of the peculiarity of speaking about "the rights of natural objects." This will be accomplished by a discussion of guardianship and the current rights of other inanimate objects. Secondly, in emphasizing the evolution of legal and ethical systems, I hope to challenge humankind to expand its moral sensitivities and to consider the possibility that the environment might have rights. Finally, although my proposal cannot be shown to be a completely desirable one, it can be shown to be appropriate within a certain world view of ourselves and nature. If we choose to adopt this world view, then we are really choosing not merely a policy for our law courts but also an ethic for our lives. Let us look at how one might develop and defend such an ethic.

Planetary Interests and Human Interests

If one admits that natural objects are endowed with a value that conveys rights or obligates us, then such an admission might be founded on an appeal either to human welfare or to the interests of all ecosystems.[4] René Dubos believes that selfish reasons alone could constitute sufficient grounds for recognizing the rights of natural objects; such rights are necessary, he says, "for the protection of humanized nature and for the preservation of mental health."[5]

Obviously, humankind cannot ignore pollution of air, water, and earth, for example, and at the same time attempt to protect itself against carcinogenic and mutagenic injuries. The well-being of the planet has a direct effect on human interests. For this reason the question of our obligations to natural objects has a clear answer, at least in one sense. To the extent that the welfare of nature affects human interests, to that same degree do persons have obligations to protect the well-being of natural objects.

What about our obligations to natural objects in themselves? Do they have rights apart from their value to human beings? This sense of the question is more difficult and its answer is less charted; resolving it will take us to the frontiers of ethics and legal practice. It is the query raised by Leopold and the subject before us now.

Christopher D. Stone, Professor of Law at the University of Southern California, and author of *Should Trees Have Standing?,* has written a classic book in response to the question of whether nature has rights, apart from its effect on human well-being. Stone is convinced that the answer to this query is "yes"; he believes that ethics ought to undergo the extension of rights to the environment. The only way to do this, he says, is to "get the courts thinking" about nature "as a jural person."[6] If cases could be pleaded on the theory that natural objects themselves are rights-holders, then society would be able to adopt a new environmental ethic heretofore recognized only in the consciences of people like Leopold.

One recent opportunity for recognition of this new ethic came in 1971, when the Sierra Club brought suit for an injunction against the US Forest Service, which had granted a permit to Walt Disney Enterprises to "develop" Mineral King Valley, a wilderness area in California's Sierra Nevada Mountains. The Sierra Club lost the suit because it did not have "standing," said the Ninth Circuit, to bring the question to the court. In other words, since the injury to the Sierra Club from development of Mineral King was said to be tenuous, the court ruled that the Club had no "standing." When the Sierra Club's appeal was reviewed by the US Supreme Court, the judges upheld the Ninth Circuit. Justice Douglas, however, wrote a dissent. He proposed a new federal rule that would allow environmental issues to be litigated in the name of the inanimate object about to be injured. Although Douglas' opinion did not prevail, three of the seven judges voting endorsed his notion and would have interpreted the Sierra Club case such that Mineral King was the party in interest and the Club, its guardian. Despite the fact that the 4-3 decision did not give nature "standing," the majority opinion in this case was not unfriendly to the notion. In fact, Garrett Hardin believes that we are near "the ridge of a watershed," and that we will soon recognize the fact that nature has standing, or legal rights.[7]

Arguments for the Rights of Natural Objects

A number of ethical and legal considerations suggest that there is a strong rational basis for affirming that natural objects have rights. According to

Professor Stone, our concept of "legal right" is a product of traditional, historical mental baggage, rather than the result of clear analysis. Since the future of the planet is dependent upon humanity's adopting a less homocentric view of legal rights, Stone proposes that full legal rights be extended to the inanimate world. Natural objects would be allowed to sue for damages, to have their injuries considered in measuring relief, and to recover damages for their benefit. Under this proposal, a friend of the natural object in question could apply to the court for guardianship, just as a friend may now do on behalf of a legal incompetent. Since the guardian could sue to protect the rights of the natural object, any damages awarded could be placed in a trust fund to be administered by the guardian on behalf of the natural object.

In building his case for the thesis that nature ought to be accorded rights, Stone points out (as did Leopold) that, historically, children, women, and aged, prisoners, aliens, Blacks, Indians, and the insane were treated as objects or as things, rather than as persons possessing rights.[8] Hence, he argues, it is not unthinkable to recognize the rights of natural objects, both because legal history reveals that rights are continually being expanded in principle and in practice, and because each time rights are conferred on some new entity, the proposal sounds odd, frightening, or laughable.

There Are Inanimate Rights-Holders

One of the most common arguments, that the environment has rights, is based on the fact that society already recognizes the existence of numerous inanimate rights-holders. These include trusts, corporations, municipalities, and ships, among others.[9] To recognize the rights of nature, however, means neither that inanimate objects would have the same rights as human beings, nor that everything in the environment would have the same rights as every other natural object. Rather, to say that nature has legal rights simply means that legal actions may be instituted at its behest, that injury to it must be taken into account by the courts, and that relief may be given to its benefit.[10]

At present natural objects have no such legal rights. Nature has no "standing" to sue for damages, for example, at its own behest; only humans able to show possible invasions of their rights have standing. Moreover, even when someone is willing to establish standing, the merits of a case (e.g., involving pollution of a natural object such as a stream) are currently decided on the basis of the conflicting interests of the (human) plaintiff and defendant. The fate of the natural object, *per se,* is ignored.[11] In cases involving injury to the environment, the beneficiary of a favorable judgment is always a human. No money goes to the benefit of the natural object to repair its damages.[12]

Such a procedure, it can be argued, is not wise because it does not enable nature to seek redress in its own behalf. Just as legal incompetents have guardians to manage their affairs, so also friends of natural objects could be recognized by the courts as their guardians. In this way, the Environmental Defense Fund, for example, might raise the land's rights in the land's name. One reason for doing so might be to prevent unsound strip-mining operations.

The Need To Calculate All Social Costs

Besides the fact that there are already inanimate rights-holders, another argument for establishing the legal rights of nature is that such recognition will further enable society to become aware of all the costs of its activities. If the legal rights of the environment are not recognized, for example, in the case of a stream polluted by a paper mill, then there is no feasible way for all the fragmented and unrepresented damage claims to be brought before the courts. For practical reasons, the claims will not be pressed. Owners of summer homes and sellers of fishing bait, for example, are unlikely to bring forward their claims. These distantly injured humans, including unborn generations, however, could all be represented by the guardian of the natural object,[13] if the object itself were made the focus of damages to it.

If injury to the environment were viewed as a violation of rights, then it would be a simple matter for a guardian to prove an economically measurable loss in court. A calculation of the damages to a natural object might be thought of as analogous to a measure of the injuries to an automobile-accident victim. In both cases, the responsible party might bear the expense of making either the person or the natural object "whole" or healthy.[14]

The Empathy and Reciprocity Between Humans and Nature

A third reason for recognizing the rights of natural objects is based on the recognition that we share fundamental needs with the nonhuman world. Plants and humans, for example, both require water, oxygen, and nutrition; both grow, reproduce, and die. With these bases for empathy and identity established, human moral evolution could conceivably result in the extension of rights to the nonhuman world. "The inner dynamic of every assault on domination is an ever broadening realization of reciprocity and identity."[15]

If we humans realized the reciprocity in our relationships with the nonhuman world, and if we identified and empathized with other beings on the planet, then we would have to give up some psychic investment in humankind's sense of separateness and specialness. Although this might be difficult, it could also be desirable, in enabling us to "free ourselves of needs for supportive illusions."[16] Extending legal rights to natural objects thus might enable us to reaffirm several truths, viz., "that the oppressor is among the first to be liberated when he lifts the yoke," and "that freedom can be realized only in fidelity to obligation."[17] In other words, recognizing the legal rights of nature might develop not only humanity's capacity for empathy and understanding but also its ability to live free from the illusion that members of only one species are the center of the universe.

In response to the various arguments that have been used to support the thesis that one ought to recognize the legal rights of the environment, a number of counterarguments have been made. Some of these responses address alleged practical problems with implementing such a thesis, while others speak to purely philosophical difficulties with the notion of extending rights to the nonhuman world. Let us examine first some of the philosophical problems.

Philosophical Problems with Recognizing Rights of Nature

Several authors have maintained that one ought not to argue that natural objects have rights on the grounds that corporations, municipalities, ships, and other inanimate objects have recognized rights. The analogy, they say, is not a valid one, because legal recognition of the rights of inanimate objects, such as corporations, is based on homocentric interest. This legal recognition is a means to promote human activity. Recognition of the rights of nature, on the other hand, would promote an interest different in kind, an interest in a particular natural object.[18] What this argument really comes down to is the notion that extending rights to natural objects, in themselves, has no legal precedents because inanimate objects have been given rights only insofar as their recognition serves some human goal. Hence, there are no legal precedents for promoting the interests of the environment.

In response to the fact that there is an absence of legal precedence for promoting nature's interests, several replies can be made. First, one can cite the reasons given in our subsection on "Empathy and Reciprocity" for recognizing environmental interests, as well as those of humanity. Secondly, one can point out that, because there are no legal precedents for an action, this does not mean that the action is not moral, rational, or legally desirable. If the absence of legal precedents did mean this, then ethical and legal systems would never be able to evolve or progress far enough to meet the demands of the times and the heightened moral sensitivities of greater numbers of people. This gradual extension of rights, for example, to Blacks, women, and children, illustrates the thesis that legal and ethical systems have not been static. And if they have not, there is no legal basis for denying rights to natural objects, purely because of the absence of certain precedents.

A related reply to the thesis that nature's rights ought to be recognized is that the monetary cost of expanding legal rights to the environment is very great, and hence that the public may be unwilling to tax itself by sustaining the increased financial burden that such a proposal would impose.[19] This argument raises the question, "What's in it [recognition of the rights of nature] for us?"

The difficulty with the argument, that the public is probably unwilling to bear the costs of extending rights to natural objects, is that it asks one to justify acknowledgment of nature's rights in anthropocentric and hedonistic terms. Yet if one ought to recognize the rights of natural objects, in themselves, then by definition this justification cannot be accomplished in purely anthropocentric or hedonistic terms. As Stone points out, perhaps the best way to respond to such a request is to counter with another question. "Couldn't a white person raise the same questions about cost when asked to compromise his preferred rights-status with Blacks?"[20] Or, could not a male, when asked to recognize the rights of a female, ask, "What's in it for me?" Raising both these questions illustrates the point that moral justification for recognition of rights need not, and indeed often cannot, be given in terms of benefits to those who already enjoy such rights. In the case of the rights of nature, however, it is clearly in the best interests of humans to recognize their existence. It is well-known by now, thanks to documents such as *The Limits to Growth,* that human welfare is inextricably bound up with planetary well-being. If authors such as Hardin, Ehrlich, Carson, and Cousteau are correct, then environmental hazards threaten our very survival, not only our health.

Even though ethics ought not to be justified solely in egoistic, anthropocentric, or hedonistic terms, there are other reasons why it is in our self-interest to recognize the rights of the environment. For one thing, such an acknowledgment might make us better persons. The underlying human qualities reinforced by our attitudes toward property, consumption, and the natural world are often qualities which obstruct our personal growth and happiness. Frequently our own desire to extend our will over things, including natural objects, to manipulate them, and to control them, bespeaks a lack of integration with nature and a stunting of our ability to love. This is why, in recognizing the rights of the environment, oppressors might be the first to be liberated when they lift the yoke from nature.

Apart from the desirability of acknowledging the rights of natural objects, there is at least one major philosophical problem that arises in connection with such a recognition. If rights are granted to the environment, then for many of the same reasons it should also bear liabilities.[21] The idea of a natural object being liable for the injuries it causes, however, has been said by some authors to be contrary to nature.[22] Although the realization, that nature ought to bear liabilities, reflects the need for symmetry in the law, it flies in the face of the philosophical principle that responsibility cannot exist where there is no freedom. How could a tree be responsible, for example, for killing a person, if the tree were not free not to fall when struck by lightening?

Even Stone admits that there is a "troublesome" ontological difficulty with affirming that natural objects ought to bear liability.[23] Although there is no clear solution to this problem, it is helpful to put it in perspective by recalling similar difficulties with accepted notions of liability. At least in a legal sense, persons are often held liable for injuries they cause, even when they are not free to avoid them. For example, if the brakes fail on the car of person A, and he unavoidably injures person B, then under strict liability, A is responsible for B's injury and is legally liable. Hence legal responsibility is often assessed in cases when the person was not free to do other than what he did. This specific case seems only slightly less problematic than the case of ascribing liability to a natural object which is also unfree to cause particular happenings, other than those causing injury. Thus, although the ontological problem posed by ascribing liability to the environment is not resolved, it may be seen as analogous to other difficulties currently faced as a result of our legal notions of liability and responsibility.

Practical Problems with Recognizing Rights of Nature

Apart from the philosophical questions raised by the proposal to grant rights to natural objects, a number of practical difficulties are also apparent. One of the most serious of these is the charge that recognition of the rights of nature will give too much power to the judicial branch of government. If standing is extended to a new category of potential litigants, then according to the Solicitor General, there will be no "legal issue ... which will not have to await one or more decisions of the court before the administrator sworn to uphold the law can take any action."[24] This could result in government by the judiciary rather than in government by all three branches.

In response to this argument, a number of considerations can be brought to bear. According to the late Justice Douglas, making governmental administrators subject to the courts could have a beneficial effect, since those sworn to uphold the law "have been notoriously derelict in protecting the environment."[25] Rather, if a guardian could sue to protect the rights of a natural object and could recover damages, then a trust fund would be available to be administered by the guardian on behalf of the natural object. This system not only would be workable, say attorneys, but also would provide for better attention to environmental injuries.[26]

According to Stone, one way of avoiding a flood of litigation, as a result of granting standing to natural objects, would be to use class action law. Another possibility would be for the courts simply to designate the guardian *de jure* representative of the natural object, with the understanding that the natural object is legally bound by an adverse decision.[27] In this way much redundant litigation could be prevented; although various groups could intervene on behalf of nature, the single-guardian concept would help protect the efficiency of legal proceedings. It would have the futher benefit of providing the natural object with continuous supervision and a stronger ground of expertise on the basis of which suits could be brought.

The single-guardian concept would also enable the courts and the public to cope with another practical problem raised by the proposal to extend rights to the environment. Some authors have argued that the guardian concept is deficient in that guardianship of a natural object will be sought only by those who value the object in question.[28] This means, it is alleged, that granting rights to nature will again be based on criteria of homocentricity, since natural objects not having persons sufficiently interested in them will not have their rights served as effectively as those things in whom humans are deeply interested.

While this argument is a sound one, the difficulties it points up can, to some extent, be remedied. If guardian *de jure* representatives of all natural objects were appointed by the courts, then the environment would have representatives on a continuing basis. Natural objects would not have to wait until persons were sufficiently interested, in their particular plights at a particular time, to sue on their behalf. Moreover, although there is no way to guarantee that every natural object will have an effective guardian, this problem is not so much the consequence of granting rights to nature as it is the result of human frailty. The same difficulty (with insuring adequate guardianship of the environment) is faced by those attempting to insure expert guardianship of all protected classes. Hence, to a large degree, recognizing the rights of nature raises no new problems not already posed by the institution of guardianship.

The adequacy of guardianship brings up another practical problem associated with granting rights to the environment. This difficulty is that, even if natural objects were said to have "standing" in court, this would not guarantee that environmental interests would be served. As one author points out, "the balancing of human interests must still remain the major consideration"; in any situation, he says, the interest of nature "could be subordinated to that of the polluter."[29] This means that granting rights to natural objects does not guarantee success to the environmental plaintiff.

While the point of the previous argument is correct, it is not clear that it provides grounds for rejecting the notion that natural objects ought to be accorded rights. Obviously, the judicial system is operated by "human chauvinists" who sometimes might disregard the rights of nature.[30] This does not mean, however, that one ought not to recognize the rights of natural objects, just as it would not mean that one ought not to recognize the rights of women, if judicial decisions were made only by sexists. Obviously if morality were limited only to ethical beliefs with which all or most people were in agreement, then there would never be reason to ask what one ought to do. One could simply find out what people were in fact doing, and then use that as a norm for behavior. Hence, to argue that rights of nature ought not be acknowledged, because they might be disregarded, would be to allow morality to become circumscribed by actual behavior, apart from whether that behavior were moral or not.

This particular argument against recognition of the rights of the environment brings to mind numerous other arguments regarding *de facto* and *de jure* resolution of issues. Just as *de jure* prohibition of segregation did not stop *de facto* segregation, because numerous "balancing actions" have enabled persons to "get around" the law, so also *de jure* granting of rights to natural objects will not stop all *de facto* violations of those rights. This does not mean, however, that one ought to argue, either against laws prohibiting segregation, or against laws recognizing rights of nature. Clearly, the morality and legality of both are determined by factors in addition to mere public acceptance. Recognizing a right "is essentially an antidemocratic process."[31]

The basic reason behind legal recognition of rights, even when the public is not wholly ready to admit them, is that such acknowledgment itself may help to bring about an expanded value system. Increased environmental awareness is an evolutionary process, as both Justice Douglas and Aldo Leopold so eloquently argued. If a right is a legitimate one, then its legal recognition can well be tutor to the spirit of environmental appreciation and protection. Moreover even though recognition of the rights of nature does not guarantee the success of environmental plaintiffs, it does increase the probability that environment-related court decisions will be more realistic and more honest. As one law professor put it: Once the court or any other decisionmaker admits that there are difficult questions involved, "it would seem less likely that a reckless decision will be made."[32]

Perhaps the most serious practical objection against recognition of the rights of natural objects, however, is that it is not needed. According to one attorney, it is not necessary to acknowledge rights of the environment, because under present procedural rules a substantial portion of a private plaintiff's recovery, in an environmental-injury suit, "could be used for the restoration of that environmental object."[33] While such an action would not change the locus of the right to sue, "it would insure that a reversal of the deterioration be initiated with any recovery from the defendant."[34]

While the preceding argument is correct, as far as it goes, it ignores the fact that recognizing rights of natural objects (as opposed to recognizing rights only of persons) brings into the legal system "a flexibility and open-endedness" that no present procedural rules permit.[35] If nature were said to have "rights," then all the force of this word could be brought to bear in investigating claims that the environment had been injured. If judges were able to speak about "the legal rights of natural objects," then they could develop, more readily, a viable body of law designed to protect the environment.[36] This means that a society which acknowledged "the legal rights of natural objects" would be likely to produce a legal system more protective of the environment than would a society in which only humans beings were said to possess such rights.

Conclusion

Although these considerations and the ones preceding them do not prove, comprehensively, that nature has rights that ought to be recognized, they do accomplish several tasks. First, these reflections remove some of the peculiarity of speaking about "the rights of the environment," by virtue of the fact that they include a discussion of guardianship and the rights of other inanimate objects. Secondly, in emphasizing the evolution of legal and ethical systems, I have challenged readers to expand their moral sensitivities and to avoid being trapped in a static view of right and wrong. Finally, although one cannot prove that environmental "interests" exist, one can develop a scenario within which recognition of certain rights seems appropriate. This is, in part, what has been attempted here. It implicitly presents two world views: one based on the metaphors of human control and domination over nature, the other based on the metaphors of human empathy and reciprocity with nature.

In choosing to recognize or not to recognize the rights of the nonhuman world, we are really choosing a specific metaphor for our lives and not merely an ethic for our law courts. Essentially, this is a choice between the injunction in Genesis 1:28, "to fill the earth and conquer it," and the call of Francis of Assisi, to embrace Brother Fire and Sister Water. Let us hope we are evolving toward an embrace, and not a conquest, of nature.

Notes

1. Aldo Leopold, *A Sand County Almanac,* New York, Oxford University Press, 1949, pp. 201-202; hereafter cited as: *Almanac.*

2. See Leopold, *Almanac,* p. 202.

3. Leopold, *Almanac,* p. 203.

4. See Holmes Rolston, "Is There an Ecological Ethic?" *Ethics,* 85 (January 1975): 103; hereafter cited as: Ethic.

5. René Dubos, *A God Within,* New York, Charles Scribner's Sons, 1972, pp. 166-67.

6. Garrett Hardin, "Foreword," in C.D. Stone, *Should Trees Have Standing?* Los Altos, William Kaufmann, 1974, p. xiii; hereafter cited as: *Trees.*

7. Hardin, "Foreword," in Stone, *Trees,* p. xvi. Background on the Sierra Club case is given on pp. xii-xvii. See also H.O. Hughes, "Who's Standing? Problems with Inanimate Plaintiffs," *Environmental Law,* 4 (Winter 1974): 315-323; hereafter cited as: Standing.

8. Stone, *Trees,* pp. 3-4.

9. This argument is made by Stone, *Trees,* p. 5 ff.; Hughes, Standing, p. 318; and L.H. Tribe, "Ways Not To Think about Plastic Trees: New Foundations for Environmental Law," *Environment Law Review—1975,* New York, Clark Boardman, 1975, p. 743; hereafter cited as: Ways. See also K.A. Manaster, "Law and the Dignity of Nature: Foundations of Environmental Law," *Environment Law Review—1978,* New York, Clark Boardman, 1978, p. 23; hereafter cited as: Law.

10. Stone, *Trees,* pp. 10-11.

11. Stone, *Trees,* pp. 13-14. See also Hughes, Standing, p. 320.

12. Stone, *Trees,* pp. 14-15.

13. This is proposed by Stone in *Trees,* pp. 27-28.

14. Stone, *Trees,* pp. 29 ff.

15. This argument is made by Tribe, Ways, pp. 746-48; the quotes in this paragraph are from Tribe, Ways, p. 747.

16. Stone, *Trees,* p. 49.

17. Tribe, Ways, p. 747.

18. Hughes, Standing, p. 319, and Tribe, Ways, p. 744, make this argument.

19. This argument is made by Hughes, Standing, pp. 319-20. See also James Huffman, "Trees as a Minority," *Environmental Law,* 5 (Fall 1974): 200; hereafter cited as: Minority.

20. Stone, *Trees,* p. 44.

21. Stone, *Trees,* p. 34, makes this point.

22. See, for example, Huffman, Minority, p. 201.

23. Stone, *Trees,* p. 34.

24. Cited in Hughes, Standing, p. 321, where this argument is made. See also J.Y. Pearson, "Toward a Constitutionally Protected Environment," *Environment Law Review—1971,* New York, Clark Boardman, 1971, pp. 72-77; hereafter cited as: Environment.

25. Cited in Hughes, Standing, p. 321.

26. See Huffman, Minority, p. 200.

27. Stone, *Trees,* pp. 23-24.

28. See, for example, Huffman, Minority, p. 201.

29. Hughes, Standing, pp. 320-21. This same argument is discussed by Huffman, Minority, p. 200; Tribe, Ways, pp. 727 ff.; and D.S. Willner, "Who Has Standing in Oregon To Defend the Environment?" *Environmental Law,* 1 (Spring 1970): 57; hereafter cited as: Oregon.

30. Tribe, Ways, p. 727, makes this same point.

31. Pearson, Environment, p. 77

32. Manaster, Law, p. 20.

33. Hughes, Standing, p. 322.

34. Hughes, Standing, pp. 322-323.

35. Stone, *Trees,* p. 41.

36. Stone, *Trees,* p. 42, makes this same point.

Further Reading

Abbott, M. 1990. The Stanislaus: A River Once More? *Sierra* 75, No. 6 (November / December): 36-44.

Aquinas, T. 1956. That Rational Creatures Are Governed for Their Own Sakes, While Others Are Governed in Subordination to Them. *On the Truth of the Catholic Faith,* trans. V.J. Bourke, pp. 115-119. New York: Doubleday.

Berry, W. 1987. Amplications: Preserving Wilderness. *Wilderness* 50 (Spring): 91-92.

Brennan, A. 1986-1987. A Moral Standing of Natural Objects. *Environmental Ethics 6 (No. 1 (Spring): 35-56.*

Foreman, D. 1986-1987. A Modest Proposal for a Wilderness System. Whole Earth Review 53 (Winter): 42-45.

Guha, R. 1989. Radical American Environmentalism and Wilderness Preservation: A Third World Critique. *Environmental Ethics* 11, No. 1 (Spring).

Hill, T. 1983. Ideals of Human Excellence and Preserving Natural Environments. *Environmental Law* 4 (Winter): 315-323.

Leopold, A. 1949. *A Sand County Almanac.* New York: Oxford University Press.

Lindbergh, C. 1971. The Wisdom of Wilderness. In *Environ/Mental,* ed. P. Shepard and D. McKinley, pp. 230-235. New York: Houghton Mifflin.

Moore, G.E. 1912. *Principia Ethica.* Cambridge: Cambridge University Press.

Naess, A. 1990. Man Apart and Deep Ecology. *Environmental Ethics* 12, No. 2 (Summer): 185-192.

Naess, A. 1973. The Shallow and the Deep, Long-Range Ecology Movement. *Inquiry* 16: 96-97.

Nash, R. 1989. The Rights of Nature. Madison: University of Wisconsin Press.

Olson, R. 1990. Wilderness International. *Wilderness* 54, No. 190 (Fall): 14-20.

Pytlak, J. 1974. In Defense of Plants. In *Earthkeeping,* ed. C. Juzek and S. Mehrtens, pp. 111-115. Pacific Grove, California: Boxwood Press.

Sax, J. 1977. Freedom: Voices from the Wilderness. *Environmental Law* 7 (Spring): 565-574.

Stillman, C. 1977. On Saving Places. *Environmental Law 7 (Spring): 485-497.*

Stone, C. 1987. Earth and Other Ethics. New York: Harper and Row.

Stone, C. 1974. *Should Trees Have Standing?* pp. 3-11, 45-54. Los Altos, California: William Kaufmann.

Taylor, P. 1986. *Respect for Nature.* Princeton: Princeton University Press.

Tribe, H. 1975. Ways Not to Think about Plastic Trees. *Environment Law Review*—1975, pp. 717-750. New York: Clark Boardman.

Vest, J. 1987. The Philosophical Significance of Wilderness Solitude. *Environmental Ethics* 9, No. 4 (Winter): 303-330.

5

THE RIGHTS OF ANIMALS

Introduction

IN 1989, housewife Karen Wood, standing in her own backyard, took a slug in her lung and died. A hunter mistook her white mittens for a fleeing white-tailed buck. The *Bangor Daily News* blamed her for not wearing orange, and the grand jury in the prohunting community refused to indict the hunter.

For groups like *Defenders of Wildlife,* the death of Karen Wood is not significantly different from the deaths of 200 million animals killed each year by hunters in the USA. Stephen Clark wrote recently that sport hunting shows an outdated bias in favor of our own species and nurtures aggressive, violent instincts that we ought to outgrow.

At the heart of the hunting controversy is a larger conflict over animal rights. Questions about whether animals have rights arise from at least three somewhat different considerations: (1)the morality of killing; (2) the social and agricultural incentives for vegetarianism; and (3) the loss of biodiversity. Animal-rights philosophers, like Regan, argue that it is immoral to kill another living, sentient being for no good cause. Vegetarians argue that it would be easier to feed the people of the world if we ate grains rather than meat, since much more grain is needed to nourish animals used for food than to feed humans directly.

Ecologists also often argue for animal rights and species protection because they are concerned with loss of biodiversity. Paul Ehrlich, a Stanford biologist, compares species to the rivets in "spaceship earth." Once the rivets go, the whole interdependent craft will fall apart. The rivets, or species, are natural resources that provide agricultural, ecological, medicinal, and economic benefits to everyone on earth. For example, fully three-fourths of medicines used in developing nations are derived from plant and animal species, some of which are already on the verge of extinction. Likewise, even in the USA, more than one-fourth of all prescription medicines are derived from plants and animals. USA agriculture, in particular, has been able to retain its productivity and resistance to disease only by developing hybrids obtained from other strains growing elsewhere. Without access to these other species and strains, USA agriculture would have faced massive failure. Yet, despite these benefits of species depletion, we are destroying our species resources. Norton estimates, for example, that 25 percent of all existing species will be caused to go extinct during the next 25 years. Most of these extinctions occur because of habitat destruction by humans, not by direct killing, such as hunting. Nevertheless, whether humans kill animals directly or merely fail to protect them and their habitats, animal deaths and extinctions are often questionable on moral, agricultural, and ecological grounds.

Arguments in favor of protecting animals and entire species generally fall into at least two categories, anthropocentric or biocentric. Anthropocentric or human-centered arguments, such as the medicinal and agricultural arguments for biodiversity and species preservation, focus on using animals or plant species to attain human goods. Biocentric or life-centered arguments, such as the ethical claim that animals ought not be harmed because they suffer pain, focus on the inherent worth of animals or species, apart from human needs. Most of the discussion in this chapter will focus on the biocentric arguments, and specifically on the ethical arguments for duties to animals, since these require us to ask fundamental and far-reaching questions about our environmental philosophy. Anthropocentric or utilitarian arguments for biodiversity, while compelling, merely require us to act in ways that serve human goods.

Whereas advocates of animal rights appeal to a heightened ethical sensitivity and to the fact that humans and animals are alike in that they are living beings capable of suffering, opponents of animal rights tend to justify their position by emphasizing the supremacy of human interests over those of animals. They stress the differences between human and nonhuman members of the animal kingdom. From their viewpoint, proponents of animal rights might be characterized as sloppy sentimentalists who care for nonhuman animals more than for people. According to Richard Neuhaus, author of the polemical book against the ecology movement, *In Defense of People,* the criticism is justified. Neuhous sees those who argue for animal rights as "zealots of the peripheral." He questions why these crusaders are not concerned instead about urban housing and world hunger. On the other hand, animal protectionists often caricature their opponents as biologically unwise tyrants who advocate a utilitarian domination of the ecosphere by one species, *Homo sapiens.*

The readings in this chapter have been chosen, not only because they represent varying perspectives on the issue of animal rights, but also because they avoid most of the irrational or dogmatic discussions often characterizing arguments on this topic. Hence these selections illustrate very well the extent to which the question of animal rights is both a meaninful and a difficult philosophical issue.

The first article is by Peter Singer, Professor of Philosophy at Monash University, Victoria, Australia. It is probably the single most acclaimed article on animal rights. Singer argues that rights are based on moral, rather than factual, equality among beings, and that since both humans and other animals are capable of suffering, they have "interests" equally deserving of recognition. In the second article Michael A. Fox, Professor of Philosophy at Queen's University, Kingston, Ontario takes issue with each of Singer's points. Fox argues that factual considerations are relevant, and that one cannot consider only moral equality when attempting to determine rights. Therefore, says Fox, it makes sense neither to say that animals have interests equal to those of humans, nor to claim that animals have rights.

Despite the fact that one of the authors represented in this chapter believes that animals ought to be said to have rights while the other maintains that this ought not to be said, both of them agree that human duties relevant to animals often have been neglected. Regardless of one's position regarding animal rights, it follows that it is important to examine this area in which there has been so much human error. Perhaps when we understand the reasons for these failings we will have arrived at a better understanding both of men and of animals.

Farley Mowat, author of *A Whale for the Killing,* has written extensively about one recent event in which human mistreatment of an animal provides a context for questioning traditional theories about rights. The situation about which he wrote occurred in 1967, when an 80-ton fin whale became trapped in a Newfoundland cove after the tide went out. Mowat and many other persons were disturbed when the gentle animal died a slow, painful death. The whale succumbed to massive infection after having been shot by numerous "sportsmen." Reminiscing on the death of the animal and the humans who caused it, Mowat asked: "who was more a beast, the sportsman or the whale?"

Readings

1. Singer, P. 1973. Animal Liberation. *The New York Review of Books,* Section 20 (April): 17-21. Reprinted by permission of Peter Singer and *The New York Review of Books.*
2. Fox, M.A. 1978. "Animal Liberation: A Critique." Reprinted from *Ethics* 88 (January) 106-118., by permission of the University of Chicago Press. ©1978 by the University of Chicago. 0014-1704/78/8802-0002801.16.

ANIMAL LIBERATION[1]

Peter Singer

I

WE ARE FAMILIAR with Black Liberation, Gay Liberation, and a variety of other movements. With Women's Liberation some thought we had come to the end of the road. Discrimination on the basis of sex, it has been said, is the last form of discrimination that is universally accepted and practiced without pretense, even in those liberal circles which have long prided themselves on their freedom from racial discrimination. But one should always be wary of talking of "the last remaining form of discrimination." If we have learned anything from the liberation movements, we should have learned how difficult it is to be aware of the ways in which we discriminate until they are forcefully pointed out to us. A liberation movement demands an expansion of our moral horizons, so that practices that were previously regarded as natural and inevitable are now seen as intolerable.

Animals, Men and Morals is a manifesto for an Animal Liberation movement. The contributers to the book may not all see the issue this way. They are a varied group. Philosophers, ranging from professors to graduate students, make up the largest contingent. There are five of them, including the three editors, and there is also an extract from the unjustly neglected German philosopher with an English name, Leonard Nelson, who died in 1927. There are essays by two novelist/critics, Brigid Brophy and Maureen Duffy, and another by Muriel the Lady Dowding, widow of Dowding of Battle of Britain fame and the founder of "Beauty without Cruelty," a movement that campaigns against the use of animals for furs and cosmetics. The other pieces are by a psychologist, a botanist, a sociologist, and Ruth Harrison, who is probably best described as a professional campaigner for animal welfare.

Whether or not these people, as individuals, would all agree that they are launching a liberation movement for animals, the book as a whole amounts to no less. It is a demand for a complete change in our attitudes to nonhumans. It is a demand that we cease to regard the exploitation of other species as natural and inevitable, and that, instead, we see it as a continuing moral outrage. Patrick Corbett, Professor of Philosophy at Sussex University, captures the spirit of the book in his closing words:

> ... we require now to extend the great principles of liberty, equality and fraternity over the lives of animals. Let animal slavery join human slavery in the graveyard of the past.

The reader is likely to be skeptical. "Animal Liberation" sounds more like a parody of liberation movements than a serious objective. The reader may think: We support the claims of blacks and women for equality because blacks and women really are equal to whites and males—equal in intelligence and in abilities, capacity for leadership, rationality, and so on. Humans and nonhumans obviously are not equal in these respects. Since justice demands only that we treat equals equally, unequal treatment of humans and nonhumans cannot be an injustice.

This is a tempting reply, but a dangerous one. It commits the non-racist and non-sexist to a dogmatic belief that blacks and women really are just as intelligent, able, etc., as whites and males—and no more. Quite possibly this happens to be the case. Certainly attempts to prove that racial or sexual differences in these respects have a genetic origin have not been conclusive. But do we really want to stake our demand for equality on the assumption that there are no genetic differences of this kind between the different races or sexes? Surely the appropriate response to those who claim to have found evidence for such genetic differences is not to stick to the belief that there are no differences, whatever the evidence to the contrary; rather one should be clear that the claim to equality does not depend on IQ. Moral equality is distinct from factual equality. Otherwise it would be nonsense to talk of the equality of human beings, since humans, as individuals, obviously differ in intelligence and almost any ability one cares to name. If possessing greater intelligence does not entitle one human to exploit another, why should it entitle humans to exploit nonhumans?

Jeremy Bentham expressed the essential basis of equality in his famous formula: "Each to count for one and none for more than one." In other words, the interests of every being that has interests are to be taken into account and treated equally with the like interests of any other being. Other moral philosophers, before and after Bentham, have made the same point in different ways. Our concern for others must not depend on whether they possess certain characteristics, though just what that concern involves may, of course, vary according to such characteristics.

Bentham, incidentally, was well aware that the logic of the demand for racial equality did not stop at the equality of humans. He wrote:

> The day *may* come when the rest of the animal creation may acquire those rights which never could have been withholden from them but by the hand of tyranny. The French have already discovered that the blackness of the skin is no reason why a human being should be abandoned without redress to the caprice of a tormentor. It may one day come to be recognized that the number of the legs, the villosity of the skin, or the termination of the *os sacrum,* are reasons equally insufficient for abandoning a sensitive being to the same fate. What else is it that should trace the insuperable line? Is it the faculty of reason, or perhaps the faculty of discourse? But a full-grown horse or dog is beyond comparison a more rational, as well as a more conversable animal, than an infant of a day, or a week, or even a month old. But suppose they were otherwise, what would it avail? The question is not, Can they *reason?* nor Can they *talk?* but, Can they *suffer?*[2]

Surely Bentham was right. If a being suffers, there can be no moral justification for refusing to take that suffering into consideration, and, indeed, to count it equally with the like suffering (if rough comparisons can be made) of any other being.

So the only question is: do animals other than man suffer? Most people agree unhesitatingly that animals like cats and dogs can and do suffer, and this seems also to be assumed by those laws that prohibit wanton cruelty to such animals. Personally, I have no doubt at all about this and find it hard to take seriously the doubts that a few people apparently do have. The editors and contributors of *Animals, Men and Morals* seem to feel the same way, for although the question is raised more than once, doubts are quickly dismissed each time. Nevertheless, because this is such a fundamental point, it is worth asking what grounds we have for attributing suffering to other animals.

It is best to begin by asking what grounds any individual human has for supposing that other humans feel pain. Since pain is a state of consciousness, a "mental event," it can never be directly observed. No observations, whether behavioral signs such as writhing or screaming or physiological or neurological recordings, are observations of pain itself. Pain is something one feels, and one can only infer that others are feeling it from various external indications. The fact that only philosophers are ever skeptical about whether other humans feel pain shows that we regard such inference as justifiable in the case of humans.

Is there any reason why the same inference should be unjustifiable for other animals? Nearly all the external signs which lead us to infer pain in other humans can be seen in other species, especially "higher" animals such as mammals and birds. Behavioral signs—writhing, yelping, or other forms of calling, attempts to avoid the source of pain, and many others—are present. We know, too, that these animals are biologically similar in the relevant respects, having nervous systems like ours which can be observed to function as ours do.

So the grounds for inferring that these animals can feel pain are nearly as good as the grounds for inferring other humans do. Only nearly, for there is one behavioral sign that humans have but nonhumans, with the exception of one or two specially raised chimpanzees, do not have. This, of course, is a developed language. As the quotation from Bentham indicates, this has long been regarded as an important distinction between man and other animals. Other animals may communicate with each other, but not in the way we do. Following Chomsky, many people now mark this distinction by saying that only humans communicate in a form that is governed by rules of syntax. (For the purposes of this argument, linguists allow those chimpanzees who have learned a syntactic sign language to rank as honorary humans.) Nevertheless, as Bentham pointed out, this distinction is not relevant to the question of how animals ought to be treated, unless it can be linked to the issue of whether animals suffer.

This link may be attempted in two ways. First, there is a hazy line of philosophical thought, stemming perhaps from some doctrines associated with Wittgenstein, which maintains that we cannot meaningfully attribute states of consciousness to beings without language. I have not seen this argument made explicit in print, though I have come across it in conversation. This position seems to me very implausible, and I doubt that it would be held at all if it were not thought to be a consequence of a broader view of the significance of language. It may be that the use of a public, rule-governed language is a precondition of conceptual thought. It may even be, although personally I doubt it, that we cannot meaningfully speak of a creature having an intention unless that creature can use a language. But states like pain, surely, are more primitive than either of these, and seem to have nothing to do with language.

Indeed, as Jane Goodall points out in her study of chimpanzees, when it comes to the expression of feelings and emotions, humans tend to fall back on nonlinguistic modes of communication which are often found among apes, such as a cheering pat on the back, an exuberant embrace, a clasp of hands, and so on.[3] Michael Peters makes a similar point in his contribution to *Animals, Men and Morals* when he notes that the basic signals we use to convey pain, fear, sexual arousal, and so on are not specific to our species. So there seems to be no reason at all to believe that a creature without language cannot suffer.

The second, and more easily appreciated way of linking language and the existence of pain is to say that the best evidence that we can have that another creature is in pain is when he tells us that he is. This is a distinct line of argument, for it is not being denied that a nonlanguage-user conceivably could suffer, but only that we could know that he is suffering. Still, this line of argument seems to me to fail, and for reasons similar to those just given. "I am in pain" is not the best possible evidence that the speaker is in pain (he might be lying) and it is certainly not the only possible evidence. Behavioral signs and knowledge of the animals's biological similarity to ourselves together provide adequate evidence that animals do suffer. After all, we would not accept linguistic evidence if it contradicted the rest of the evidence. If a man was severely burned, and behaved as if he were in pain, writhing, groaning, being very careful not to let his burned skin touch anything, and so on, but later said he had not been in pain at all, we would be more likely to conclude that he was lying or suffering from amnesia than that he had not been in pain.

Even if there were stronger grounds for refusing to attribute pain to those who do not have a language, the consequences of this refusal might lead us to examine these grounds unusually critically. Human infants, as well as some adults, are unable to use language. Are we to deny that a year-old infant can suffer? If not, how can language be crucial? Of course, most parents can understand the responses of even very young infants better than they understand the responses of other animals, and sometimes infant responses can be understood in the light of later development.

This, however, is just a fact about the relative knowledge we have of our own species and other species, and most of this knowledge is simply derived from closer contact. Those who have studied the behavior of other animals soon learn to understand their responses at least as well as we understand those of an infant. (I am not just referring to Jane Goodall's and other well-known studies of apes. Consider, for example, the degree of understanding achieved by Tinbergen from watching herring gulls.)[4] Just as we can understand infant human behavior in the light of adult human behavior, so we can understand the behavior of other species in the light of our own behavior (and sometimes we can understand our own behavior better in the light of the behavior of other species).

The grounds we have for believing that other mammals and birds suffer are, then, closely analogous to the grounds we have for believing that other humans suffer. It remains to consider how far down the evolutionary scale this analogy holds. Obviously it becomes poorer when we get further away from man. To be more precise would require a detailed examination of all that we know about other forms of life. With fish, reptiles, and other vertebrates the analogy still seems strong, with molluscs like oysters it is much weaker. Insects are more difficult, and it may be that in our present state of knowledge we must be agnostic about whether they are capable of suffering.

If there is no moral justification for ignoring suffering when it occurs, and it does occur in other species, what are we to say of our attitudes toward these other species? Richard Ryder, one of the contributors to *Animals, Men and Morals,* uses the term "speciesism" to describe the belief that we are entitled to treat members of other species in a way in which it would be wrong to treat members of our own species. The term is not euphonious, but it neatly makes the anology with racism. The non-racist would do well to bear the analogy in mind when he is inclined to defend human behavior toward nonhumans. "Shouldn't we worry about improving the lot of our own species before we concern ourselves with other species?" he may ask. If we substitute "race" for "species" we shall see that the question is better not asked.

"Is a vegetarian diet nutritionally adequate?" resembles the slave-owner's claim that he and the whole economy of the South would be ruined without slave labor. There is even a parallel with skeptical doubts about whether animals suffer, for some defenders of slavery professed to doubt whether blacks really suffer in the way that whites do.

I do not want to give the impression, however, that the case for Animal Liberation is based on the analogy with racism and no more. On the contrary, *Animals, Men and Morals* describes the various ways in which humans exploit nonhumans, and several contributors consider the defenses that have been offered, including the defense of meat-eating mentioned in the last paragraph. Sometimes the rebuttals are scornfully dismissive, rather than carefully designed to convince the detached critic. This may be a fault, but it is a fault that is inevitable, given the kind of book this is. The issue is not one on which one can remain detached. As the editors state in their Intro duction:

> Once the full force of moral assessment has been made explicit there can be no rational excuse left for killing animals, be they killed for food, science, or sheer personal indulgence. We have not assembled this book to provide the reader with yet another manual on how to make brutalities less brutal. Compromise, in the traditional sense of the term, is simple unthinking weakness when one considers the actual reasons for our crude relationships with the other animals.

The point is that on this issue there are few critics who are genuinely detached. People who eat pieces of slaughtered nonhumans every day find it hard to believe that they are doing wrong; and they also find it hard to imagine what else they could eat. So for those who do not place nonhumans beyond the pale of morality, there comes a stage when further argument seems pointless, a stage at which one can only accuse one's opponent of hypocrisy and reach for the sort of sociological account of our practices and the way we defend them that is attempted by David Wood in his contribution to this book. On the other hand, to those unconvinced by the arguments, and unable to accept that they are rationalizing their dietary preferences and their fear of being thought peculiar, such sociological explanations can only seem insultingly arrogant.

II

The logic of speciesism is most apparent in the practice of experimenting on nonhumans in order to benefit humans. This is because the issue is rarely obscured by allegations that nonhumans are so different from humans that we cannot know anything about whether they suffer. The defender of vivisection cannot use this argument because he needs to stress the similarities between man and other animals in order to justify the usefulness to the former of experiments on the latter. The researcher who makes rats choose between starvation and electric shocks to see if they develop ulcers (they do) does so because he knows that the rat has a nervous system very similar to man's, and presumably feels an electric shock in a similar way.

Richard Ryder's restrained account of experiments on animals made me angrier with my fellow men than anything else in this book. Ryder, a clinical psychologist by profession, himself experimented on animals before he came to hold the view he puts forward in his essay. Experimenting on animals is now a large industry, both academic and commercial. In 1969, more than 5 million experiments were performed in Britain, the vast majority without anesthetic (though how many of these involved pain is not known). There are no accurate US figures, since there is no federal law on the subject, and in many cases no state law either. Estimates vary from 20 million to 200 million. Ryder suggests that 80 million may be the best guess. We tend to think that this is all for vital medical research, but of course it is not. Huge numbers of animals are used in university departments from Forestry to Psychology, and even more are used for commercial purposes, to test whether cosmetics can cause skin damage, or shampoos eye damage, or to test food additives or laxatives or sleeping pills or anything else.

A standard test for foodstuffs is the "LD50." The object of this test is to find the dosage level at which 50 percent of the test animals will die. This means that nearly all of them will become very sick before finally succumbing or surviving. When the substance is a harmless one, it may be necessary to force huge doses down the animals, until in some cases sheer volume or concentration causes death.

Ryder gives a selection of experiments, taken from recent scientific journals. I will quote two, not for the sake of indulging in gory details, but in order to give an idea of what normal researchers think they may legitimately do to other species. The point is not that the individual researchers are cruel men, but that they are behaving in a way that is allowed by our speciesist attitudes. As Ryder points out, even if only 1 percent of the experiments involve severe pain, that is 50,000 experiments in Britain each year, or nearly 150 every day (and about fifteen times as many in the United States, if Ryder's guess is right). Here then are two experiments:

> O.S. Ray and R.J. Barrett of Pittsburgh gave electric shocks to the feet of 1,042 mice. They then caused convulsions by giving more intense shocks through cup-shaped electrodes applied to the animals' eyes or through pressure spring clips attached to their ears. Unfortunately some of the mice who "successfully completed Day One training were found sick or dead prior to testing on Day Two." [*Journal of Comparative and Physiological Psychology,* 1969, Vol. 67, pp. 110-116]
>
> At the National Institute for Medical Research, Mill Hill, London, W. Feldberg and S.L. Sherwood injected chemicals into the brains of cats—"with a number of widely different substances, recurrent patterns of reaction were obtained. Retching, vomiting, defaecation, increased salivation and greatly accelerated respiration leading to panting were common features." ...
>
> The injection into the brain of a large dose of Tubocuraine caused the cat to jump "from the table to the floor and then straight into its cage, where it started calling more and more noisily whilst moving about restlessly and jerkily ... finally the cat fell with legs and neck flexed, jerking in rapid clonic movements, the condition being that of a major [epileptic] convulsion ... within a few seconds the cat got up, ran for a few yards at high speed and fell in another fit. The whole process was repeated several times within the next ten minutes, during which the cat lost faeces and foamed at the mouth."
>
> This animal finally died thirty-five minutes after the brain injection. [*Journal of Physiology,* 1954, Vol. 123, pp. 148-167]

There is nothing secret about these experiments. One has only to open any recent volume of a learned journal, such as the *Journal of Comparative and Physiological Psychology,* to find full descriptions of experiments of this sort, together with the results obtained—results that are frequently trivial and obvious. The experiments are often supported by public funds.

It is a significant indication of the level of acceptability of these practices that, although these experiments are taking place at this moment on university campuses throughout the country, there has, so far as I know, not been the slightest protest from the student movement. Students have been rightly concerned that their universities should not discriminate on grounds of race or sex, and that they should not serve the purposes of the military or big business. Speciesism continues undisturbed, and many students participate in it. There may be a few qualms at first, but since everyone regards it as normal, and it may even be a required part of a course, the student soon becomes hardened and, dismissing his earlier feelings as "mere sentiment," comes to regard animals as statistics rather than sentient beings with interests that warrant consideration.

Argument about vivisection has often missed the point because it has been put in absolutist terms: would the abolitionist be prepared to let thousands die if they could be saved by experimenting on a single animal? The way to reply to this purely hypothetical question is to pose another: Would the experimenter be prepared to experiment on a human orphan under six months old, if it were the only way to save many lives? (I say "orphan" to avoid the complication of parental feelings, although in doing so I am being overfair to the experimenter, since the nonhuman subjects of experiments are not orphans.) A negative answer to this question indicates that the experimenter's readiness to use nonhumans is simple discrimination, for adult apes, cats, mice, and other mammals are more conscious of what is happening to them, more self-directing, and, so far as we can tell, just as sensitive to pain as a human infant. There is no characteristic that human infants possess that adult mammals do not have to the same or a higher degree.

(It might be possible to hold that what makes it wrong to experiment on a human infant is that the infant will in time develop into more than the nonhuman, but one would then, to be consistent, have to oppose abortion, and perhaps contraception, too, for the fetus and the egg and sperm have the same potential as the infant. Moreover, one would still have no reason for experimenting on a nonhuman rather than a human with brain damage severe enought to make it impossible for him to rise above infant level.)

The experimenter, then, shows a bias for his own species whenever he carries out an experiment on a nonhuman for a purpose that he would not think justified him in using a human being at an equal or lower level of sentience, awareness, ability to be self-directing, etc. No one familiar with the kind of results yielded by these experiments can have the slightest doubt that if this bias were eliminated the number of experiments performed would be zero or very close to it.

III

If it is vivisection that shows the logic of speciesism most clearly, it is the use of other species for food that is at the heart of our attitudes toward them. Most of *Animals, Men and Morals* is an attack on meat-eating—an attack which is based solely on concern for nonhumans, without reference to arguments derived from considerations of ecology, macrobiotics, health, or religion.

The idea that nonhumans are utilities, means to our ends, pervades our thought. Even conservationists who are concerned about the slaughter of wild fowl but not about the vastly greater slaughter of chickens for our tables are thinking in this way—they are worried about what we would lose if there were less wildlife. Stanley Godlovitch, pursuing the Marxist idea that our thinking is formed by the activities we undertake in satisfying our needs, suggests that man's first classification of his environment was into Edibles and Inedibles. Most animals came into the first category, and there they have remained.

Man may always have killed other species for food, but he has never exploited them so ruthlessly as he does today. Farming has succumbed to business methods, the objective being to get the highest possible ratio of output (meat, eggs, milk) to input (fodder, labor costs, etc.). Ruth Harrison's essay "On Factory

Farming" gives an account of some aspects of modern methods, and of the unsuccessful British campaign for effective controls, a campaign which was sparked off by her *Animal Machines* (Stuart: London, 1964).

Her article is in no way a substitute for her earlier book. This is a pity since, as she says, "Farm produce is still associated with mental pictures of animals browsing in the fields, ... of hens having a last forage before going to roost. ..." Yet neither in her article nor elsewhere in *Animals, Men and Morals* is this false image replaced by a clear idea of the nature and extent of factory farming. We learn of this only indirectly, when we hear of the code of reform proposed by an advisory committee set up by the British government.

Among the proposals, which the government refused to implement on the grounds that they were too idealistic, were: *"Any animal should at least have room to turn around freely."*

Factory farm animals need liberation in the most literal sense. Veal calves are kept in stalls five feet by two feet. They are usually slaughtered when about four months old, and have been too big to turn in their stalls for at least a month. Intensive beef herds, kept in stalls only proportionately larger for much longer periods, account for a growing percentage of beef production. Sows are often similarly confined when pregnant, which, because of artificial methods of increasing fertility, can be most of the time. Animals confined in this way do not waste food by exercising, nor do they develop unpalatable muscle.

"A dry bedded area should be provided for all stock." Intensively kept animals usually have to stand and sleep on slatted floors without straw, because this makes cleaning easier.

"Palatable roughage must be readily available to all calves after one week of age." In order to produce the pale veal housewives are said to prefer, calves are fed on an all-liquid diet until slaughter, even though they are long past the age at which they would normally eat grass. They develop a craving for roughage, evidenced by attempts to gnaw wood from their stalls. (For the same reason, their diet is deficient in iron.)

"Battery cages for poultry should be large enough for a bird to be able to stretch one wing at a time." Under current British practice, a cage for four or five laying hens has a floor area of twenty inches by eighteen inches, scarcely larger than a double page of the *New York Review of Books*. In this space, on a sloping wire floor (sloping so the eggs roll down, wire so the dung drops through) the birds live for a year or eighteen months while artificial lighting and temperature conditions combine with drugs in their food to squeeze the maximum number of eggs out of them. Table birds are also sometimes kept in cages. More often they are reared in sheds, no less crowded. Under these conditions all the birds' natural activities are frustrated, and they develop "vices" such as pecking each other to death. To prevent this, beaks are often cut off, and the sheds kept dark.

How many of those whose support factory farming by buying its produce know anything about the way it is produced? How many have heard something about it, but are reluctant to check up for fear that it will make them uncomfortable? To nonspeciesists, the typical consumer's mixture of ignorance, reluctance to find out the truth, and vague belief that nothing really bad could be allowed seems analogous to the attitudes of "decent Germans" to the death camps.

There are, of course, some defenders of factory farming. Their arguments are considered, though again rather sketchily, by John Harris. Among the most common: "Since they have never known anything else, they don't suffer." This argument will not be put by anyone who knows anything about animal behavior, since he will know that not all behavior has to be learned. Chickens attempt to stretch wings, walk around, scratch, and even dustbathe or build a nest, even though they have never lived under conditions that allowed these activities. Calves can suffer from maternal deprivation no matter at what age they were taken from their mothers. "We need these intensive methods to provide protein for a growing population." As ecologists and famine relief organizations know, we can produce far more protein per acre if we grow the right vegetable crop, soy beans for instance, than if we use the land to grow crops to be converted into protein by animals who use nearly 90 percent of the protein themselves, even when unable to exercise.

There will be many readers of this book who will agree that factory farming involves an unjustifiable degree of exploitation of sentient creatures, and yet will want to say that there is nothing wrong with rearing animals for food, provided it is done "humanely." These people are saying, in effect, that although we should not cause animals to suffer, there is nothing wrong with killing them.

There are two possible replies to this view. One is to attempt to show that this combination of attitudes is absurd. Roslind Godlovitch takes this course in her essay, which is an examination of some common attitudes to animals. She argues that from the combination of "animal suffering is to be avoided" and "there is nothing wrong with killing animals" it follows that all animal life ought to be exterminated (since all sentient creatures will suffer to some degree at some point in their lives). Euthanasia is a contentious issue only because we place some value on living. If we did not, the least amount of suffering would justify it. Accordingly, if we deny that we have a duty to exterminate all animal life, we must concede that we are placing some value on animal life.

This argument seems to me valid, although one could still reply that the value of animal life is to be derived from the pleasures that life can have for them, so that, provided their lives have a balance of pleasure over pain, we are justified in rearing them. But this would imply that we ought to produce animals and let them live as pleasantly as possible, without suffering.

At this point, one can make the second of the two possible replies to the view that rearing and killing animals for food is all right so long as it is done humanely. This second reply is that so long as we think that a nonhuman may be killed simply so that a human can satisfy his taste for meat, we are still thinking of nonhumans as means rather than as ends in themselves. The factory farm is nothing more than the application of technology to this concept. Even traditional methods involve castration, the separation of mothers and their young, the breaking up of herds, branding or ear-punching, and of course transportation to the abattoirs and the final moments of terror when the animal smells blood and senses danger. If we were to try rearing animals so that they lived and died without suffering, we should find that to do so on anything like the scale of today's meat industry would be a sheer impossibility. Meat would become the prerogative of the rich.

I have been able to discuss only some of the contributions to this book, saying nothing about, for instance, the essays on killing for furs and for sport. Nor have I considered all the detailed questions that need to be asked once we start thinking about other species in the radically different way presented by this book. What, for instance, are we to do about genuine conflicts of interest like rats biting slum children? I am not sure of the answer, but the essential point is just that we *do* see this as a conflict of interests, that we recognize that rats have interests too. Then we may begin to think about other ways of resolving the conflict—perhaps by leaving out rat baits that sterilize the rats instead of killing them.

I have not discussed such problems because they are side issues compared with the exploitation of other species for food and for experimental purposes. On these central matters, I hope that I have said enough to show that this book, despite its flaws, is a challenge to every human to recognize his attitudes to nonhumans as a form of prejudice no less objectionable than racism or sexism. It is a challenge that demands not just a change of attitudes, but a change in our way of life, for it requires us to become vegetarians.

Can a purely moral demand of this kind succeed? The odds are certainly against it. The book holds out no inducements. It does not tell us that we will become healthier, or enjoy life more, if we cease exploiting animals. Animal Liberation will require greater altruism on the part of mankind than any other liberation movement, since animals are incapable of demanding it for themselves, or of protesting against their exploitation by votes, demonstrations, or bombs. Is man capable of such genuine altruism? Who knows? If this book does have a significant effect, however, it will be a vindication of all those who have believed that man has within himself the potential for more than cruelty and selfishness.

Notes

1. This article originally appeared as a book review of *Animals, Men and Morals,* edited by Stanley and Roslind Godlovitch and John Harris.

2. *The Principles of Morals and Legislation,* Ch. XVII, Sec. 1, footnote to paragraph 4. (Italics in original.)

3. Jane van Lawick-Goodall, *In the Shadow of Man* (Houghton Mifflin, 1971), p. 225.

4. N. Tinbergen, *The Herring Gull's World* (Basic Books, 1961).

"ANIMAL LIBERATION": A CRITIQUE[1]

Michael Fox — Queen's University, Kingston, Ontario

IN THE PAST few years, philosophers have actively engaged in a (long overdue) discussion of racism and sexism. The growing body of literature on these subjects has just begun to generate a new controversy—or perhaps more accurately, to rekindle an old one—in which some of those interested in applied ethics or "current moral issues" are taking part: that of so-called animal liberation. I wish to examine here two very recent attempts to provide the cause of humane treatment for animals with a solid philosophical foundation. These are Peter Singer's *Animal Liberation: A New Ethics for Our Treatment of Animals* and Tom Regan's "The Moral Basis of Vegetarianism."[2] The authors of both works propound a doctrine of animal liberation, that is, liberation from being discriminated against and used at pleasure by human beings—Singer's treatment of the subject being both wider (supporting antivivisectionism as well as vegetarianism) and developed in more detail. They do so by advancing a case for granting moral rights to animals (the right to equal consideration of interests, the right to life).

Now, I am prepared to concede that animals may have interests, in the sense that they are capable of distinguishing between states of consciousness which are painful and those that are pleasurable or accompany physical well-being, and that they seek the latter and avoid the former as much as possible.[3] It does not seem objectionable to say that because animals are capable of pleasure as well as suffering a pleasurable existence is "in their best interest." Both Singer and Regan assert that animals' painful and pleasurable experiences are qualitatively and quantitatively the same as those of humans and that, hence, animals have a capacity to enjoy life equal to that of humans. But even leaving aside this peculiar claim (which I must confess I have no idea how to interpret or evaluate), it is very difficult to see how animals' having interests per se entails their having equal interests with human beings and, as a consequence, the associated moral rights that the latter possess. Singer and Regan, in other words, take animals' capacity to enjoy and suffer as the sole fact that is morally relevant to these alleged entailments.

Regan goes further than Singer, however, contending that animals have, in addition to the right, *ceteris paribus,* to equal consideration of interests with humans in the matter of treatment, a natural right to life, which cannot be overridden except by the most stringent utilitarian considerations. He maintains that just as no amount of human pleasure equal to or greater than a given amount of "non-trivial" (R, p. 198) animal suffering caused by man can ever neutralize the moral condemnation engendered by the infliction of that suffering, so, too, the death of an animal cannot, in general, be justified by the amount of human pleasure which is consequent upon it. His reason for saying this is that any argument which purports to show that humans have a right not to be maltreated or unjustly caused to suffer to a degree equal to or greater than the level of someone else's gain in pleasure, or that human beings have a right to life, will also hold in the case of animals (at least higher sentient animals).

I shall argue, against both of these views, that the concept of moral rights cannot be extended to include animals, and that the question of animals' rights is therefore a bogus issue. We may and ought to be concerned about the welfare of animals and their present exploitation by man because they are sentient beings. But this concern and this simple fact neither license nor entail the postulation of animal rights.

Singer and Regan approach the question of animals' rights by focusing on a condemnation of what Singer calls "speciesism,"[4] which is defined as "a prejudice or attitude of bias toward the interests of members of one's own species and against those of members of other species" (S, p. 7). The main thesis advanced by them is that if we cannot morally justify discriminating against other human beings on grounds of race or sex (for example) then, for exactly the same sorts of reasons, we cannot morally justify discriminatory treatment of animals. Just as there are no morally relevant considerations which warrant exploiting other humans for our own ends, so there are none to warrant the exploitation of nonhumans. Just as there is no difference between human groups (such as those of different races, sexes, or intelligence levels) with respect to their capacity to suffer, so there is none between humans and animals—at least those animals with highly complex nervous systems like our own. From this standpoint humans are animals *tout court* and, as such, do not stand in any position of natural superiority. This is not, of course, to deny any significant differences between humans and animals, and neither Singer nor Regan commits himself to this absurdity. Rather, it is to deny any moral superiority on the part of *Homo sapiens.*

How can this view be sustained? It is important to note that Singer and Regan adopt the principle, first enunciated clearly by Jeremy Bentham in his *Introduction to the Principles of Morals and Legislation* (1789), that the only capacity which counts in assigning moral rights is the capacity to suffer. Color of skin, sex, rationality, intelligence, and ability to communicate are not relevant in justifying unequal treatment. And if they do not count in one instance (discrimination against other humans), then they likewise fail in every other case (including the discriminatory treatment of animals by humans). Everyone recognizes that a mature, healthy horse, cat, or pig is more intelligent than a newborn infant or a severely retarded child and communicates at least as well. Furthermore, if the ability to reason is to be attributed to any of these, it will be to the animals and not the infant. But although humans are supposed to exercise rightful dominion over all of nature because they are more intelligent, rational, capable of communicating, and so on, no one suggests that we may eat, experiment upon, hunt, or make shoes and soap out of infants and retarded children because they are less endowed in these same respects (and, indeed, less so than some animals). Nor does anyone believe that a mongoloid baby may be used as a mere means to someone else's ends because it has less potentiality for developing those capacities that are most characteristically human, though this "lesser potentiality" argument is often used as the justification for our treatment of animals—in spite of the fact that some animals have greater potentiality than many defective infants for developing or exhibiting the same valued characteristics. Now why are these things so? Singer and Regan hold that we have no answer to this question, and this merely shows the hypocrisy and logical inconsistency in our dealings with the animal kingdom.

The strategy of both authors throughout is to force the reader into the uncomfortable position of either (*a*) having to refute their claims—many of which are presented quite convincingly—or (*b*) confessing that he is a speciesist, and that since speciesism is morally indefensible and, hence, reprehensible, he must either change his behavior toward animals or be a hypocrite. Thus Regan argues, for example, that if pain is truly an intrinsic evil—regardless of whose pain it is—if causing pain is therefore prima facie morally wrong and must always be justified, and if today's large-scale rearing and slaughtering methods do genuinely involve the unjustified infliction of (a great deal of) pain and suffering on undeserving animals, then these methods are immoral and should be prohibited. It follows that anyone who benefits from the rearing and slaughtering practices in question tacitly condones and helps sustain and encourage them and is hence in a morally untenable position.[5] For both Regan and Singer, we "are rationally compelled to regard animals as beings who count for something when we attempt to determine what we morally ought or ought not to do" (R, p. 186).

Now there is a great deal in the Singer-Regan position that merits extended discussion, and I can only hope to consider some main points here. I shall start with the question of the nature of rights itself. On this vital matter, Singer has surprisingly little to day. He accepts without reservation Bentham's line of reasoning, which may be reconstructed as follows: (1) "equal consideration of interests ... [is] a basic moral principle" (S, p. 8); (2) "the capacity for suffering and enjoyment is a *prerequisite for having interests at all*" (S, p. 9; Singer's italics); (3) this capacity is the condition that requires us to grant a right to equal consideration of interests; (4) any being which possesses this capacity has such a right and should be treated accordingly; (5) some animals possess this capacity; (6) therefore, some animals have a right to equal consideration of interests when we decide how to treat them (i.e., when suffering and enjoyment are possible consequences of our behavior toward them). (Regan's grounds for assigning this same right to animals closely parallel those stated here and do not require independent consideration. He asserts the additional right to life, which Singer does not explicitly do, because he clearly recognizes that the Benthamite argument above entails that if animals were reared and slaughtered painlessly, and their natural needs allowed to be satisfied during their rearing period, there would be no moral objection to meat eating.) Singer and Regan both acknowledge that, as Singer points out, "the basic principle of equality does not require equal or identical *treatment;* it requires equal *consideration.* Equal consideration for different beings may lead to different treatment and different rights" (S, p. 3; Singer's italics). So it might be said that differential treatment and granting of rights can be justified by empirically determined differences among individuals or groups—but only in the light of this general moral principle. (For example, it is not society's obligation to send hopelessly brain-damaged children through elementary school, but giving them costly medical care may well be.) What is important to maintain, then, is the principle of equitable or fair treatment. Hence (known and possibly discoverable) empirical differences between races or sexes cannot by themselves justify differential treatment, for "equality is a moral idea, not an assertion of fact" (S, p. 5). Humans are, in many ways, unequal in fact. So any attempt to make equality rest on characteristics that all human beings share, both authors agree, has to seek the lowest common denominator. But this will not enable us to separate human beings from animals as beings whose interests deserve equal consideration, for the lowest common denominator will have to be nothing other than the capacity to enjoy and suffer.

There are numerous flaws in this argument. First of all, if all talk about interests (in the moral sense) is meaningful only in relation to the capacity to enjoy and suffer, then it now appears that the ethical idea of equality does rest upon an assertion of factual equality after all—namely, the fact that all human beings have this capacity. But if one factual consideration is relevant to assigning rights to beings, then others may be as well, and the question of the other capacities which beings must have to be proper subjects of morality becomes important to consider.

Singer and Regan insist that any characteristic which is used as a basis for assigning moral rights to human beings must be universal, that is, possessed by all humans without exception. This is why they fasten onto the capacity to enjoy and suffer, with the totally unsurprising result that we cannot find anything else that fits this extreme requirement. But even if we play by their rules, it can be doubted whether any characteristic is really universal in so strong a sense, the capacity to enjoy and suffer included. To begin with, as physiologists well know, there is a rare but thoroughly documented condition called "congenital universal indifference (or insensitivity) to pain,"[6] which is characterized by complete absence, throughout life, of any pain-sensing capability. But if the capacity to experience pain is missing, any rights predicated on it must vanish as well. In addition, completely anesthetized, hypnotized, or deeply comatose human beings lack the capacity in question and hence, too, any corresponding rights. If this is an unacceptable conclusion, however (as I think everyone would agree it is), it is instructive to see why. The reasons are: (1) that basic moral rights arise from other criteria than the capacity to enjoy and suffer (this capacity being a necessary but not a sufficient condition for the granting of the rights in question);[7] and (2) that what counts in establishing rights are the characteristics that a certain class of beings share in general, even if not universally.

The search for attributes that all humans, without exception, share in common and which are supposed to furnish the grounds for the assigning of moral rights to them, as well as to any sufficiently similar beings, is bound to be futile; for even the capacity of humans to experience pain and pleasure falls short of complete universality, as we have just seen. But then if we shift our attention instead to capacities that are nearly or virtually universal among humans, as we are forced to do, it will be seen that humans generally possess them and (probably) no animals do and, hence, that the concept of a moral right to equitable treatment makes no sense except as applied to humans.

Regan challenges the assertion that humans are different from animals in morally relevant ways by declaring that the opponents of his position bear the onus of providing adequate empirical evidence to support their claim and that such evidence does not at present exist. It seems to me, however, that (as I think most people would agree on the basis of experience) all animals—whatever their place on the evolutionary scale—are prima facie significantly different kinds of creatures from humans, in morally relevant as well as other ways, and that the onus of proof is therefore on those who would hold otherwise. Further, though experimental psychology, comparative anatomy and physiology, and the biological and ecological sciences are far from being able to yield all the evidence Regan demands, it is surely naive in the extreme to blithely brush aside as of no consequence (R, p. 191) all the data on the important differences between animals and humans which have been gathered to date. Let us assume, for the sake of argument, that it could be shown to everyone's satisfaction that animals experience pleasure and pain in the same way and to the same degree as humans and, further, that many also reason, have emotions, use some form of symbolic communication, and have a sense of self-identity. It still would not follow that these facts would qualify such animals to be recipients of moral rights. For, as H.J. McCloskey has recently pointed out,[8] to appreciate (1) that the existence of certain higher animals is intrinsically valuable because they possess some capacities (like sentience, intelligence, emotionality), the exercise of which enables them to enjoy a quality of life that humans can recognize as of value, (2) that they are capable of suffering psychologically as well as physically, and, (3) that as a consequence of 1 and 2, good reasons are required to be given for killing such beings is not tantamount to, and does not entail, assigning animals moral rights.

What other characteristics, then, that humans share in general should be cited in order to give an adequate account of the reasons why they have, and animals lack, moral rights? A complete list of these would have to include at least the following: the capacities to be critically self-aware, manipulate concepts, use a sophisticated language,[9] reflect, plan, deliberate, choose, and accept responsibility for acting. In a similar vein, McCloskey suggests that the crucial morally relevant characteristics of humans which we are seeking here are those which manifest the attributes of truly autonomous beings, where this entails being capable of acting freely, choosing and deciding rationally in the fullest sense, creating, and self-making (self-realizing).[10] I have drawn attention to certain cognitive capacities (critical self-awareness, concept manipulation, and the use of a sophisticated language) because these are the essential tools or vehicles by means of which an agent's autonomy is evolved, made known to himself reflexively, and manifested or expressed. The possession of these cognitive capacities, therefore, is a necessary pre-prequisite for autonomy, which is the capacity for self-conscious, voluntary, and deliberate action, in the fullest sense of these words. Autonomy, which thus entails certain cognitive capacities, is necessary (and, together with the capacity to enjoy and suffer, sufficient) for the possession of moral rights. It follows that all (and only) those beings which are members of a species of which it is true in general (i.e., typically the case at maturity, assuming normal development) that members of the species in question can be considered autonomous agents are beings endowed with moral rights.[11]

Now how can the above entailments be defended? I cannot give full treatment to this important topic here, but I should like to suggest that only autonomous beings, as just described, can and do belong to a moral community, which is the sort of social group within which (and only within which) such concepts as those of rights and duties have any meaning and application. For it is only in a community of interacting autonomous beings of this sort that there can be the kind of mutual recognition required for these concepts to evolve and be understood. Obligations and rights, as well as the moral discourse generated by these and ancillary notions, are functions of mutual recognition and accountability and are, consequently, inapplicable outside the context specified. It should be made clear that the foregoing is not an attempt merely to legislate concerning the kinds of beings which qualify as possessors of moral rights. Rather, my analysis is meant to suggest that, since the only species we know of that has developed the concepts of rights and obligations (and the institutions associated with them) is *Homo sapiens,* there must be something about this peculiar sort of social being that accounts for the phenomenon in question. And my argument is that the relevant features of humans (other than their capacity to suffer and enjoy) that explain why they have rights are their possession of a certain kind of consciousness, particular cognitive and linguistic abilities, and the capacity to comprehend, undertake, and carry out obligations and to expect the same of like beings.

The considerations taken up briefly here should suffice to show that regarding the cognitive capacities of human beings as relevant to the question of possessing moral rights is not tantamount to invoking some simplistic notion of humans' rationality to settle a vastly more complex set of issues, as proponents of animal rights frequently suppose. Singer and Regan just conveniently leave the capacities I have mentioned out of the picuture or else systematically misunderstand and underrate their significance.

I conclude, then, that it is difficult to see how an argument for ascribing specifically moral rights to animals can get started. And if it cannot get off the ground, then there also appears to be no case for saying either that animals ought not to be treated as means to human ends, provided that they are treated in as humane a manner as possible in the process, or that they have a right to life. But it seems to me that the overall obligation to prevent or minimize animal suffering should suffice as a moral basis for prohibiting the atrocious conditions of crowding and confinement that prevail on modern "factory farms," for drastically curtailing the use of animals in excruciating but pointless experiments in product testing, and for ending other inhumane practices (in slaughtering, trapping, the keeping of pets, hunting, racing, and so on). Undoubtedly, animals should not be maltreated. They should not be made to suffer needlessly or excessively.[12] Singer and Regan are surely correct to single out animals' capacity to suffer as the reason why we should treat them humanely. But it is no more clear how this extends moral rights to them than how our dawning ecological sense that we ought not to waste natural resources and systematically ravage the environment would establish moral rights for trees, lakes, or mineral deposits. What should be said is that we have an obligation to avoid mistreating animals, but that this is an obligation without a corresponding right on the part of the beings affected by our behavior.[13]

The argument presented thus far undercuts Singer's surely exaggerated claim that philosophers have felt the need to posit "some basis for the moral gulf that is commonly thought to separate humans and animals, but can find no concrete difference that will do this without undermining the equality of humans ... " (S, pp. 266-67). It is difficult to see how Singer can maintain the position that there is no "moral gulf" separating humans from animals when he also makes the following (clearly speciesist) remark: "It is not arbitrary to hold that the life of a self-aware being, capable of abstract thought, of planning for the future, of complex acts of communication, and so on, is more valuable than the life of a being without these capacities" (S., p. 23). For once it is admitted that certain forms of life are inherently more valuable than others (valuable to whom, incidentally, if not to humans?), then it has already been conceded that the allegedly "more valuable" beings have a greater claim to life, pleasure, and freedom from suffering than those lacking the capacities in question. And it becomes highly problematic how Singer can go on from there to defend such views as that animal pleasure and pain are both qualitatively and quantitatively the same as those of humans and that their capacity for enjoying life is the same.

There is another important weakness in Singer's central argument, arising from his own (otherwise carefully suppressed) speciesist commitments. At one point, he considers the objection that if it were shown that plants, too, are capable of suffering (though there seems to be no reason at all for thinking they are), then it would follow that humans would be morally obliged to starve themselves to death rather than cause suffering to sentient beings. Singer replies, "If we must inflict pain or starve, we would then have to choose the lesser evil. Presumably it would still be true that plants suffer less than animals, and therefore it would still be better to eat plants than to eat animals" (p. 263). But if human beings have this kind of (absolute?) right to live and are thereby licensed to "choose the lesser evil" in this way, then they are in a position of "moral superiority" after all vis-à-vis the rest of nature. The same point can be made in regard to Singer's claim (p. 260) that animal populations may legitimately be controlled if they threaten our food supply. How can this be so if their right to life is on a par with our own? No answer is given.

It would seem to follow that if "lesser evil" arguments are to be admitted into this discussion, then the use of animals in research may be justified by a similar (and, in fact, the usual) appeal: that it is a lesser evil to subject some animals to suffering and possible death than to allow many humans (including those yet unborn) to suffer and perhaps die for lack of the knowledge that could be attained by such research. Singer does document in vivid and sordid detail the extraordinary and often apparently pointless suffering to which many laboratory animals have been routinely forced to submit. One can agree that this situation is shameful and intolerable, if accurately reported. However, the reader is never provided with descriptions of experiments that have proved beneficial to mankind—or to animals, for that matter. Nor is he ever given a larger context into which to place the experiments depicted in order to be able to judge whether they are, indeed, pointless and whether the number of pointless experiments is within the margin of error that might be acceptable when so many experiments are being performed by so many different investigators. But without this larger context, and in light of the general failure of the case for animal rights, and for the consequent principle that animals may not justifiably be used for human ends where some degree of discomfort or suffering is caused, no grounds remain for Singer's assertion that "if the experimenter would not be prepared to use a human infant then his readiness to use nonhuman animals reveals an unjustifiable form of discrimination on the basis of species ..." (S, p. 79).

Singer's entire case against animal experiments is buttressed by carefully selected and one-sided accounts, partial information, and outright misinformation. Of course, there are some insensitive persons engaged in animal research, just as there are in research on humans. But why one should regard these as typical remains a mystery—unless one deliberately adopts an a priori assumption about the fiendish qualities of laboratory personnel. Singer conveniently neglects to acknowledge the degree to which researchers have come to realize that better and more consistent experimental results are obtained when their animals' total environments are controlled and made to simulate natural conditions as closely as possible. It has also been found that when this is done fewer animals are needed for a given experiment than would otherwise be used. This shows that even aside from humane considerations, intelligent researchers have a practical interest in treating their animals with the best of care. Though it is, of course, difficult to say how many have learned these lessons, there is reason to believe that there is a growing awareness of the need for better animal care in the laboratory.

Singer points out that there are alternatives to experiments on animals, such as the use of tissue cultures and computer simulations. No doubt further advances will be made in these areas. But he misleads the reader seriously when he suggest that virtually all animal experiments could be eliminated by such surrogates. For the biomedical researcher and the teacher there is no substitute for a complete and healthy cardiovascular or central nervous system.

It might be of interest to note here that in Canada (about which Singer says nothing), most animal experiments for research purposes are done under guidelines set down by the Canadian Council on Animal Care. Under its experiment assessment scheme, responsibility for appraising the amount of pain an animal will be likely to suffer, as well as its environment, housing, procurement and transportation, anesthesia and euthanasia, is assigned to local committees at each research facility. These guidelines, though voluntary rather than legislated, have proved highly effective in preventing the abuse of animals. In Ontario, these matters are the subject of broad-ranging legislation, covering everything from anticipated pain levels in experiments to all aspects of hygiene in the research facility; from the animal's total artificial environment to postoperative care. Ontario Regulation 139/71 frequently states that conditions must be "suitable for the health, comfort, and welfare" of the animals concerned.[14] Maybe the lot of most research animals is not so fortunate, but at least we can see that the situation is not so hopeless as Singer would have us believe and is, in fact, improving.

Again, Singer paints a shocking and lamentable picture of the competitive and virtually unregulated large-scale animal-rearing methods utilized in the United States and Britain. Something clearly must be done to prevent and eventually ban the abuses he documents. But this issue has little to do with the "justification" for eating meat, if one is required. Singer gives a very distorted view of humans' position in nature. If humans are part of the larger ecological balance and are responsible for maintaining it, then why may they not be viewed as part of the carnivorous, as well as the herbivorous, food chain as well? How can these eating habits be regarded as antinature, so long as humans replenish (or allow to replenish) what they take away? And if intervening in the natural course of events to regulate the rampant population of certain species through fertility control (as we saw above Singer suggests) is not somehow immoral, then in what way is farming livestock or commercial fishing inherently wrong? These questions are never realistically or fully examined by Singer, Regan, or any other animal-rights spokesman, as far as I know. It is true that unlike other animals, humans have the capacity to weight alternatives and make a conscious moral choice whether or not to kill animals for food, as Singer points out (S, p. 250). But what follows from this fact? Simply that meat-eating humans are morally obligated to address themselves to the problem of animal suffering caused by intensive farming methods and to demand that the rearing (and slaughtering) of animals be brought under strict regulation.

Singer admits that there is no logical inconsistency in thinking that animals should be guarded against cruel treatment and simultaneously including meat in one's diet. However, he tries to base part of his argument for becoming a vegetarian on the claim that

> ... practically and psychologically it is impossible to be consistent in one's concern for nonhuman animals while continuing to dine on them. If we are prepared to take the life of another being merely in order to satisfy our taste for a particular type of food, then that being is no more than a means to our end. In time we will come to regard pigs, cattle and chickens as things for us to use, no matter how strong our compassion may be; and when we find that to continue to obtain supplies of the bodies of these animals at a price we are able to pay it is necessary to change their living conditions a little, we will be unlikely to regard these changes too critically. The factory farm is nothing more than the application of technology to the idea that animals are means to our ends. [S, p. 172]

Here we have a classical "slippery slope" argument. Such arguments should always be considered suspect, since careful scrutiny usually shows them to be meretricious at best. In this case, it is not at all clear that people would fail, out of narrow self-interest, to challenge the abominable farming methods Singer describes, if they were confronted with the full facts. To declare otherwise is to assume an unwarrantably low estimate of human decency (a typical feature of slippery slope arguments, incidentally). To go beyond this and assert that "no one in the habit of eating an animal can be completely without bias in judging whether the conditions in which that animal is reared cause suffering" (S, p. 172) is plainly absurd (cf., "No radical can get a fair and impartial trial by jury"). Singer's own conversion from meat eater to vegetarian itself falsifies this extreme dictum. Finally, factory farms are only a technological triumph in the eyes of their greedy and unfeeling proprietors.

The strongest part of Singer's case against meat eating is his brief discussion of the world food crisis. It is a patent truth that by any conceivable health standards most North Americans are overfed. More specifically, they eat far more meat than is necessary to maintain adequate nutrition. Surely some of the excess food they consume should be distributed, in some form, to the starving millions of the world. One can only agree. Modern livestock farming on a grand scale also wastes a colossal amount of feed grains on animals which, in times past, would simply have fed off the land. Even if, contrary to fact, none of this feed grain could be used to nourish humans elsewhere in the world, at least the land which yields the grain could be sown with high-protein-yielding crops, such as soybeans, according to Singer. There is no doubt a good deal of truth in this last point as well, and we are here presented with a serious moral problem concerning the world food supply. But even this fails to establish a case for vegetarianism. All it establishes is that we should eat far less meat so that factory farms become obsolete and that, in conjuction with this, arable land should be turned over to the production of high-protein crops, where possible, so that world hunger can be alleviated somewhat.

We are given to believe that meat is merely a wasteful luxury and that vegetarians are probably healthier than meat eaters. But meat is not a wasteful luxury per se: ruminants can (whereas humans cannot) utilize cellulose to produce carbohydrates and, from these, synthesize certain essential amino acids, the constituents of protein. Not only this, but as Singer himself notes (S, p. 196), "animal foods ... have a very well-balanced amino acid composition" (which a diet that excludes all animal products can easily lack); they are also a natural source of vitamin B_{12}, which pure vegetarians (or "vegans") have to take as a dietary supplement for complete nutrition. Nor is it clear that being a vegetarian is inherently healthier, unless one is impressed by such statistics as that concerning "the 'mean transit time' of food through the digestive system" (76-83 hours for nonvegetarians, 42 hours for vegetarians), and the highly speculative inferences based on them (S, p. 193 n.).[15]

In closing his book, Singer emphasizes the rational approach he has taken toward the discussion of speciesism: "I have *argued* for it, appealing to reason rather than to emotion or sentiment" (p. 270; Singer's italics). In general this is true, and his arguments are often well constructed. But he likewise often falls considerably short of such objective detachment: "Flesh taints our meals. Disguise it as we may, the fact remains that the centerpiece of our dinner has come to us from the slaughterhouse, dripping blood. Untreated and unrefrigerated, it soons begins to putrefy and stink. When we eat it, it sits heavily in our stomachs, blocking our digestive processes until, days later, we struggle to excrete it" (p. 193). There are also repeated examples, both in Singer's book (pp. x, 81-82, 240-41) and in Regan's article (pp. 182, 213-14), of a disturbing penchant for equating experiments on animals with Nazi death-camp experiments performed on hapless, unanesthetized human beings. The overall impression one gains from such lurid passages, despite the legitimate points Singer and Regan may have to make regarding unjustified cruelty, is that in their zeal to help launch a new and popular movement for animal rights they cast their usual caution to the breeze. Or are we, instead, merely being subjected to the self-righteousness of recent converts?

Is speciesism immoral, then? The only sensible verdict, I think, is "not proven." The effort to establish speciesism, on the one hand, and racism and sexism, on the other, as identical forms of unjust discrimination which flout basic moral rights cannot succeed because neither Singer nor Regan has shown any meaningful sense in which rights can and should be ascribed to animals to begin with.

It would seem, therefore, that while the issue of the infliction of unnecessary and excessive pain and suffering upon animals, which is not offset by a significant long-term gain in pleasure for humans or for animals, is a matter that ought to concern every thoughtful and caring person, the question of animals' rights in which it has unfortunately become embroiled—and hence, that of "animal liberation"—is a nonstarter. But Regan and Singer have an important moral to teach. As Regan rightly notes, "The onus of justification is always on anyone who supports a practice that is known to inflict nontrivial, undeserved pain on a sentient creature to show that, in doing so, he is not doing anything wrong" (R, p. 202). The point implicit here, it seems to me, is not that everyone who finds great animal suffering odious to contemplate should rush to dump the contents of his frozen meat locker and medicine cabinet or makeup kit in the garbage pail and don the nearest available (synthetic) hair shirt. Rather, it is that each concerned person should consider carefully the amount of meat a sensible diet, the world food crisis, and the cost of living really should allow him or her, and what sorts of drugs and cosmetics are really essential, and begin lobbying for the elimination of factory farms and for more stringent regulation of the use of animals in experiments and product testing.

Singer and Regan confine themselves to a consideration of the rights and wrongs of killing animals for purely human ends. If I may be permitted to append my own moral to this discussion, a somewhat wider perspective on the issues they have raised would suggest that political involvement aimed at effecting the redistribution of world food supplies and the control of proliferating consumer goods is also morally obligatory.

Notes

1. An earlier version of this article was presented at the Philosophy Colloquium, Queen's University, Kingston, and subsequently published as the Editor's Column in *Queen's Quarterly* 83, no. 1 (Spring 1976): 178-87, under the title "The Use and Abuse of Animals." I would like to express my appreciation to the following for advice and helpful criticism: R. Greenwood, director of the Queen's University Animal Care Service, the Office of the Dietician, Kingston General Hospital, Jonathan Mallov, and Warner Wick, editor of *Ethics.*

2. Peter Singer, *Animal Liberation: A New Ethics for Our Treatment of Animals* (New York: New York Review, 1975) (hereafter cited as S); and Tom Regan, "The Moral Basis of Vegetarianism," *Canadian Journal of Philosophy* 5, no. 2 (October 1975): 181-214 (hereafter cited as R).

3. For the purposes of this paper, I shall take "pain" and "suffering" to be equivalent in meaning. "Suffering" is probably a preferable term, since it embraces both physical and psychological distress more readily. Be this as it may; I shall assume that animals, like humans, can and do experience both kinds of distress.

4. A term which he borrows from Richard Ryder, author of *Victims of Science* (London: Davis-Poynter, 1975).

5. Though I do not wish to digress too far here, it might be argued (and a Marxist surely would) that almost all the benefits which North Americans enjoy routinely, as part of their exorbitantly high standard of living relative to the rest of the world, depend upon the correlated and disproportionate suffering and deprivation caused others elsewhere in the world (e.g., in those countries which supply the raw materials that North American industry and consumerism devour at a staggering rate). From this perspective, the animal-rights debate seems considerably less urgent and a relatively "safe" area of controversy. One wonders why here (as elsewhere) there is so much concern for the plight of animals and evidently so little for that of humans.

6. See, for example, D.W. Baxter and J. Olszewski, "Congenital Universal Insensitivity to Pain," *Brain* 83 (1960); 381-93.

7. Singer actually appears to acknowledge 1 when he says, "The capacity for suffering and enjoyment is *a prerequisite for having interests at all* (p. 9; Singer's italics). But if so, then he owes us an account of the other prerequisites.

8. H.J. McCloskey, "The Right to Life," *Mind* 84 (July 1975): 410-13.

9. Regan assumes that the use of language is an uncomplicated phenomenon and that granting animals the same language capacities as humans is unproblematic. This is certainly empirically false, but it is also philosophically naive. As McCloskey points out (ibid., p. 413), it is not just the capacity to use language that is involved when we refer to humans' linguistic endowment as a criterion for the assigning of rights; it is their capacity to use language "to express thoughts, decisions, wishes, choices."

10. Ibid., pp. 413-17.

11. To anticipate a possible objection, individual beings (say one or more extraterrestrials) may be granted moral rights on the same basis (i.e., if they show evidence of autonomy, etc.), without our knowing the general characteristics of their species.

12. Singer often talks as if we have an obligation to avoid ever deliberately causing an animal to suffer. Thus, for instance, we may not eat any organism, however rudimentary a form of life, if it shows any sign whatever that it has the capacity to suffer (S, pp. 185 ff.). But even our treatment of other human beings does not rest upon so unrealistic and stringent an attitude toward suffering. Punishment and the infliction of pain are often required in the pursuit of a greater long-range good for the individual concerned or sometimes for society as a whole.

13. I realize that denying the universal correlation of rights and obligations is controversial. It seems to me, however, that there is at least one clear case of an obligation which each of

us has but for which it makes no sense to speak of anyone else possessing a corresponding right. The case I have in mind is the general duty of benevolence, where there is no individual or group that can justifiably claim a right to benevolent treatment from a given person. Another example would be duties toward oneself (supposing there to be such).

14. Ontario Regulation 139/71, "Research and Supply Facilities," *Regulations Made under the Animals for Research Act, 1968-69* (Toronto: William Kinmond, Queen's Printer and Publisher, 1971), passim.

15. If a shorter "mean transit time" indicates a healthier diet, it follows that an alcoholic should be healthier than a vegetarian, since alcohol passes through the digestive system much faster than a vegetarian's diet. The reasons why a vegetarian diet is processed faster by the body than a nonvegetarian diet are *(a)* roughage stimulates peristalsis and *(b)* the body can utilize less of the bulk of vegetable matter (cellulose) than that of meat and, hence, discards it more rapidly. This is an example of the tendency that many philosophers have of failing to get their facts straight when borrowing empirical data from other disciplines.

Further Reading

Burr, S.I. 1975. Toward Legal Rights for Animals. *Environmental Affairs* 4 (Spring): 205-254.
Caras, R.A. 1966. *Last Chance on Earth: A Requiem for Wildlife.* New York: Chilton.
Causey, A. 1989. On the Morality of Hunting. *Environmental Ethics* 11, No. 4 (Winter): 327-344.
Clark, S. 1982. *The Nature of the Beast.* Oxford: Oxford University Press.
Clark, S. 1984. *The Moral Status of Animals.* Oxford: Oxford University Press.
Cobb, J.B., Jr. 1974. The Population Explosion and the Rights of the Subhuman World. In *Environment and Society,* ed. R.T. Roelofs, Jr. Crowley, and D.L. Hardesty. Englewood Cliffs, New Jersey: Prentice-Hall.
Cooper, T. 1978. Nature—of the People, by the People, for the People. *Frontiers* 42 (Summer) 10-15.
Crowe, P.K. 1967. *The Empty Ark.* New York: Scribners.
Cutler, M.R. 1978. Perspectives on a Bill of Rights for Living Things. *Frontiers* 42 (Summer) 109-15/
Dembeck, H. 1965. *Animals and Men,* trans. R. Winston and C. Winston. Garden City, New York: Natural History Press.
Dodge, V. 1978. Editorial: Rights of Living Things. *Frontiers* 42 (Summer): 9.
Dombrowski, D. 1984. *The Philosophy of Vegetarianism.* Amherst, Massachusetts: University of Massachusetts Press.
Dombrowski, D. 1988. *Hartshorne and the Metaphysics of Animal Rights.* Albany, New York: SUNY Press.
Etling, C.D. 1973. Who Owns the Wildlife? *Environmental Law* 3 (Spring): 23-31.
Feinberg, J. 1974. The Rights of Animals and Unborn Generations. In *Philosophy and Environmental Crisis,* ed. William T. Blackstone, pp. 43-68. Athens, Georgia: University of Georgia Press.
Fisher, J., and H.R.H. The Prince Philip, 1970. *Wildlife Crisis.* New York: Cowles Book Company.
———., Simon, N., and Vincent, J. 1969. *Wildlife in Danger.* New York: Viking.
Fox, M.A. 1978. Animal Suffering and Rights: A Reply to Singer and Regan. *Ethics* 88 (January): 134-138.
Fox, M.W. 1980. *Returning to Eden: Animal Rights and Human Responsibility.* New York: Viking Press.
Frome, M. 1969. *The Varmints: Our Unwanted Wildlife.* New York: Coward.
Godlovitch, S., Godlovitch, R., and Harris, J., eds. 1972. *Animals, Men, and Morals.* New York: Toplinger.
Griffin, D. 1984. *Animal Thinking.* Cambridge: Harvard University Press.
Haworth, L. 1978. Rights, Wrongs and Animals. *Ethics* 88 (January): 95-105.
Gunn, A. 1983. Traditional Ethics and the Moral Status of Animals. *Environmental Ethics.* 5, No. 2. (Summer): 133-154.
Laycock, G. 1966. *The Alien Animals.* New York: The Natural History Press.
Livingston, J. 1981. *The Fallacy of Wildlife Conservation.* Toronto: McClelland and Stewart.
Louise, M., Grossman, S., and Hamlet, J. 1969. *Our Vanishing Wilderness.* New York: Grosset and Dunlap.
Lowry, J.W. 1975. Natural Rights: Men and Animals. *Southwestern Journal of Philosophy* 6 (Summer): 109-122.
McCloskey, H.J. 1965. Rights. *The Philosophical Quarterly* 15: 115-127.
———. 1976. Rights—Some Conceptual Issues. *Australasian Journal of Philosophy,* 54 (August): 99-115.
Midgley, M. 1983. *Animals and Why They Matter.* Athens: University of Georgia Press.
Miller, H., and Williams, W., eds. 1983. *Ethics and Animals.* Clifton, New Jersey:' Humana Press.
Murphy, E.F. 1970. A Law for Life. *Environment Law Review—1970,* pp. 3-20. New York: Clark Boardman.
———. 1971. Has Nature Any Right to Life? *The Nash, R. 1989. The Rights of Nature.* Madison, Wisconsin: University of Wisconsin.
Hastings Law Journal 22 (February): 467-484.
Neuhaus, R. 1971. *In Defense of People,* pp. 128-129; 181-189. New York: Macmillan.
Norton, G.B., ed. 1986. *The Preservation of Species: The Value of Biological Diversity.* Princeton University Press.
Nozick, R. 1974. *Anarchy, State and Utopia.* New York: Basic Books.

Olsen, J. 1971. *Slaughter the Animals, Poison the Earth.* New York: Simon and Shuster.
Palmer, W.D. 1975. Endangered Species Protection. *Environmental Affairs* 4 (Spring): 255-295.
Passmore, J. 1974. *Man's Responsibility for Nature,* pp. 111-126. New York, Scribners.
______. 1975. The Treatment of Animals. *Journal of the History of Ideas* 36 (April-June): 195-218.
Pyle, E. 1978. The Tiger. *Frontiers* 42 (Summer): 48.
Rachels, J. 1976. Do Animals Have a Right to Liberty? In *Animal Rights and Human Obligations,* ed. T. Regan and P. Singer, pp. 205-219. Englewood Cliffs, New Jersey: Prentice-Hall.
Regan, T. 1975. The Moral Basis of Vegetarianism. *Canadian Journal of Philosophy* 5 (October): 181-214.
______. 1983. *The Care for Animal Rights. Los Angeles, California: University of California Press.*
______. 1986. *Animal Sacrifices.* Philadelphia: Temple University Press.
______. 1978. Fox's Critique of Animal Liberation. *Ethics* 88 (January): 126-133.
______. 1976. McCloskey on Why Animals Cannot Have Rights. *The Philosophical Quarterly* 26 (July): 251-257.
______., and Singer, P., eds. 1976. *Animal Rights and Human Obligations.* Englewood Cliffs, New Jersey: Prentice-Hall.
Reisner, M. 1978. What Are Species Good For? *Frontiers* 42 (Summer): 24-27.
Rollin, B. 1981. *Animal Rights and Human Morality.* Buffalo, New York: Prometheus.
Rolston, H. 1986. *Philosophy Gone Wild.* Buffalo, New York: Prometheus.
______.1988. *Environmental Ethics.* Philadelphia, Pennsylvania: Temple University Press.
Rood, R. 1971. *Animals Nobody Loves.* New York: Stephen Green Press.
Ruggieri, S. 1990. Why I Don't Hunt. *Sierra* 75, No. 6 (November/December): 52-53.
Sauer, C.O. 1971. Theme of Plant and Animal Destruction in Economic History. In *Environ/Mental,* ed. P. Shepard and D. McKinley, pp. 52-60. Boston: Houghton Mifflin.
Searles, H.F. The Role of the Nonhuman Environment. In *Environ/Mental,* ed. P. Shepard and D. McKinley, pp. 80-86. Boston: Houghton Mifflin.
Singer, P. 1975. *Animal Liberation.* New York: New York Review Books.
______. 1978. "The Fable of the Fox and the Unliberated Animals," *Ethics* 88 (January): 119-125.
Sisson, D. 1990. Why I hunt. *Sierra.* 75, No. 6 (November/December): 50-51.
Stevens, C. 1978. Our Obligations Toward Animal Species and Their Individual Members. *Frontiers* 42 (Summer): 33-34.
Talbot, L.M. 1978. "Science Court"; A Possible Way to Obtain Scientific Certainty for Decisions Based on Scientific "Fact"? *Environmental Law* 8 (Spring) 827-850.
Taylor, P.W. 1986. *Respect for Nature.* Princeton, New Jersey: Princeton University Press.
Train, R.E. 1978. Reverence for Life. *Frontiers* 42 (Summer): 38-39.
Vandeveer, D. 1976. Defending Animals by Appeal to Rights. In *Animal Rights and Human Obligations,* ed. T. Regan and P. Singer, pp. 224-229. Englewood Cliffs, New Jersey: Prentice-Hall.
Varner, G. 1987. Do Species Have Standing? Environmental Ethics 9, No. 1 (Spring): 57-72.
Vitali, T. 1990. Do Species Have Standing? *Environmental Ethics* 12, No. 1 (Sporing): 69-82.
Wilson, E.O. 1988. *Biodiversity.* Washington: National Academy Press.

6

THE RIGHT TO A LIVEABLE ENVIRONMENT

Introduction

EVERY DAY throughout the USA, persons directly and intentionally kill other persons, and we call their acts "murder." USA murders average about 9 per 100,000 annually, and total approximately 22,000 each year. Even greater numbers of persons die prematurely from preventable, pollution-induced diseases, however, but we choose not to call these deaths "murder." For example, according to the US Office of Technology Assessment, up to 90 percent of all cancers—accounting for 500,000 deaths each year in the USA—are "environmentally induced and theoretically preventable." The US Office of Technology Assessment likewise estimates that air pollution is responsible for 50,000 premature deaths each year in the USA. It kills more Americans annually than do murderers. Wordwide, the World Heal Organization (WHO) and the United Nations Environment Programme estimated in 1990 that 625 million persons are exposed to unhealthy levels of sulfur dioxide and more than a billion persons breathe unhealthy levels of particulates.

The effects of air pollution are particularly acute for children, because their body cells divide more rapidly, causing a greater probability that mutations will initiate cancer, and because children inhale several times more pollutant, per body weight, than do adults. In the USA, from 3 to 6 million children suffer the effects of lead pollution, causing mental retardation and poor academic performance. In Mexico City, 7 out of 10 newborns have lead levels in their blood that are in excess of WHO standards. How do such children have a right to life when they have no right to breathe clean air necessary for life—when they have no legal right to a liveable environment?

We realize that murder violates our inalienable right to life, but we have been much slower to recognize that air, water, or soil pollutants also violate this same right. Our slowness arises not merely from moral blindness, but because the victims of most environmental hazards are "statistical casualties." Epidemiologists and risk assessors, using a dose-response curve, can predict precise numbers of statistical casualtics attributable to a particular dose of a specific pollutant. Because the victims are "statistical casualties," and because the individual causal chains of harm are very complex, often we cannot identify each such victim. As a result, our pollution-induced deaths are soaring, and few environmental criminals are punished. Some of the culprits are ourselves, of course, as Pogo realized. We drive the cars, for example, that expose half of all Americans to unhealthy levels of air pollution and more than 100,0000 Los-Angeles children to ozone levels that typically triple the federal standard.

A 700-page law, the amended 1990 Clean Air Act, addresses some of our problems with air pollution. It mandates a phase-out of chemicals harmful to the ozone layer, for example, and requires chemical companies to inform the public about accident risks at their plants. Nevertheless, the 1990 law also contains shortcomings: slow deadlines for auto-pollution reductions and exemptions for utilities and other major polluters.

Dirty water, as well as dirty air, also jeopardizes our rights to life and to a liveable environment. In the USA, cancer rates in cities using the Ohio and Mississippi Rivers for drinking water—full of industrial and agricultural effluents—are among the highest in the world. In China, one-third of the water in major rivers is polluted beyond safe health levels, and in most of Latin America, municipal sewage and industrial effluents are discharged, untreated, into the nearest rivers and streams.

Moreover, global freshwater supply—polluted or not—will have declined, by the end of the century, by 24 percent. Per capital supplies in certain nations, such as Kenya and Nigeria, will decline by as much as 50 and 42 percent, respectively. To meet their water needs, China and the Soviet Union are planning river diversion projects of an unprecedented scale, costing billions of dollars. Water tables, especialy in the USA, are declining rapidly. In areas of Dallas-Fort Worth, water tables have fallen more than 120 meters over the last 25 years, and over 50 meters in parts of Tucson. Groundwater withdrawals routinely exceed recharge in many regions. The Ogallala aquifer is half depleted, for example; Mono Lake has shrunk by a third; seawater has intruded because of heavy pumping of coastal aquifers; and water-rights sales in certain areas, such as the Colorado Front Range, are $6,000 per acre-foot (325,850 gallons).

Our rights to life and therefore to a liveable environment are not threatened merely by air and water pollution. Our land is also being harmed. In fact, the USA produces more than one ton of hazardous waste, each year, for every man, woman, and child. As a result, some of this waste shows up in our food, our water, and our air; the US Environmental Protection Agency (EPA) claims that 90 percent of all Americans carry measurable concentrations of dioxins, furans, chlorobenzene, dichlorobenzene, benzene, and styrene in their fatty tissues. All 6 of these hazardous chemicals are thought to be carcinogens.

As the previous remarks indicate, the most frequently cited grounds for acceptance of the right to a liveable environment are probably medical. No one today is able to deny the carcinogenic nature of many substances and the hazardous effects of numerous environmental pollutants to which we are subject. Recently the President's Commission on Chronic Illness defined "medicine" as merely a branch of human ecology; this definition bears testimony to the numerous connections between environmental well-being and human well-being. A toxicogenic environment, however, is not merely one in which there is air and water pollution. Deleterious psychological and sociological effects also result from various forms of environmental degradation. People who have been exposed to high-decibel sound near airports, for example, have exhibited several ill effects of this noise pollution. The clamor has been shown to produce stress and nervous disorders, as well as inner ear changes leading to destruction of hearing tissue.

Despite such medical evidence, however, not everyone is willing to grant that there is a right to a liveable environment. In the case of airport noise, for example, producing a more desirable environment is almost impossible. As many writers have argued, it is unlikely that airlines could continue to operate if they were forced to provide adequate compensation to all victims of noise. Even when airports could be located in zones of low population density, it is likely that society would trade this environmental advantage for the convenience of having the facilties nearby.

Other authors maintain that "greater good" considerations argue against recognition of the right to a liveable environment. Perhaps the most significant of the "goods" they cite are economical. Maintaining quality surroundings costs money better spent on other items, they claim. Moreover, they say, granting the right to a liveable environment might have consequences equally as undesirable as the economic ones. An example often cited in this regard is that of air pollution. According to the Council on Environmental Quality, roughly 40 to 60 percent of all air pollution (80 percent in Los Angeles), in terms of tonage of emissions, is caused by automobiles. Reduction of auto use, and/or escalation of vehicle price, to provide more pollution-control equipment, are viewed by many people as undesirable costs of the right to a liveable environment. This particular argument assumes even more significance when one realizes that the consumer is likely to bear the ultimate costs of nearly all types of planetary cleanup.

Still other writers oppose recognition of the right to a decent environment on the grounds that it would result in greater restrictions of freedom, both of industry and of the individual. For these persons, liberty is judged to be a greater good than a clean environment, and the gains wrought by pollution exceed its losses.

According to Ezra J. Mishan, Professor of Economics at the London School of Economics and at American University in Washington, D.C., however, one cannot simply use a "gains exceed losses" argument. This mode of reasoning, he says, justifies neither environmental damages nor neglect of the right to liveable habitat. He maintains that such an approach ignores basic considerations of logic and equity. This is because "what *are* costs under the law and what *ought* to be counted as costs is just what is in issue." Mishan argues that violations of the right to a decent environment (he calls these infringements of "amenity rights") should be, but are not, counted as a "cost" of industrialization and planetary degradation. Were all relevant parameters considered, says Mishan, a quite different calculation of "the greatest good" would result that which has been assumed by society thus far. Mishan is convinced that the ethical and economic principles served by such amenity rights can be implemented legally in an easy manner.

These are outlines of only a few arguments regarding the right to a liveable environment. The readings in this chapter reflect the diversity of considerations relevant to the topic, and each addresses a different aspect of the problem.

The first selection is by William T. Blackstone, Professor of Philosophy at the University of Georgia. Editor of *Philosophy and Environment Crisis,* Blackstone argues that there is a right to liveable surroundings. Although he discusses many philosophical difficulties with this conception, his basic justification for his position is that such a right "is essential for one to fulfill his human capacities."

In the second essay, Richard B. Stewart, Professor of Law at Harvard University, cites a number of controversial court cases in arguing that the *legal* aspects of this "right" are not so simple as Blackstone suggests. Stewart maintains that it is problematic, in any particular case, to establish that such a right has been violated or that it ought to be protected. "The collective nature of the environmental goods and bads involved is the root of the difficulty," says Stewart. Because such goods are not traded in any market, he claims, and because there are no adequate political mechanisms for controlling bureaucratic environmental regulators, it is difficult, both theoretically and practically, to deal with such a right. This means it also is problematic to assess whether environmental "rights" or individual liberty represent either a higher ethical norm or a greater good.

Whether one takes the approach of a philosopher, an attorney, or an economist, the question of the right to liveable surroundings raises some important ethical concerns. Like other issues discussed in this volume, its practical resolution depends on more than abstract philosophical speculation regarding the relationship between individual liberty and environmental regulation. In the end, this question comes down to a very basic ethical point: What ought we pay to secure our well being?

Readings

1. Blackstone, W.T. 1974. Ecology and Rights. In *Philosophy and Environmental Crisis,* ed. W.T. Blackstone, pp. 29-38, Atlanta: University of Georgia Press. Reprinted by permission of the University of Georgia Press; ©1974 by the University of Georgia Press.

2. Stewart, R.B. 1977. Paradoxes of Liberty, Integrity, and Fraternity: The Collective Nature of Environmental Quality and Judicial Review of Administrative Action. *Environmental Law* 7 (Spring): 464-484. Reprinted by permission of *Environmental Law.*

ECOLOGY AND RIGHTS

William Blackstone

MUCH HAS BEEN SAID about the right to a decent or livable environment. In his 22 January 1970 state of the union address, President Nixon stated: "The great question of the seventies is, shall we surrender to our surroundings, or shall we make our peace with Nature and begin to make the reparations for the damage we have done to our air, our land, and our water? ... Clean air, clean water, open spaces—these would once again be the birthright of every American; if we act now, they can be." It seems, though, that the use of the term *right* by President Nixon, under the rubric of a "birthright" to a decent environment, is not a strict sense of the term. That is, he does not use this term to indicate that one has or should have either a legal right or a moral right to a decent environment. Rather he is pointing to the fact that in the past our environmental resources have been so abundant that all Americans did in fact inherit a livable environment, and it would be *desirable* that this state of affairs again be the case. Pollution and the exploitation of our environment is precluding this kind of inheritance.

Few would challenge the desirability of such a state of affairs or of such a "birthright." What we want to ask is whether the right to a decent environment can or ought to be considered a right in a stricter sense, either in a legal or moral sense. In contrast to a merely desirable state of affairs, a right entails a correlative duty or obligation on the part of someone or some group to accord one a certain mode of treatment or to act in a certain way.[1] Desirable states of affairs do not entail such correlative duties or obligations.

The Right to a Livable Environment as a Human Right

Let us first ask whether the right to a livable environment can properly be considered to be a human right. For the purposes of this paper, however, I want to avoid raising the more general question of whether there are any human rights at all. Some philosophers do deny that any human rights exist.[2] In two recent papers I have argued that human rights do exist (even though such rights may properly be overridden on occasion by other morally relevant reasons) and that they are universal and inalienable (although the actual exercise of such rights on a given occasion is alienable).[3] My argument for the existence of universal human rights rests, in the final analysis, on a theory of what it means to be human, which specifies the capacities for rationality and freedom as essential, and on the fact that there are no relevant grounds for excluding any human from the opportunity to develop and fulfill his capacities (rationality and freedom) as a human. This is not to deny that there are criteria which justify according human rights in quite different ways or with quite different modes of treatment for different persons, depending upon the nature and degree of such capacities and the existing historical and environmental circumstances.

If the right to a livable environment were seen as a basic and inalienable human right, this could be a valuable tool (both inside and outside of legalistic frameworks) for solving some of our environmental problems, both on a national and on an international basis.

Are there any philosophical and conceptual difficulties in treating this right as an inalienable human right? Traditionally we have not looked upon the right to a decent environment as a human right or as an inalienable right. Rather, inalienable human or natural rights have been conceived in somewhat different terms; equality, liberty, happiness, life, and property. However, might it not be possible to view the right to a livable environment as being entailed by, or as constitutive of, these basic human or natural rights recognized in our political tradition? If human rights, in other words, are those rights which each human possesses in virtue of the fact that he is human and in virtue of the fact that those rights are essential in permitting him to live a human life (that is, in permitting him to fulfill his capacities as a rational and free being), then might not the right to a decent environment be properly categorized as such a human right? Might it not be conceived as a right which has emerged as a result of changing environmental conditions and the impact of those conditions on the very possibility of human life and on the possibility of the realization of other rights such as liberty and equality?[4] Let us explore how this might be the case.

Given man's great and increasing ability to manipulate the environment, and the devastating effect this is having, it is plain that new social institutions and new regulative agencies and procedures must be initiated on both national and international levels to make sure that the manipulation is in the public interest. It will be necessary, in other words, to restrict or stop some practices and the freedom to engage in those practices. Some look upon such additional state planning, whether national or international, as unnecessary further intrusion on man's freedom. Freedom is, of course, one of our basic values, and few would deny that excessive state control of human action is to be avoided. But such restrictions on individual freedom now appear to be necessary in the interest of overall human welfare and the rights and freedoms of *all* men. Even John Locke with his stress on freedom as an inalienable right recognized that this right must be construed so that it is consistent with the equal right to freedom of others. The whole point of the state is to restrict unlicensed freedom and to provide the conditions for equality of rights for all. Thus it seems to be perfectly consistent with Locke's view and, in general, with the views of the founding fathers of this country to restrict certain rights or freedoms when it can be shown that such restriction is necessary to insure the equal rights of others. If this is so, it has very important implications for the rights to freedom and to property. These rights, perhaps properly seen as inalienable (though this is a controversial philosophical question), are not properly seen as unlimited or unrestricted. When values which we hold dear conflict (for example, individual or group freedom and the freedom of all, individual or group rights and the rights of all, and individual or group welfare and the welfare of the general public) something has to give; some priority must be established. In the case of the abuse and waste of environmental resources, less individual freedom and fewer individual rights for the sake of greater public welfare and equality of rights seem justified. What in the past had been properly regarded as freedoms and rights (given what seemed to be unlimited natural resources and no serious pollution problems) can no longer be so construed, at least not without additional restrictions. We must recognize both the need for such restrictions and the fact that none of our rights can be realized without a livable environment. Both public welfare and equality of rights now require that natural resources not be used simply according to the whim and caprice of individuals or simply for personal profit. This is not to say that all property rights must be denied and that the state must own all productive property, as the Marxist argues. It is to insist that those rights be qualified or restricted in the light of new ecological data and in the interest of the freedom, rights, and welfare of all.

The answer then to the question, Is the right to a livable environment a human right? is yes. Each person has this right qua being human and because a livable environment is essential for one to fulfill his human capacities. And given the danger to our environment today and hence the danger to the very possibility of human existence, access to a livable environment must be conceived as a right which imposes upon everyone a correlative moral obligation to respect.[5]

The Right to a Livable Environment As a Legal Right

If the right to a decent environment is to be treated as a legal right, then obviously what is required is some sort of legal framework which gives this right a legal status. Such legal frameworks have been proposed. Sen. Gaylord Nelson, originator of Earth Day, recently proposed a Constitutional Amendment guaranteeing every American an inalienable right to a decent environment.[6] Others want to formulate an entire "environmental bill of rights" to assist in solving our pollution problems. Such a bill of rights or a constitutional revision would provide a legal framework for the enforcement of certain policies bearing on environmental issues. It would also involve the concept of "legal responsibility" for acts which violate those rights. Such legal responsibility is beginning to be enforced in the United States. President Nixon on 23 December 1970 signed an executive order requiring industries to obtain federal permits before dumping pollutants. He issued the order not under the authority of new legislation but under the Refuse Act of 1899, which was originally designed to control discharges in connection with dredging and water construction operations but now has been broadened by Supreme Court decisions to cover pollution resulting from industrial operations. (The extension of this act is similar to my suggestion above, namely, the extension of the consitutional rights to equality, liberty, and property to include the right to a livable environment.)

Others propose that the right to a decent environment also be a cardinal tenet of international law. Pollution is not merely a national problem but an international one. The population of the entire world is affected by it, and a body of international law, which includes the right to a decent environment and the accompanying policies to save and preserve our environmental resources, would be an even more effective tool than such a framework at the national level. Of course, one does not have to be reminded of the problems involved in establishing international law and in eliciting obedience to it. Conflicts between nations are still settled more by force than by law or persuasion. The record of the United Nations attests to this fact. In the case of international conflict over environmental interests and the use of the environment, the possibility of international legal resolution, at least at this stage of history, is somewhat remote; for the body of enforceable international law on this topic is meager indeed. This is not to deny that this is the direction in which we should (and must) move.

A good case can be made for the view that not all moral or human rights should be legal rights and that not all moral rules should be legal rules. It may be argued that any society which covers the whole spectrum of man's activities with legally enforceable rules minimizes his freedom and approaches totalitarianism. There is this danger. But just as we argued that certain traditional rights and freedoms are properly restricted in order to insure the equal rights and welfare of all, so also it can plausibly be argued that the human right to a livable environment should become a legal one in order to assure that it is properly respected. Given the magnitude of the present dangers to the environment and to the welfare of all humans, and the ingrained habits and rules, or lack of rules, which permit continued waste, pollution, and destruction of our environmental resources, the legalized status of the right to a livable environment seems both desirable and necessary.

Such a legal right would provide a tool for pressing environmental transgressions in the courts. At the present the right to a livable environment, even if recognized as a human right, is not generally recognized as a legal one. One cannot sue individuals or corporations for polluting the environment, if the pollution harms equally every member of a community. One can sue such individuals or corporations if they damage one's private property but not if they damage the public environment. It is true that public officials have a legal standing in cases of generalized pollution, but unfortunately they have done little to exercise that standing.

Since public officials are failing to take steps to protect the environment, Joseph Sax, professor of law at the University of Michigan, argues that the right to take environmental disputes to court must be obtained as a right of private citizens: "The No. 1 legal priority of those concerned with governmental protection now is that the old restraints of the public nuisance doctrine and other archaic rules be rooted out of the law and be replaced with the recognition of every citizen's opportunity to enforce at law the right to a decent environment."[7] Sax himself drafted a model environmental law which was presented to the Michigan legislature, which "empowered any person or organization to sue any private or public body and to obtain a court order restraining conduct that is likely to pollute, impair, or destroy the air, water, or other material resources or the public trust therein."[8] This bill has now become law, and a similar bill is pending in the U.S. Congress, having been introduced by Senators McGovern and Hart and Representative Udall. I concur with Sax that this law, which provides an enforceable right to a decent environment and which places the burden of proof on a polluter or would-be polluter to show that his action affecting the environment is consistent with public health and welfare, "offers the promise of a dramatic legal break-through in the effort to protect environmental quality."[9] Although it may add to the clogged conditions of the courts, it should also have the effect of encouraging careful planning of activities which affect the environment.

The history of government, in this country and elsewhere, has been that of the gradual demise of a laissez-faire philosophy of government. Few deny that there are areas of our lives where government should not and must not intrude. In fact, what we mean by a totalitarian government is one which exceeds its proper bounds and attempts to control nearly all human activities. But in some areas of human life, it has been seen that the "keep-government-out-of-it" attitude just will not work. The entire quality of life in a society is determined by the availability and distribution of goods and services in such vital areas as education, housing, medical treatment, legal treatment, and so on. In the field of education, for example, we have seen the need for compulsory education and, more recently, for unitary school systems in order to provide equality of educational opportunity. In the area of medical treatment it is plain that we need "new systems of health service delivery that stress better distribution, accessibility, cost containment and increased quality without sacrificing quality."[10] Dr. John Knowles, former president of the American Medical Association argues that we cannot expect this from a system in which the free market economy sets the fees of physicians. Nor can we expect it from the AMA which some characterize as an "ogre in the form of a vast and powerful trade union interested only in economic advancement of its members."[11] Government must step in to provide these vital health services.

In the same way, it is essential that government step in to prevent the potentially dire consequences of industrial pollution and the waste of environmental resources. Such government regulations need not mean the death of the free enterprise system. The right to private property can be made compatible with the right to a livable environment, for if uniform antipollution laws were applied to all industries, then both competition and private ownership could surely continue. But they would continue within a quite different set of rules and attitudes toward the environment. This extension of government would not be equivalent to totalitarianism. In fact it is necessary to insure equality of rights and freedom, which is essential to a democracy.

Ecology and Economic Rights

We suggested above that it is necessary to qualify or restrict economic or property rights in the light of new ecological data and in the interest of the freedom, rights, and welfare of all. In part, this suggested restriction is predicated on the assumption that we cannot expect private business to provide solutions to the multiple pollution problems for which they themselves are responsible. Some companies have taken measures to limit the polluting effect of their operations, and this is an important move. But we are deluding ourselves if we think that private business can function as its own pollution police. This is so for several reasons: the primary objective of private business is economic profit. Stockholders do not ask of a company, "Have you polluted the environment and lowered the quality of the environment for the general public and for future generations?" Rather they ask, "How high is the annual dividend and how much higher is it than the year before?" One can hardly expect organizations whose basic norm is economic profit to be concerned in any great depth with the long-range effects of their operations upon society and future generations or concerned with the hidden cost of their operations in terms of environmental quality to society as a whole. Second, within a free enterprise system companies compete to produce what the public wants at the lowest possible cost. Such competition would preclude the spending of adequate funds to prevent environmental pollution, since this would add tremendously to the cost of the product—unless all other companies would also conform to such antipollution policies. But in a free enterprise economy such policies are not likely to be self-imposed by businessmen. Third, the basic response of the free enterprise system to our economic problems is that we must have greater economic growth or an increase in gross national product. But such growth many ecologists look upon with great alarm, for it can have devastating long-range effects upon our environment. Many of the products of uncontrolled growth are based on artificial needs and actually detract from, rather than contribute to, the quality of our lives. A stationary economy, some economists and ecologists suggest, may well be best for the quality of man's environment and of his life in the long run. Higher GNP does not automatically result in an increase in social well-being, and it should not be used as a measuring rod for assessing economic welfare. This becomes clear when one realizes that the GNP

> ... aggregates the dollar value of all goods and services produced—the cigarettes as well as the medical treatment of lung cancer, the petroleum from offshore wells as well as the detergents required to clean up after oil spills, the electrical energy produced and the medical and cleaning bills resulting from the air-pollution fuel used for generating the electricity. The GNP allows no deduction for negative production, such as lives lost from unsafe cars or environmental destruction perpetrated by telephone, electric and gas utilities, lumber companies, and speculative builders.[12]

To many persons, of course, this kind of talk is not only blasphemy but subversive. This is especially true when it is extended in the direction of additional controls over corporate capitalism. (Some ecologists and economists go further and challenge whether corporate capitalism can accommodate a stationary state and still retain its major features.)[13] The fact of the matter is that the ecological attitude forces one to reconsider a host of values which have been held dear in the past, and it forces one to reconsider the appropriateness of the social and economic systems which embodied and implemented those values. Given the crisis of our environment, there must be certain fundamental changes in attitudes toward nature, man's use of nature, and man himself. Such changes in attitudes undoubtedly will have far-reaching implications for the institutions of private property and private enterprise and the values embodied in these institutions. Given that crisis we can no longer look upon water and air as free commodities to be exploited at will. Nor can the private ownership of land be seen as a lease to use that land in any way which conforms merely to the personal desires of the owner. In other words, the environmental crisis is forcing us to challange what had in the past been taken to be certain basic rights of man or at least to restrict those rights. And it is forcing us to challenge institutions which embodied those rights.

Notes

1. This is a dogmatic assertion in this context. I am aware that some philosophers deny that rights and duties are correlative. Strictly interpreted this correlativity thesis is false, I believe. There are duties for which there are no correlative rights. But space does not permit discussion of this question here.

2. See Kai Nielsen's "Scepticism and Human Rights," *Monist* 52, no. 4 (1968): 571-594.

3. See my "Equality and Human Rights," *Monist* 52, no. 4 (1968): 616-639 and my "Human Rights and Human Dignity," in Laszlo and Gotesky, eds., *Human Dignity*.

4. Almost forty years ago, Aldo Leopold stated that "there is as yet no ethic dealing with man's relationship to land and to the non-human animals and plants which grow upon it. Land, like Odysseus' slave girls, is still property. The land relation is still strictly economic entailing privileges but not obligations." (See Leopold's "The Conservation Ethic," *Journal of Forestry* 31, no. 6 (October 1933): 634-643. Although some important changes have occurred since he wrote this, no systematic ethic or legal structure has been developed to socialize or institutionalize the obligation to use land properly.

5. The right to a livable environment might itself entail other rights, for example, the right to population control. Population control is obviously essential for quality human existence. This issue is complex and deserves a separate essay, but I believe that the moral framework explicated above provides the grounds for treating population control both as beneficial and as moral.

6. *Newsweek*, 4 May 1970, p. 26.

7. Joseph Sax, "Environment in the Courtroom," *Saturday Review*, 3 October 1970, p. 56.

8. Ibid.

9. Ibid.

10. John Knowles, "Where Doctors Fail," *Saturday Review*, 27 August 1970, p. 63.

11. Ibid.

12. See Mellville J. Ulmer, "More then Marxist," *New Republic*, 26 December 1970, p. 14.

13. See Murdock and Connell, "All About Ecology," p. 63.

PARADOXES OF LIBERTY, INTEGRITY AND FRATERNITY: THE COLLECTIVE NATURE OF ENVIRONMENTAL QUALITY AND JUDICIAL REVIEW OF ADMINISTRATIVE ACTION

Richard B. Stewart[1]

THE JUXTAPOSITION of environmental regulation and individual liberty reflects a lawyer's view of environmental problems. Economists, on the other hand, typically view issues of environmental policy as problems of collective or public goods, for reasons that may briefly be rehearsed here.[2] Clean air, unpolluted oceans, scenic vistas and other forms of environmental quality are typically enjoyed in common by many individuals in ways that make it impractical to exclude some individuals from enjoying such goods while admitting others. This lack of excludability represents a form of market imperfection that deprives potential suppliers of environmental quality of incentives to provide it. Since a polluter that installs controls cannot limit enjoyment of the resulting clean air to those who pay for it, it will be unable to recover the costs of such controls and will fail to voluntarily undertake them.

The non-excludability feature also inhibits individual beneficiaries from contributing voluntarily to the costs of cleanup, since each individual may reason that if others make such contributions, he can enjoy the resulting clean air for free. But if all reason in the same way, none will contribute, and controls will not be undertaken.[3]

The typical solution to the "public goods" problem in other contexts is for the government to underwrite the costs of providing the goods by paying suppliers to produce them and recouping its outlays through taxation. If we followed this approach with respect to environmental problems, then the government should pay firms and individuals to produce environmental quality in the same way that it pays for the production of other collective goods and services, such as bomber fleets and police protection. Thus steel mills would be paid by the government to reduce pollution, and landowners paid not to develop pristine areas by erecting residential housing tracts or high-rise buildings. If a practice of purchasing environmental quality as a "collective good" were consistently followed, there would be relatively little conflict between environmental quality and individual liberty, although there would be a sharp clash between government and taxpayers.[4]

We have, however, largely viewed the "problem" of environmental degradation not as a problem of collective goods but collective bads. Certain classes of actors —industrial firms, developers—have been identified as "causing" environmental harm to others, and are subjected to coercive regulation.[5] It is this characterization of the "problem," and the corresponding remedy, that creates the conflict between environmental quality and individual liberty. Many economists in practice share the view that the problem of environmental degradation is one of collective bads rather than collective goods, but would employ, in lieu of regulation, a fee or tax on those causing collective bads or limit their activity through transferrable pollution or development permits.[6]

Why is it that we largely perceive the problem of environmental degradation as one of collective bads rather than collective goods, and identify certain classes of actors as "causing" that degradation? Concern with a more equitable overall distribution of income is not an adequate explanation, for the net impact of a "collective bad" characterization is probably more regressive than a "collective good" prescription.[7]

Considerations of utility (or efficiency) in a world of transaction costs, on the other hand, do seem to go a fair way towards explaining the "collective bad" treatment of the problem.[8] But not far enough. For our characterization of the environmental "problem" also reflects fundamental moral concerns that go to the heart of property right allocations.

If environmental quality (degradation) were merely a commodity (negative commodity or "bad") on a par with sofas or whiskey, then environmental controls would be a function of our preferences for environmental quality versus our preferences for other goods and services. Such analysis would largely dissolve the surface conflict between environmental regulation and liberty.

Following Hayek, *liberty* may be defined as comprising: (a) freedom to do what one pleases; and (b) autonomy (all constraints on freedom to do what one pleases are rule-governed).[9] While environmental controls may contract our freedom to do what we wish in some respects (*e.g.*, operation of high-polluting automobiles may be prohibited or heavily taxed), they may enlarge our opportunities in other respects (*e.g.*, enable us to see distant mountains or live in Los Angeles without health risk). Moreover, as demonstrated by the example of emission fees, personal autonomy can be maintained through use of rule-governed forms of environmental control that are impersonal, predictable, and facilitate individual adaptation. Since liberty consists of well-ordered opportunities, there are no grounds for supposing that environmental controls will necessarily or even typically curtail liberty. The issue then becomes whether we value opportunities for environmental quality more than other sorts of opportunities.[10]

But the problem is more complicated. Environmental quality, or the reciprocal avoidance of environmental degradation, is not your run-of-the-mill commodity. It is "special" in several ways. Avoidance of environmental degradation is closely tied with human health and the physical integrity of the body. It is also essential to preserving the integrity of our legacy to future generations, which includes maintenance of natural environments of high quality as well as learning, arts, ethics, and the other fruits of culture. Maintenance of diverse physical environments (some of exceptionally high quality) and associated diverse cultures and styles of life is central to the potential for experiential diversity that nurtures the moral development of a society.[11] These various aspects of the material and moral health of the individual and the civilization are attributes of *integrity*.

Integrity is logically prior to liberty, for liberty presupposes a choosing actor who can only exist with physical integrity, with a set of preferences reflecting the heritage of her culture, and (if the actor is to be anything more than an automaton) a capacity for diversity of experience, changes in preferences, and the other ingredients of moral growth.[12] It also marks out a subset of liberties—here, opportunities generated by maintenance of high quality environments—that have a special claim to be maintained.[13]

Because the problem of environmental quality is one of collective goods and bads, it is one that can be resolved only by the collective endeavors of people in political society. This circumstance immediately raises two questions of collective choice, or *fraternity*. First, by what mechanisms shall collective choices about the environment be made? Ideally, each should participate in a choice which affects and concerns all. Second, how should the benefits of environmental quality and the burdens of securing it be distributed? Ideally, both the benefits and burdens should be shared equally by those who enjoy or suffer the collective goods or bads in question. Thus the ideal of fraternity embraces an equal responsibility in decisionmaking and an equal sharing of benefits and burdens.[14]

It would be wonderful indeed if liberty, integrity, and fraternity fit harmoniously together in one reinforcing analytic and institutional whole. But I hope to persuade you in this essay that there are profound, paradoxical conflicts among these central values in the realm of environmental policy. Each is at war with the others. Even more paradoxically, each is at war with itself—liberty with liberty, integrity with integrity, fraternity with fraternity.

To exemplify my argument, I have selected four court decisions relating to judicial review of administrative action:

Ethyl Corp. v. EPA[15]
Brown v. EPA[16]
Sierra Club v. Ruckleshaus[17]
Alyeska Pipeline Service Co. v. Wilderness Society[18]

This emphasis on administrative law reflects not only my own interests but the important role of judicial review of administrative action respecting environmental quality.

The common law can be viewed as an exercise in management of market exchange by the judiciary in order to produce collective goods or avoid collective bads.[19] But the collective nature of goods and bads in a developed industrial society has become so complex and far reaching that the efficacy of private litigation to deal with them has been outstripped. Accordingly, administrative agencies were created to secure collective interests in matters such as environmental quality. While we in the United States have long relied upon judicial review to check administrative power in order to protect liberty, more recently the courts have also attempted to correct agency insensitivity to the full range of social interests and values implicated in administrative decisions.[20] As the environmental decisions discussed below illustrate, the broadened scope of judicial review has forced the courts to confront the conflicts and paradoxes that inhere in liberty, integrity and fraternity.

I. *Ethyl Corp. v. EPA—The Paradoxes of Uncertainty and Risk.*

In its *Ethyl* decision, a closely-divided Court of Appeals for the D.C. Circuit sustained an order by the Federal Environmental Protection Agency to phase out use of lead additives in gasoline because of health hazards assertedly posed by airborne lead. The decision dealt at length with specific clinical and epidemiological evidence which did not establish the extent or magnitude of adverse health effects attributable to such use of lead and its subsequent atmospheric diffusion, but did indicate a potential, undefined risk to health, particularly in the case of children.

The basic issue was far more general: what leeway should agencies be allowed in imposing restrictions to guard against uncertain risks to health or the natural environment posed by various forms of pollution and environmental degradation? In common law adjudication there existed (at least in principle) relatively clear points of demarcation determining whether a defendant was the "cause" of "harm" to a plaintiff such that liability should arise.[21] And environmental statutes are often framed on the supposition that a bright line can be drawn between pollution levels that cause adverse health effects and those that are "safe."[22] Such assumptions about thresholds and unambiguous causal links may lack scientific validity, but they have served a socially useful purpose by providing an accepted basis for demarcation between the need for regulation to protect the physical integrity of human individuals and of the ecosystem, and the countervailing claims of liberty threatened by regulation.[23]

As *Ethyl* exemplifies, however, this useful naivete cannot survive the modern era of environmental decisionmaking, where exhaustive documentary records make clear that our information and understanding of environmental degradation and its effects is partial at best, that harms and causes are seldom clear and unambiguous, and that dramatic thresholds are rare. In most problems of environmental decisionmaking, we are confronted with gradations of risks of harm that must be evaluated under conditions of considerable uncertainty.[24] The data and analysis rarely exhibit an obvious plateau or breaking point of severity sufficient to fix where governmental control may begin.[25]

How then are the competing claims of integrity and liberty to be reconciled? It will not do to regulate activity so as to eliminate all risk of harm, which could only be achieved through perfect paralysis and the extinction of all opportunity to do what we please.[26] It is equally unsatisfactory to dispense with government controls altogether. The demarcation between integrity and liberty comes down to an essentially subjective assessment of risk under conditions of uncertainty.[27] Statutes typically fail to provide clear directions for such determinations.[28] The only practical choice in most cases is either to leave this assessment to the agency—as the majority did in *Ethyl*—or for the judges to indulge their own subjective preferences by fixing an essentially arbitrary point at which government intervention becomes permissible.[29] Either solution violates the first norm of fraternity, because each relegates a judgment in which all are concerned to the essentially arbitrary judgment of an elite, be it bureaucratic or judicial. Yet this result seems inevitable. Most individuals have little capacity or time to evaluate and respond to the myriad potential hazards posed by environmental degradation. The position of legislators is similar. The decision must be made by those immersed in the problem at hand.

Yet the nature of the problem makes its resolution essentially ad hoc, precluding the development of consistent principles that would permit some outside review of decisions by agencies or courts. Liberty and integrity can accordingly be reconciled only at the sacrifice of fraternity. Moreover, preservation of some freedom of choice is achieved by the imposition by the judge or administrator of ad hoc subjective limits on regulation. Autonomy is eroded because one's liability to regulation becomes a function not of ordered rules, but of another's will. Thus freedom from constraint is purchased only at the sacrifice of autonomy; liberty is thrown at war with itself.

II. *Brown v. EPA: The Paradoxes of Centralization and Decentralization.*

American political philosophy has traditionally reflected a presumption in favor of decentralized decisionmaking, particularly in environmental matters. Federal displacement of state and local decision bears a heavy burden of justification. The presumption of decentralization has rested on several important values that decentralized decision has been thought to serve:

Liberty. Decentralization promotes a diffusion of power that tends to enlarge individual opportunities for choice.

Utility. Local decisionmakers are better attuned to the preferences of local constituents and can tailor policies which provide such constituents with opportunities (liberties) that better enable them to satisfy their preferences.

Fraternity. Small scale decisionmaking facilitates participation in, or at least identification with, the political life of the society by each affected individual.

Diversity. Decentralized decision encourages diversity in social policies, which not only has utilitarian justifications in better enabling individuals with diverse tastes to satisfy them and facilitating experimentation in social policies, but promotes a corresponding diversity of experience that makes possible the moral growth of the individual and the civilization.

Nonetheless, rising public concern with environmental quality has translated into a vast increase in dictation by the federal government of environmental policies through detailed legislation and far-reaching administrative programs. This shift is attributable to a number of structural factors which inhibit the states from fully satisfying the heightened desire for environmental quality:

Federal Economics of Scale. States often face a "commons" dilemma in which each would prefer higher environmental quality than it now has but none dares take unilateral action to achieve that quality for fear of losing economic growth to other states. Because of its ability to minimize the transaction cost barriers to national agreement, the federal government can alleviate commons problems by imposing uniform standards for all. The federal government also enjoys other economies of scale in research, data gathering and analysis.

Spillovers. Decisions about environmental quality can generate spillover "bads" and "goods" in other states—physical, economic, ideological. These spillovers create "market imperfections" in decentralized decisionmaking and conflicts which can only be resolved by federal authority.

Imbalance in Representation. Because of transaction costs obstacles to effective organization by environmental interests, such interests probably enjoy (on the average) more effective influence where environmental decisions are made in Washington than in multiple state and local forums. With an increase in the relative political power of environmental interests, the shift to federal determination of policy is not surprising.

Politics of Sacrifice. Control measures requiring material sacrifices may be more palatable if undertaken by all on a national basis, aided by fiscal mechanisms (such as federal waste water treatment grants) that mask some of the costs involved, and by implementation through federal bureaucracies less sensitive to sacrifice-generated local or short-term opposition.

Despite the shift of policymaking to Washington, however, it is increasingly clear that federal agencies neither have nor are likely to acquire the resources needed to implement federal policies themselves. This is because the federal agencies are dependent upon state and local officials for administration and enforcement and such officials are often reluctant to carry out federal programs, both because of the structural factors that initially precipitated resort to federal legislation, and because of justified criticisms of the federal programs that have emerged: their rigid uniformities, their insensitivity to local conditions, their imposition of unwarranted sacrifices on particular locales.[30]

May federal officials compel the states to implement federal programs? Such compulsion may be essential if adequate environmental controls are to be achieved. *Brown v. EPA* found that the federal EPA lacked statutory authority to compel state officials to implement federal transportation control plans to curb automobile pollution, the court adverting to constitutional difficulties that might be presented if Congress attempted to grant such authority.[31] Other courts of appeals have ruled differently.[32] The Supreme Court granted certiorari to resolve the question but vacated and remanded when the government conceded that its regulations would have to be modified substantially.[34]

This is not the occasion to explain the details of the statutory and constitutional issues presented.[35] Here it is sufficient to point out that a serious conflict had arisen between integrity on the one hand, and liberty and fraternity on the other. Federal conscription of state authority may be essential to effective implementation of federal policies to protect health and the integrity of our natural endowment. Promotion of environmental quality is an ingredient of moral integrity that may only be achievable through a national "politics of sacrifice."

Yet the costs of federal conscription are substantial. Liberty—the maintenance of diverse opportunities of choice—is threatened by the effective decentralization of power, and the levelling of self-determination and political diversity.[36] Fraternity is more grievously threatened in the subversion of local decisionmaking mechanisms to the will of federal officials who bear slight, if any, accountability to the local citizens upon whom they impose sacrifices. The burdens are no longer self-imposed, the benefits too often attenuated. The integrity of the environment, of human health, and of collective moral aspiration is thus profoundly at odds with the values of liberty and meaningful fraternity as realized within a federal system of government.

III. *Sierra Club: The Paradoxes of Uniformity and Diversity.*

The federal dialectic of centralization and decentralization also plays strange tricks with environmental diversity. As already sketched, diversity in national environments, together with an associated diversity in urbanization and modes of life and thought, serves more than utilitarian uses. It comprises a vital part of the media through which individual and collective moral growth goes on.

Traditionally, diversity has been secured by decentralized decisionmaking in a federal system of states with differing economies, geographies and cultures. In more recent times, however, the homogenizing effects of industrial and commercial development, and the "commons" dilemmas thus created by the states have eroded the capacity of decentralized decisions to produce and preserve diversity.[37] Concomitantly, the centralization of policy determinations in Washington has exacerbated the problem by mandating national uniformities in environmental standards and programs.[38]

In the *Sierra Club* "non-degradation" decision, the federal judiciary intervened to protect diversity values by reading the Clean Air Act to require federal EPA measures to protect and preserve air quality in undeveloped areas. This intervention, to the extent that it is ultimately successful, is well calculated to preserve what I have elsewhere celebrated as "quasi-constitutional" integrity interests in environmental diversity.[39] But this vindication of integrity conflicts with liberty and fraternity, and with aspects of integrity itself.

Dictation by the federal judiciary that a state—say, Utah—may not pollute, and therefore may not develop to the same extent as other states robs local decisionmaking of its efficacy. The resulting insult to liberty and fraternity is particularly acute because imposed by the judiciary, clogging political mechanisms of decisionmaking potentially more responsive to the preferences of affected citizens. Moreover, the citizens of states denied an equal opportunity for development are required to sacrifice for the benefit of those in other states.

This geographical disparity in benefits and burdens is exacerbated by analogous horizontal disparities among income groups. It is the wealthy and educated that will primarily benefit from the maintenance of pristine environments. It is the poor who will bear much of the burden—the rural poor denied the fruits of development, the urban poor choking on pollution that cannot be exported to more thinly populated areas.[40] Where diversity is the good sought, the fraternal ideal of equal distribution of benefits and burdens is unlikely to be feasible. Moreover, integrity itself is divided when the price of wilderness preservation is the health of urban dwellers.

These sacrifices may be judged worthwhile. Art, culture, and the "higher" manifestations of civilizations' integral development have often been at odds with fraternal ideals and (in some measure) with liberty as well.[41] But this does not lessen the conflicts' bitterness.

IV. *Alyeska: The Paradoxes of the Few and the Many.*

The problem of collective goods (or bads) is generated by transaction costs. The scattered, unorganized individuals with an interest in environmental quality cannot organize effectively to bribe or coerce those who release pollutants or develop pristine environments in order to make them cease their activities.[42] The difficulties in effectively "pooling" collective interests and redressing collective wrongs similarly explain the inadequacies of private litigation as a solution, and the consequent creation of administrative agencies to secure collective goods or prevent collective bads. The mission, however, is rarely self-defining, and there are political and practical obstacles to the legislature's providing administrators with precise and workable directions in so uncharted and volatile a field.[43] The agencies accordingly employ considerable discretion, the exercise of which may be unduly influenced by already organized economic interests such as corporations or unions, that have a sufficient individual stake in given decisions to underwrite legal representation and other forms of advocacy.[44] Thus the existence of transaction costs constitutes a political and economic version of original sin, both creating the collective problem of environmental degradation and undermining the efficacy of institutional response to the problem.

Judicial review of administrative action provides a potential archimedean point to leverage a solution by requiring the agencies to protect collective interests in environmental quality. The courts have the potential to become a forum in which transaction costs could be shortcircuited and the interests of an unorganized environmental fraternity vindicated.

But what is this fraternity, and who speaks for it? Resources are the first requisite of access to the courts, but the scattered character of environmental interests precludes a fully effective pooling of the collective stake. Given the limitations of foundation funding and voluntary contributions, environmental advocates have asked for court-awarded reimbursement of the costs of advocacy.[45] In the *Alyeska* decision, however, the Supreme Court denied the federal courts general power to award litigation costs to successful litigants assertedly representing unorganized "public" interests.

The decision exposes profound contradictions in the effort to give flesh to the values of fraternity in a realm of collective goods or bads. Given the realities and limits of legislative and administrative politics, agency policies will often fail to mirror the collective interests they purportedly serve. Judicial protection of such interests may be no more valid or efficacious. Unorganized collective interests are by their very nature not self-defining; their hypostasis is not revealed in any market or election. In court, they are represented by spokesmen who are inevitably self-appointed and formally unaccountable. The effective lack of participation by all in what concerns them is further diminished if judges are able to decide which self-appointed representatives shall be funded, and which not.[46]

Effective advocacy, of one sort or another, of inchoate interests is probably essential to the capacity for self-transformation in which a civilization's moral integrity must reside.[47] It is equally vital to the fraternal ideal of participation. Yet these ideals are necessarily flawed by reliance on the isolated decisions of a few advocates or judges whose only badge of fraternity is personal earnestness. Those ostensibly represented can neither speak for themselves not repudiate the advocacy asserted on their behalf.[48] Moreover, it is clear that many decisions to represent or sustain environmentalists' positions are hostile to the immediate interests of the poor or less educated. Thus the effort to effectively realize shared interests is necessarily at odds with the fraternal ideal of participation, and threatens the associated ideal of equal benefit and burden.

V. *The Collective Nature of Environmental Problems and the Role of Reviewing Courts*

I do not propose to offer solutions to these paradoxes and contradictions in the ideals of liberty, integrity and fraternity. I do hope to have convinced you of the profound difficulty of the questions posed by environmental problems—a difficulty far more intricate than the simple opposition of regulation and liberty might indicate. Under such a bipolar model, we might rate the decisions we have reviewed an even tie—two favoring more effective regulation *(Ethyl Corp.* and *Sierra Club)* and two less regulation *(Alyeska* and *Brown).*[49] But our analysis indicates that the opposition between regulation and liberty is in large measure false, and that the problems posed by the decisions under review are far more complex than any such accounting would suggest.

The collective nature of the environment goods and bads involved is the root of the difficulty. Economists such as Samuelson,[50] Baumol,[51] and Olson[52] have elucidated the inadequacies of market exchange in dealing with such goods and bads, and the consequent need for collective action. More recently, we have come to realize that our existing political mechanisms for dealing with these collective problems are also gravely flawed. Because the collective goods (or avoidance of collective bads) secured by administrative action are not traded in any markets, and because political mechanisms for controlling bureaucracies through elected legislatures are highly imperfect, agencies exercise considerable discretion in policies with respect to collective bads and goods. There is a corresponding danger that this discretion will not be exercised in the interests of all those affected, but may instead be skewed in favor of well-organized and well-financed private interests, or in favor of the interests of the administrators themselves.[53]

The courts, which have traditionally been accorded a large role in controlling administrative action, have perceived this danger and sought to alleviate it through the development of an interest-representation model of administrative law.[54] Under this model, a range of affected interests broader than regulated or client firms is entitled to participate in agency decisions and to secure judicial review. Courts have also attempted to correct the perceived biases in agencies' exercise of discretion by demanding fuller explanations for their decisions and requiring the production of records reflecting consideration of all pertinent issues and interests.[55] This development responds to the fraternal ideal by ensuring a broader (albeit indirect) representation of individuals in the process of governmental decision and (to the extent that these procedural innovations shape policy outcomes) a more equal sharing of benefits and burdens.

This judicial development of essentially procedural solutions to agency discretion is by no means a "neutral" mechanism for ensuring that agency policies with respect to collective goods and bads mirror the preferences of all affected individuals. The interests entitled to participate in agency decisions and secure judicial review on a particular matter are defined by reference to the purposes of the relevant statute.[56] Given the opaque quality of statutory purposes,[57] judges can exercise a measure of creativity in selecting and defining those purposes that will convey corresponding rights of access to private interests. Moreover, in the process of specifying the interests protected by statutory delegations of administrative power, the courts are at the same time redefining the terms of the agency's mission.[58]

In discharging this role, the courts may take part in the moral growth of the society by stimulating bureaucratic response to new and emergent values—such as the value of environmental quality. In addition, the courts can become a forum in which the administrative and political exigencies of the day can be mediated with more general conceptions of the good society and of human excellence. As Paul Freund once put it:

> The courts are the substations which transform the high-tension charge of the philosophers into the reduced voltage of a serviceable current. ... [J]udicial review is not merely a derivative from a society in agreement on fundamentals; in itself it is an educative and formative influence which ... may have consequences beyond its immediate applications for the mind of a people.[59]

The courts in both of these ways contribute to society's capacity for critical self-examination and renewal—in a word, to its integrity.

These contributions are not without cost. Considerable resource burdens, direct and indirect, are involved in time-consuming procedural formalities and judicial review. There are attendant dangers that the judicial apparatus of interest representation will break the capacity and incentive for decisive or innovative administrative action.[60] The reluctance of the Supreme Court in *Alyeska* to take the initiative in financing the further implementation of the interest-representation model is accordingly understandable. Moreover, the efficacy of the interest representation model in securing its goal is open to question. While I believe that expanded standing, increased rights of access to agency decisionmaking, and judicial insistence on more complete records and better agency explanations of decision have had salutary impacts on agency decisionmaking,[61] I share some of Professor Sax' pessimism about the value of additional doses of procedural requirements.[62]

Finally, to the extent that judicially-imposed procedural requirements have been effective in altering administrative decisions with respect to collective goods or bads, the courts might be charged with improperly intervening in resource allocation decisions that should be reserved to legislators or administrators. It is no answer to this criticism that courts have an independent role in determining fair procedures, and that any impacts on resource allocation are indirect and irrelevant.[63] For it is precisely the aim of these procedural requirements to influence agency decision in favor of judically-selected purposes and interests.

Nonetheless, in my view these criticisms dictate judicial advance rather than retreat. Rather than resort to costly and indirect procedural stratagems, courts could well direct agencies to accord greater weight to emergent interests or enduring values through innovative statutory interpretation. The very lack of clarity in statutory directives that enables courts to define the interests entitled to procedural protection also enables them to set constraints on agency discretion. In decisions such as *Citizens to Preserve Overton Park, Inc. v. Volpe,*[64] the *Sierra Club* nonsignificant deterioration cases previously discussed,[65] and cases curtailing clearcutting in National Forests,[66] the courts have effectively reformed statutes to prevent environmental degradation. Such interventions are more efficacious and less costly than procedural stratagems. They also make more explicit the impact of the courts' intervention and facilitate legislative correction in cases where the courts' initiatives are too flagrantly at odds with the dominant political consensus.[67]

Such judicial initiatives may be condemned as intermeddling in resource decisions that belong to the political branches. But the post-New Deal conventional wisdom that courts should concern themselves with procedural fairness and individual liberties, leaving to the political branches responsibility for resource allocation and income distribution, is simply untenable. It denies the role assumed by the courts under the common law and throughout most of our history.[68] Its weaknesses are particularly acute in the environmental field, where protection of individual interests in bodily integrity and experiential diversity necessarily have far reaching allocational and distributional implications. The maintenance of environmental liberties requires more than staying the coercive powers of the state. If the courts are to play an important role, as I think they should, in the maintenance of those environmental liberties central to integrity values, then they must be prepared to accept a substantial role in resource allocation decisions.[69]

This is not to argue that judges should feel entirely free to rewrite statutes according to their own notions of sound policy, taking comfort in the availability of legislative correction if they err. The weight of legislative inertia is considerable, and even where the legislature intervenes, the end result may be substantially different because of statutory initiatives by the court.[70] Yet that same legislative inertia and the character of bureaucratic politics demand a substantial role for the courts in shaping administrative policies in the light of emerging concerns and longer run societal interests. The crucial task is defining a lawmaking role for the courts in an era of statutes and administrative agencies that is at least analogous to their role at common law. Professor Calabresi, in his recent Holmes Lectures, outlined a general doctrine of judicial "second look" to deal with problems such as statutory obsolescence.[71] Elsewhere I have suggested a realm of "quasiconstitutional" law in which courts would depart from conventional principles of statutory interpretation in order to promote enduring values frustrated by structural defects in the political process.[72] These efforts are simply the first beginnings on a large and difficult endeavor.

The task of constructing a substantial lawmaking role for courts in the administrative implementation of statutes is further complicated by the technical complexity that infests environmental and other regulatory questions, a complexity well illustrated in the *Ethyl* decision. If environmental problems had technical solutions, then judges could be properly remitted to a minimal role of determining whether the agency was operating within the grant of statutory authority. Judges can contribute little to the design of computerized diffusion models, the statistical analysis of epidemiological data, or the construction of dynamically efficient incentive systems.[73] But the burden of this essay is that environmental problems are not solely or even primarily technical. Such problems demonstrate that a responsible society cannot ignore the conflicts between liberty, integrity and fraternity. Rather it must keep those conflicts alive in creative balance.

Notes

1. Professor of Law, Harvard University, B.A., Yale, 1961; A.B. Oxford, 1963; LL.B. Harvard, 1966.

2. For comprehensive discussion of the economic analysis of environmental problems, *see* ECONOMICS OF THE ENVIRONMENT (R. Dorfman & N. Dorfman ed., 1972; O. HERFINDAHL & A. KNEESE, THE ECONOMIC THEORY OF NATURAL RESOURCES, 1974); A. KNEESE, R. AYRES & R. D'ARGE, ECONOMICS AND THE ENVIRONMENT: A MATERIALS BALANCE APPROACH (1970).

3. *See generally* M. OLSON, THE LOGIC OF COLLECTIVE ACTION: PUBLIC GOODS AND THE THEORY OF GROUPS (1971).

4. Financing improvements in environmental quality with federal tax revenues would have a less regressive distributional impact than requiring providers of goods and services to install controls without compensation. *See* note 6 *infra.* On the other hand, the total social costs of pollution control are likely to be greater under a policy of paying persons not to pollute. *See* note 7 *infra.* But the current emphasis on regulatory solutions to problems of environmental degradation is not so much attributable to these considerations as to two other factors. First, pollution is widely viewed as in some measure immoral, and there is repugnance to paying people to cease activities which they shouldn't undertake in any event. (By contrast, few believe that it is immoral for defense contractors to demand payment for fighter planes). Second, regulatory controls tend to obscure both the costs of providing environmental quality and the responsibility of legislators for the imposition of such costs.

5. The achievement of Ronald Coase's famous article, *The Problem of Social Cost,* 3 J.L. & ECON. 1 (1960), is to make clear that when conflicts in resource use occur, there is no "natural" or "objective" ground for assigning responsibility for the losses caused by the conflict to one of the persons causing the conflict rather than the other. Rather, the assignment is a social choice about property rights and liability rules. *See* Calabresi & Malamed, *Property Rules, Liability Rules and Inalienability: One View of the Cathedral,* 85 HARV. L. REV. 1089 (1972). Viewing environmental degradation as a problem in collective goods implicitly assumes a right in the polluter to use natural resources, whereas viewing it as a problem in collective bads assumes that the right inheres in the receptors. The characterization of the environmental "problem" thus necessarily implies social choices. *Cf.* Ludwig Wittgenstein's discussion of the "duck-rabbit", PHILOSOPHICAL INVESTIGATIONS 194-200 (2d ed. 1958).

6. *See* J. Dales, POLLUTION, PROPERTY AND PRICES (1968); A. KNEESE & C. SCHULTZE, POLLUTION, PRICES AND PUBLIC POLICY (1975).

7. *See* 4 CEQ Annual Report 101-109 (1973); Stewart, *The Development of Administrative and Quasi-Constitutional Law in Judicial Review of Environmental Decisionmaking: Lessons from the Clean Air Act,* 62 IOWA L. REV. 713, 715-716 & n. 14 (1977) [hereinafter cited as Stewart, *Lessons from the Clean Air Act*].

8. Assuming that pollution sources are the cheapest cost-avoiders, it might seem that paying producers of such sources not to pollute

would be preferable because it would achieve a desired level of emission reductions at least cost as polluters vied for payments. Least cost achievement of a given level of emissions through regulatory controls cannot be a-cheived without extensive information costs. Emission fees cannot guarantee achievement of a given level of reductions without incurring similar costs, and transferrable permit systems may be more readily thwarted by the presence of market imperfections than may a payment scheme.

But this conclusion ignores dynamic effects. Paying firms not to pollute will, over the long run, encourage firms to create (or threaten to create) more rather than less pollution. Even in the short run, payments not to pollute may maintain pollution sources that would otherwise have suffered losses and ceased operations. From the viewpoint of dynamic efficiency it is accordingly preferable, as a general matter, to deal with environmental degradation as a problem in collective bads rather than collective goods. For discussion, *see* Polinsky, *Notes on the Symmetry of Taxes and Subsidies in Pollution Control* (Harvard Institute of Economic Research Discussion Paper No. 515, 1976).

9. *See* F. Hayek, THE CONSTITUTION OF LIBERTY 12-13, 205-09 (1960).

10. It may be objected that the polluter's loss of freedom to drive her car is not qualitatively comparable to the receptor's enhanced opportunity to view distant mountains and breathe clean air. The car driver's liberty is being interfered with by coercive state intervention, while the receptor is simply enjoying an increment in his welfare. We are not trading off liberty with liberty, but liberty with welfare—a characterization that enables the polluter publicly to identify her interests with the Long Parliament, Patrick Henry, and resistance to tyranny.

But consider the obverse case. Assume that nuisance law, which formerly forbade pollution causing substantial injury to others, is altered to permit such pollution, free of liability, whenever it is judged to advance the net welfare of society. As a result, a receptor's orchard is ruined and he can no longer enjoy the view of distant mountains. Has not the receptor's liberty been diminished by the shift in tort rules just as much as the polluter's liberty has been infringed by enforcement of regulations against driving polluting automobiles?

There is, to be sure, a difference in the agency by which the respective liberties of the polluter and receptor have been infringed in the two cases. The polluter is liable to fine or arrest by state officials if she drives her car, whereas the infringement of the receptor's opportunity to grow fruit and enjoy the view is accomplished by a private actor.

We continue to so far differentiate the state and civil society that the polluter who rains sulfates on a receptor under the imprimatur of tort law is not regarded as an agent of the government in the same measure as a policeman. Psychologically, at least, this may be an important distinction. Suffering a diminishment of opportunities because of another private actor's pursuit of her own self interest may be viewed as less threatening to one's integrity than being subjected to the official power of the state. But fixing on the identity of the immediate agent tends to obscure the similarity between the two cases. It is the coercive power of the state which ultimately restrains the opportunities of the receptor, who might otherwise resort to self-help to destroy the polluting facility. On the other hand, the polluter need not incur coercive sanctions if she acquiesces in the limitation of her opportunities in the same way that a receptor who eschews self-help acquiesces in the loss of opportunity to see distant mountains and grow apples. The fundamental choice in either case is the relative allocation of opportunities between polluter and receptor. *Cf.* Note 4 *supra.*

For argument that there is no general "right to liberty" (in the sense of being free to do what one pleases), *see* R. DWORKIN, TAKING RIGHTS SERIOUSLY 266-72 (1977).

11. *See* Stewart, *Lessons from the Clean Air Act, supra* note 6 at 750-54.

12. For analysis of the logical priority of the essential constituents of individual integrity, *see* C. FRIED, RIGHT AND WRONG (forthcoming).

13. If, as previously agreed, liberty consists of rich and well-ordered opportunity sets, then we generally cannot enjoy any presumptive or special immunity against governmental action, which is in large measure responsible for the creation of those opportunity sets. Liberties entitled to special protection would have to consist of some sub-class of all possible liberties defined by criteria other than liberty. *See* R. Dworkin, Taking Rights Seriously, 266-72 (1977). The principles of integrity sketched in the text would constitute such criteria. Of

course, they would have to be elaborated and defended at much greater length in order to merit widespread acceptance and application.

14. There is no necessary or inevitable conflict between these fraternal ideals and the general notion of liberty as we have defined it. *Cf.* R. Dworkin, Taking Rights Seriously 272-74 (1977). However, I argue below that there are serious conflicts in the context of environmental policy, between fraternal ideals and the maintenance of certain liberties essential to integrity. *Compare* C. FRIED, RIGHT AND WRONG (forthcoming).

15. 541 F.2d 1 (D.C. CIR.) *(en banc), cert, denied,* 96 S. Ct. 2663 (1976).

16. 521 F.2d 827 (9th CIR. 1975), *vacated and remanded* 45 U.S.L.W. 4445 (May 2, 1977).

17. 344 F. Supp. 253 (D.D.C.), *aff'd by mem.,* 4 ERC 1205 (D.C. Cir. 1972), *aff'd by an equally divided court, sub nom.* Fri v. Sierra Club, 412 U.S. 541 (1973). The basic conclusion in the *Sierra Club* decision—that the federal Clean Air Act precludes significant deterioration of air quality anywhere in the United States—was reaffirmed in Sierra Club v. EPA, 540 F. 2d 1114 (D.C. Cir. 1976), sustaining EPA regulations implementing the earlier decision.

18. 421 U.S. 240 (1975).

19. *See generally* R. POSNER, ECONOMIC ANALYSIS OF LAW 10-102 (1973).

20. *See* Stewart, *The Reformation of American Administrative Law,* 88 HARV. L. REV. 1667 (1975) [hereinafter cited as Stewart, *Reformation of Administrative Law*].

21. *See* Gelpe & Tarlock, *The Uses of Scientific Information in Environmental Decisionmaking,* 48 SO. CALIF. L. REV. 371 (1974).

22. *See, e.g.,* Clean Air Act § 109 (b)(1), 42 U.S.C. § 1857c-4(b)(1) (1970) requiring the EPA to establish ambient air quality standards "requisite to protect the public health" after allowance for "an adequate margin of safety."

23. *See* Gelpe & Tarlock, note 20 *supra.*

24. *See, e.g.,* U.S. NUCLEAR REGULATORY COMMISSION, WASH-1400, Reactor Safety Study (1975); Yellin, *The Nuclear Regulatory Commission's Reactor Safety Study,* 7 Bell J. of Econ. 317 (1976).

25. Borrowing on Schelling's analysis of bargaining, *see* T. SCHELLING, THE STRATEGY OF CONFLICT 60-66 (paper ed. 1970), we may say that the data on adverse effects rarely afford "clues" as to the point at which curtailment of liberty becomes justifiable.

26. *See* C. FRIED, AN ANATOMY OF VALUES 183-87 (1970).

27. *Compare* Ethyl Corp. v. EPA, 541 F.2d 1 (D.C. Cir.) *(en banc) cert. denied,* 96 S. Ct. 2663 (1976) (evidence of "potential harm" sufficient to justify regulatory action) *with id.* at 70 (dissenting opinion of Wilkey, J.). Also *compare* Reserve Mining Co. v. United States, 498 F.2d 1073 (8th Cir. 1974) (requiring proof of "demonstrable health hazard" as precondition to relief against polluting facility) *with* Reserve Mining Co. v. United States, 514 F.2d 492 (8th Cir. 1975) *(en banc)* (relief may issue as a "precautionary and preventive measure" if possibility of harm gives rise to "concern for the public health").

28. *See* note 21 *supra. See also* the discussion in the majority opinion and the principal dissenting opinion as to the significance of variations in statutory language in different sections of the Clean Air Act authorizing regulatory action to prevent adverse health effects. 541 F.2d at 13-17, 96-97. Given Congress' inability to foresee the terms of choice in particular regulatory contexts and the general difficulty of securing legislative agreement on specific directives in areas of regulation where risks are uncertain and the social costs of control high, reliance on such variations in statutory language is largely makeweight.

29. Of course, were judges to refuse to sustain environmental controls on the ground that the risks of harm presented do not justify curtailment of liberty, they would do so not on constitutional grounds but through statutory interpretation, reliance on the "substantial evidence" test, or else remanding for a further explanation by the agency of its action. *See, e.g.,* South Terminal Corp. v. EPA, 504 F.2d 646 (1st Cir. 1974); International Harverster Corp. v. Ruckleshaus, 478 F.2d 615 (D.C. Cir. 1973).

30. *See* Krier, *The Irrational National Air Quality Standards: Macro- and Micro-Mistakes,* 22 U.C.L.A. L. REV. 323 (1974).

31. *See* 521 F.2d 827, 837-42 (9th Cir. 1975).

32. Pennsylvania v. EPA, 500 F.2d 246 (3d Cir. 1974) found that the Clean Air act authorized the EPA to compel state officials to implement EPA-drafted transportation control measures, and that such authority was

constitutional. In District of Columbia v. Train, 521 F.2d 971 (D.C. Cir. 1975), *cert. granted,* 96 S.Ct. 2224 (1976), the court sustained such authority in part and in part denied it. Maryland v. EPA, 530 F.2d 215 (4th Cir. 1975) followed Brown v. EPA in finding that the Clean Air Act did not grant such authority, relying on asserted constitutional difficulties that would be presented if the statute were construed otherwise.

In Friends of the Earth v. Carey, 552 F.2d 25 (2d Cir. 1977) the court sustained the EPA's authority to require state implementation of transportation control measures drafted by the state and subsequently approved by the EPA as part of the state's implementation plan.

33. *See* 96 S.Ct. 2224, *granting certiorari* in Brown v. EPA, District of Columbia v. Train and Maryland v. EPA.

34. EPA v. Brown, 45 U.S.L.W. 4445 (May 2, 1977).

35. For more detailed analysis, *see* Stewart, Pyramids of Sacrifice?—Federalism Problems in Mandating State Implementation of National Environmental Policies, 86 YALE L.J. ______ (1977) (forthcoming); *see also* R. Hostetter and J. Sale. *Protection of the Environment and Protection of the States: The Constitutional Issue Raised by EPA Action under the Clean Air Act,* 7 ENV. L. 383 (1977).

36. *See* Elazar, *The New Federalism: Can the States Be Trusted?* 35 PUBLIC INTEREST 89, 102 (1974).

37. *See* Stewart, *Lessons from the Clean Air Act, supra* note 6, at 747-48, 750.

38. *See id,* at 748; Krier, *The Irrational National Air Quality Standards; Macro- and Micro-Mistakes,* 22 U.C.L.A. L. REV. 323 (1974); Tripp, *Conflicts and Dilemmas in Recent Federal Water Legislation,* 14 HARV. J. LEGISL. ______ *(1977) (forthcoming).*

39. *See* Stewart, *Lessons from the Clean Air Act, supra* note 6, at 750-58.

40. *See id.,* at 741, 744-45 & n. 159, 755.

41. *See* Wildavsky, Aesthetic Power or the Triumph of the Sensitive Minority over the Vulgar Mass, in POLLUTION AND PUBLIC POLICY 37 (D. Paulsen & R. Denhardt et. 1973). For discussion of the potential conflict between cultural or aesthetic ideals and fraternal norms, *see* B. BARRY, POLITICAL ARGUMENT 220-21 (1965); R. DWORKIN, TAKING RIGHTS SERIOUSLY 274-75 (1977).

42. For detailed analysis of the ways in which transaction costs and free rider effects hinder the mobilization of widely shared, unorganized interests in which any given individual's stake is small, *see* M. OLSON, THE LOGIC OF COLLECTIVE ACTION (1965).

43. *See* Stewart, *Reformation of Administrative Law, supra* note 19, at 1693-97.

44. *See id,* at 1713-15. For a sophisticated and illuminating analysis by an economist of the political economy of regulation, *see* Peltzman, *Toward a More General Theory of Regulation,* 19 J.L. & ECON. 211 (1976).

45. *See* COUNCIL FOR PUBLIC INTEREST LAW, BALANCING THE SCALES OF JUSTICE 312-21 (1976).

46. *See* Stewart, *Reformation of Administrative Law, supra* note 19, at 1762-70; *see also,* Note, *Alyeska Pipeline Service Co. v. Wilderness Society: Demise of the "Private Attorney General",* 6 ENV. L. 243 (1975).

47. *See* R. WOLFF, THE POVERTY OF LIBERALISM (1968).

48. Borrowing from O. HIRSCHMAN, EXIT, VOICE AND LOYALTY (1970), we may say that unorganized individuals assertedly represented by public interest advocates can, as a practical matter, neither "exit" by repudiating such representation nor exercise "voice" by speaking for themselves.

49. *Alyeska* was not, of course, a decision whether to improve or sustain regulation, but approval by the Supreme Court of judicial fee awards for environmental litigants would surely have resulted over the longer run in greater restraints on environmentally disruptive activities.

50. P. Samuelson, *The Pure Theory of Public Expenditure,* 36 REV. OF ECON. & STATISTICS 387 (1954).

51. W. BAUMOL, WELFARE ECONOMICS AND THE THEORY OF THE STATE (2d ed. 1965).

52. M. OLSON, THE LOGIC OF COLLECTIVE ACTION (1965).

53. *See, e.g.,* Peltzman, *Toward a More General Theory of Regulation,* 19 J. L. & ECON. 211 (1976); Stewart, *Reformation of Administrative Law, supra* note 19, at 1681-88.

54. *See* Stewart, *Reformation of Administrative Law, supra* note 19, at 1711-89.

55. *See* Stewart, *Lessons from the Clean Air Act, supra* note 6, at 727-40.

56. *See* Stewart, *Reformation of Administrative Law, supra* note 19, at 1725-30.

57. *See* Chayes, *Some Further Last Words on Erie: The Bead Game,* 87 HARV. L. REV. 741, 753 (1974).

58. *See, e.g.,* Scenic Hudson Preservation Conf. v. FPC, 354 F.2d 608 (2d Cir. 1965), *cert. denied,* 384 U.S. 941 (1966); Citizens Committee for the Hudson Valley v. Volpe, 425 F.2d 97 (2d Cir. 1970).

59. Freund, *Umpiring the Federal System,* 54 COLUM. L. REV. 561, 578 (1954).

60. *See* Stewart, *Reformation of Administrative Law, supra* note 19, at 1770-76.

61. *See* Stewart, *Lessons from the Clean Air Act, supra* note 6, at 727-40.

62. *See* Sax, *The (Unhappy) Truth About NEPA,* 26 OKLA. L. REV. 239 (1973).

63. *Cf.* Mathews v. Eldridge, 96 S. Ct. 893, 909 (1976) (Costs to the government of providing procedural formalities should be considered by courts in determining what procedures are required by constitutional due process in administrative decisionmaking.)

64. 401 U.S. 402 (1971).

65. *See* note 16 *supra.*

66. Izaak Walton League v. Butz, 522 F.2d 945, 949 (4th Cir. 1975); Zieske v. Butz, 9 ERC 1061, 1063 (D. Alas. 1976); Sierra Club v. Department of Interior, 376 F. Supp. 90, 95 (N.D. Cal. 1974).

67. For an example of legislative "correction" *see* P.L. No. 93-153, 87 Stat. 576 (1973), authorizing construction of the Trans-Alaska Pipeline and effectively overruling Wilderness Society v. Morton, 479 F.2d 842 (D.C. Cir.) *cert. denied,* 411 U.S. 917 (1973).

68. *See, e.g.,* M. HORWITZ, THE TRANSFORMATION OF AMERICAN LAW (1977).

69. To the extent that the appropriate goal of resource allocation in the provision of collective goods is economic efficiency, the case for judicial intervention is weak. A crucial problem in determining the levels of collective goods that will maximize consumer satisfaction as measured by willingness to pay is to determine consumer preferences for such goods, which is difficult because such goods are not traded on markets and often lack close market surrogates. The political branches of government are more likely to respect consumers' existing preferences for such goods than is the judicial process. *See* Stewart, *Lessons from the Clean Air Act, supra* note 6 at 749 & n. 174. But to the extent that considerations of integrity call for maintenance of certain environmental opportunities, maximizing the satisfaction of existing consumer preferences is no longer the sole criterion of decision, and the judiciary may properly play a larger role, particularly where maintenance of such opportunities is hindered by structural defects in the political process. *See id.* at 747-58, 767-69.

70. *See id.* at 757-58.

71. G. Calabresi, *The Common Law Function in the Age of Statutes* (Holmes lectures delivered at Harvard Law School, March 8-10, 1977).

72. *See* Stewart *supra* note 6, at 747-58, 767-69.

73. *See* International Harvester Co. v. Ruckleshaus, 478 F.2d 615, 650 (D.C. Cir. 1973) (Bazelon, J., concurring) (judges in technically complex cases should adopt the "wisdom of Socrates" in acknowledging "how much one does not know").

Further Reading

Bird, C. 1972. A New Bill of Rights. In *The Crowding Syndrome,* pp. 262-265. New York: McKay.

Blackstone, W.T. 1968. Equality and Human Rights. *Monist* 52: 616-369.

Environmental Protection Agency. 1989. EPA Lists Places Failing to Meet Ozone or Carbon Monoxide Standards. Press release July 27. Washington, D.C.: Environmental Protection Agency.

French, H. 1990. Clearing the Air. In *State of the World* 1990, ed. L. Brown, C. Flavin, and S. Postel, pp. 98-118. New York: Norton.

Herfindahl, O.C. 1975. Defining the Problem of Environmental Quality. In *Environmental Quality and Society,* ed. R.A. Tybout, pp. 219-238. Columbus, Ohio: Ohio State University Press.

Iltis, H.H., Loucks, O.L., and Andrews, P. 1970. Criteria for an Optimum Human Environment. *Bulletin of the Atomic Scientists* 26 (January): 2-6.

Kozlovsky, D.G. 1974. Look Where I've Got to Now. In *An Ecological and Evolutionary Ethic.* Englewood Cliffs, New Jersey: Prentice-Hall.

Lashof, J., et al. Health and Life Sciences Division of the US Office of Technology Assessment (OTA). 1981. *Assessment of Technologies for Determining Cancer Risks from the Environment.* Washington, D.C.: OTA.

Manchester, K.A. 1978. Law and the Dignity of Nature: Foundations of Environmental Law. *Environment Law Review* 1978, pp. 3-26. New York: Clark Boardman.

Mishan, B. 1970. Property Rights and Amenity Rights. In *Technology and Growth: The Price We Pay.* pp. 36-42. New York: Praeger.

Moyers, B. 1990. *Global Dumping Ground.* Washington, D.C.: Seven Locks Press.

Murphy, E.F. 1970. A Law for Life. *Environment Law Review 1970.* pp. 3-20. New York: Clark Boardman.

Neuhaus, R. 1971. *In Defense of People.* pp. 274-303. New York: Macmillan.

North, D.C. 1977. Political Economy and Environmental Policies. *Environmental Law* 7 (Spring): 449-462.

Passmore, J. 1974. *Man's Responsibility for Nature.* Scribners, New York.

Pearson, J.Y. 1971. Toward a Constitutionally Protected Environment. *Environment Law Review—1971,* pp. 53-83. New York: Clark Boardman.

Peffer, R. 1978. A Defense of Rights to Well-Being. *Philosophy and Public Affairs.* 8 (Fall): 75-87.

Pettigrew, H.W. 1971. A Constitutional Right of Freedom from Ecocide. *Environmental Law* 2 (Winter): 1-41.

Postel, S. 1986. *Altering the Earth's Chemistry.* Washington, D.C.: Worldwatch Institute.

Postel, S. 1985. *Conserving Water: the Untapped Alternative.*Washington, D.C.: Worldwatch Institute.

Postel, S. 1986. Increasing Water Efficiency. In *State of the World 1986,* ed. L. Brown and E. Wolf, pp. 40-61. New York: Norton.

Postel, S. 1985. Managing Freshwater Supplies. In *State of the World,* ed. L. Brown, pp. 42-72.New York: Norton.

Postel, S. 1989. *Water for Agriculture: Facing the Limits.* Washington, D.C.: Worldwatch Institute.

Postel, S. 1984. *Water: Rethinking Management in an Age of Scarcity.* Washington, D.C.: Worldwatch Institute.

Reynolds, S.H. (ed.). 1991. *Why Johnny Can't Breathe.* NRDC Newsline 8 (April): 1, 5.

Shaw, A. 1977. Defining the Quality of Life. *The Hastings Center Report* 7 (October): 11.

Shrader-Frechette, K.S. 1985a. *Science Policy, Ethics, and Economic Methodology. Boston, Massachusetts: Reidel.*

Shrader-Frechette, K.S. 1985b. Risk Analysis and Scientific Method. Boston, Massachusetts: Reidel.

Shrader-Frechette, K.S. 1991. *Risk and Rationality.* Berkeley, California: University of California Press.

Stainbrook, E. 1975. Mental Health and the Environment: Do We Need Nature? In *Environmental Quality and Society,* ed. R.A. Tybout, pp. 186-202. Columbus, Ohio: Ohio State University.

Udall, S.L. 1963. *The Quiet Crisis.* New York: Holt, Rinehart and Winston.

United Nations Environment Programme (UNEP) and World Health Organization (WHO). 1988. *Assessment of Urban Air Quality.* Nairobi: Global Environment Monitoring System.

US Office of Technology Assessment. 1984. *Acid Rain and Transported Air Pollutants.* Washington, D.C.: US Government Printing Office.

Young, G.L. 1975. Environmental Law: Perspectives from Human Ecology. *Environmental Law,* 6 (Winter): 289-308.

7

THE DUTY TO LIMIT CONSUMPTION

Introduction

NEVER BEFORE have the American people enjoyed as much prosperity and, at the same time, been beset by so many problems. As Charles Dickens said in beginning *A Tale of Two Cities:* "It was the best of times, it was the worst of times." In fact, it might well be said that today our increasing material wealth has brought most of our spiritual and environmental problems. We have used pesticides and fertilizers to fill our food coffers and to create the "Green Revolution"; yet, these chemicals are poisoning fish and birds and polluting our food and water. We have harnessed nuclear fission to provide an abundance of electrical energy, but our children, and their children, and their children's children, for the next one hundred thousand years, must learn to protect themselves from the radioactive wastes we have bequeathed them.

With all the technological innovations that have brought, at once, both material abundance and greater technological problems, humankind has been faced with some gnawing questions. What is our current style of life doing to nonrenewable resources? Will enough be left for generations to come? Do we waste too many goods? Is there a moral obligation to consume only what we need? Or, on the other hand, do we have a right to consume as much as we can "legally" obtain? Most importantly, what have our patterns of use and waste done to nurture our most important resource, the human spirit?

These are only some of the issues that might be raised when one examines the morality of our current levels of resource use. Specifically, this chapter will address the question of whether we have a duty to limit our consumption and possession of goods. If we do, then not only must we understand the reasons for this obligation, but we also need to know what moral considerations are relevant to establishing ethical standards for consumption. If we do *not* have this duty, then we ought to know why not, as well as how to work toward conditions of equal opportunity among all those desiring goods and resources.

In determining the answer to this question, a number of distinctions are important. Ordinarily, one speaks of resources as renewable or nonrenewable. Renewable resources are those which either maintain themselves or are able to be replenished. Most ecologists classify food crops, animals, forests, air, water and soil as renewable resources. Nonrenewable resources, on the other hand, are those whose supplies either are finite, or take so long to be renewed, that one can speak of them as being depleted. Coal, oil, phosphates, aluminum, tin, copper, and iron, for example, are in this category.

It is easy to see that the issue, of whether we have a duty to limit our consumption, might be resolved differently depending on whether one is talking about our use of renewable or nonrenewable resources. The problems raised by consumption of goods in the first category are that they might be used faster than they can be regenerated and that environmental side effects of their employment might harm the planet. The difficulties raised by consumption of nonrenewable resources are even more serious. Their use raises not only the problems already mentioned but also the question of whether supplies are being depleted at such a rate that the well-being of future generations will be jeopardized.

The renewable / nonrenewable distinction, however, does not always provide a neat classification for the gravity of problems of consumption and resource depletion. This is because materials are always coming into, or going out of, the "important resource" category. Aluminum, for example, has become a more significant resource, for many purposes, and has replaced a number of uses of iron and steel. On the other hand, if nuclear fission becomes more prominent, uranium may become more valuable than coal. Such tradeoffs depend in part on cost-benefit considerations as well as on the quality and availability of alternative materials. Hence, although it is easy to determine which resources are renewable or nonrenewable, it is not always possible to know which ones will become more important than others, and whether technology will be able to extend the life of a particular commodity.

Another factual concern is the extent to which resource consumption (as opposed to population, the profit motive, and technological innovation, for example) is responsible for environmental degradation or for threats to human survival. Clearly, the greater the threat produced by depletion of valuable raw materials, the greater the probability that we have the obligation to limit our consumption of resources.

As the previous remarks indicate, disagreement concerning the existence of this duty stems from both factual and ethical considerations. Experts such as Mesarovic and Pestel, authors of the second report to the Club of Rome, *Mankind at the Turning Point,* believe that there is no technological or economic means capable of delivering us from the consequences of finite resources. They are convinced that our own survival, as well as justice to future generations, demands that we not engage in inordinate consumption. Hence they argue that " ... A new ethic in the use of material resources must be developed which will result in a style of life compatible with the oncoming age of scarcity."

Those who disagree with the notion, that the world's resources are irrevocably finite, usually do so because they believe that technology will somehow rid us of our problems of scarcity. They argue that this will be accomplished either by developing new substances or by discovering new ways to use fewer traditional resources. Apart from what assumptions they share about the ability of technology to rescue us from material shortages, nearly all those, who oppose any attempt to limit consumption, do so because they believe that individual liberty is a greater good than conservation of resources. M.J. Grayson and T.R.

Shepard, for example, authors of *The Disaster Lobby,* point out some of the dangers to human freedom. Their reasoning is that, as one author put it, only the individual is able to determine what he needs and what he is wasting.

Those who believe, on the other hand, that there is an obligation to limit one's consumption, often argue that this duty is not a matter of individual freedom, but of justice. They maintain, for example, that it is unjust for citizens of the US, with 6 percent of the world's population, to use 60 percent of its minerals and energy. For them, our right to consume ends where another's right to his fair share begins. Why is it, they ask, that western countries with nearly the *same* per capita Gross National Product as the US use only one half or one third as much per capita energy as the US?

In response to such remarks about America's high rate of consumption, economists sometimes point out that resources are not really taken away from underdeveloped countries. Instead, they argue, the US helps Third World nations by buying their raw materials; since such countries are unable to produce sophisticated manufactured goods, their survival depends upon US markets for their natural resources.

A number of authors argue for the duty to limit consumption, neither on grounds of our survival and the depletion of resources, nor in terms of justice to underdeveloped nations. Instead they maintain that this duty can be debated purely on the basis of personal morality. For these people, such as Frederick Elder, author of *Crisis in Eden,* denying this obligation is a hallmark of materialism. Elder proposes that we need a new asceticism, a return to a Puritan ideal, and a turning away from pleonexia—wanting more and more. For Elder, a theologian, unlimited freedom to consume is irresponsible and spiritually debilitating. He explains how Americans typically place importance on being free *from* restraint rather than on being free *for* what is good.

Richard Neuhaus, author of *In Defense of People,* is critical of those who, like Elder, argue that the environmental crisis demands a new asceticism. Neuhaus reasons that use of the term, 'asceticism', implies that one has a free decision about whether to use restraint in consuming. This is unrealistic, he says, because most of the people of the world cannot make such a decision; they are too busy with life-and-death matters of attempting to survive. As Neuhaus puts it: asceticism, as a response to the world crisis of resource shortage, "is not socially relevant."

The authors represented in this chapter hold a wide variety of views on the topic of whether man has a duty to limit his consumption. They provide opposite points of view on the issues of the alleged scarcity of valuable raw materials, the obligation of the US to consume only its fair share, and the ability of technology to handle the problem of resource shortage. Ansley J. Coale, an economist and Director of the Office of Population Research at Princeton University, disagrees with the MIT report, *The Limits to Growth.* In the first reading, he claims that there is no shortage of most resources, that new substitutes for scarce substances will be found, and that US consumption patterns are actually beneficial to poor nations. Regarding the last point, he explains how American buyers provide a much-needed market for those who must sell their raw materials.

E.C. Pasour, Professor of Economics at North Carolina State University, is the author of the second reading. Like Coale, he provides reasons which can be used to show that current patterns of resource consumption are desirable. He argues why the market, with its controls on price and commodity, is not only the most efficient means of allocating scarce resources, but also the most desirable way. Ethical or voluntary ascetical incentives, such as those provided by Schumacher, are unworkable, says Pasour; without the market, it is impossible to specify what is an authethic need and what is not. For this same reason, he maintains that involuntary rationing also presents an unworkable solution to resource consumption. According to Pasour, these schemes have the additional problem of being either wittingly or unwittingly discriminatory and therefore antisocial. Hence he views both mandatory and voluntary restriction of consumption as less good than allowing the market to set such levels. For Pasour, any scheme except that of the market is counterproductive as well as unethical. Austerity militates against people's desire to work, says Pasour, and forced austerity threatens the tenets supporting a free society.

I disagree with both Coale and Pasour. In the third selection in this chapter, I use the MIT study, *The Limits to Growth,* to argue that—unless resource consumption is limited in the future—planetary environmental and political crisis will result. Other reasons for such limitation are that it will help to achieve equity in distribution of world resources and, on a personal level, will enable individuals to become freer from the need to engage in conspicuous consumption. Agreeing with E.F. Schumacher, that limiting consumption helps one's growth in spirit, I propose following an ethic based on voluntary simplicity or austerity.

Although the three readings in this chapter represent only a first attempt to address the question of whether we have an obligation to limit our consumption, they do provide some useful perspectives on this issue. Much of what remains to be said can be found in some classical treatments of distributive justice, such as those by Nicholas Rescher, Robert Nozick, or John Rawls. What is interesting about the selections presented in this chapter, however, is what they reveal about the dimensions of the environmental crisis. Clearly, the shortage of natural resources is not just a problem for international allocation or for technological answers. If theorists like Kozlovsky and Schumacher are at least partially correct, it is also a crisis of spirit, and one that touches our most basic perceptions about who we are and what we need.

Readings

1. Coale, A.J. 1970. An Economist's Review of Resource Exhaustion. *Science* 170 (October 9): 132-133. ©1970 by the American Association for the Advancement of Science. Reprinted by permission of Ansley J. Coale and *Science* magazine.

2. Pasour, E.C. 1978. Austerity, Waste, and Need. *The Intercollegiate Review* 13 (Winter-Spring): 79-85. Reprinted by permission of E.C. Pasour and *The Intercollegiate Review*.

3. Shrader-Frechette, K.S. 1979. Voluntary Simplicity and the Duty To Limit Consumption. Written especially for inclusion in this volume.

AN ECONOMIST'S REVIEW OF RESOURCE EXHAUSTION

Ansley J. Coale

ONE OF THE questions most frequently raised about the environmental effects of modern life is the rapid and rising rate of extraction of raw materials. Are we running out of resources?

I would first like to note that the distinction between renewable and nonrenewable resources is not a clear one. There are, of course, instances of nonrenewable resources in the form of concentrated sources of energy, such as the fossil fuels. These are reservoirs of reduced carbon embodying radiant energy from the sun that accumulated over many thousands of years. When these fuels are used, the energy that is released is to a large extent radiated into space, and we have no way of reclaiming it. The geological processes that are constantly renewing the fossil deposits of carbon are so slow compared to the rate at which we are burning the fuels that the designation "nonrenewable" is appropriate.

On the other hand, when we think of our resources of such useful materials as the metallic elements of iron, copper, nickel, lead, and so forth, we should realize that spaceship Earth has the same amount of each element as it had a million years ago, and will have the same amount a million years from now. All we do with these resources is to move them around. The energy we use is lost, but the minerals we find useful are still with us. It does not pay to recycle these minerals (that is, to use them repeatedly by reclaiming scrap) because the deposits of minerals in the ground or in the ocean are still such a cheap source. It must be noted that the mining of fresh ore is cheaper than the use of scrap in part because miners are not charged for their "externalities." If harmful by-products of mining could not be discharged into streams, if mine tailings were regulated, and erosion-producing or even unesthetic practices forbidden, minerals would be more expensive and recycling more attractive. In the production of any metallic element, the easier sources are exploited first. As mining gets more difficult, the ore gets more expensive, and recycling becomes more nearly competitive. It seems wholly probable that the technology of recycling will be improved.

The surprising fact is that raw materials are not at the moment very costly, and moreover their cost relative to the cost of finished goods has not been increasing. The gross national product in the United States is more than $4,500 per capita and the raw materials component per capita is less than $100. The price of raw materials relative to the price of finished goods is no higher now than at the beginning of the century, and if we were running out of raw materials, they would surely be rising in relative expensiveness. A prominent exception is saw lumber, which is substantially more expensive relative to the cost of finished wooden products than it used to be.

The reason that the future of our resource situation always seems so bleak and the past seems quite comfortable is that we can readily construct a plausible sounding estimate of the future demand for a particular raw material, but cannot form such a plausible picture of the future supply. To estimate the future demand, we need merely note the recent trends in the per capita consumption of whatever it is we are concerned about, utilize whatever plausible projection of population we are prepared to accept, multiply the two together, and project an astonishingly high rate of usage 50 years in the future. If this demand does not seem overwhelming, we need only make a projection 100 years in the future. What we cannot so readily foresee is the discovery of new sources and of new techniques of extraction, and, in particular, the substitution of other raw materials or the substitution of other industrial processes which change the demand away from the raw material we are considering. Hence it can always be made to appear that in the future we are going to run out of any given material, but that at present we never have.

It is possible to set plausible limits to the stores of fossil fuels that we are likely to discover, and with the very rapid rise in the use of these fuels they will surely become more expensive in some not too distant time. It should be noted, however, that we will not suddenly "run out" of fossil fuels. Long before the last drop of oil is used, oil will have become much more expensive. If gasoline were $5 or $10 a gallon, we would utilize it much more sparingly, with small economical automobile engines, or perhaps the substitution of some non-petroleum-based fuel altogether. In fact, the principal user of our petroleum deposits may be the petrochemical industries. I have given this special attention to fossil fuels because there is no substitute in prospect for such fuels in small mobile units such as automobiles. On the other hand, the supply of overall energy seems to pose no problem. There seems to be ample fissionable material to supply rising energy needs for many centuries, if breeding reactors are perfected. If fusion proves a practical source, the supply of energy can properly be considered limitless.

Another aspect of the relation of the United States economy to resources that is much publicized today is the fact that we are consuming such a large fraction of the current annual extraction of raw materials in the world. A much quoted figure is that 6 percent of the world's population is using 30 percent of the resources. It is concluded from figures such as these that we are robbing the low-income countries of the world of the basis of their future prosperity—that we are using up not only our resources, but theirs as well. Most economists would find this a very erroneous picture of the effect of our demand for the raw materials extracted in the less developed parts of the world. The spokesmen for the less developed countries themselves constantly complain about the adverse terms of trade that they face on world markets. The principal source of their concern is the low price of raw materials and the high price of finished goods. The most effective forms of assistance that the developed countries (including the United States) give to the less developed countries are the purchases they make from the less developed countries in international trade. A developing country needs receipts from exports in order to finance the purchase of the things they need for economic development. For example, in order to industrialize, a non-industrialized country must for a long time purchase capital equipment from more advanced countries, and the funds for such purchases come from exports—principally of raw materials. Economists in the developing countries feel that the demand for raw materials is inadequate. Perhaps the most important adverse effect of slowing down the growth of the gross national product in the United States would be that it would diminish the demand for primary products that we would otherwise import from the less developed countries. After all, if a developing country wants to retain its raw materials at home, it can always place an embargo on their export. However, it would be a policy very damaging to economic progress of that very country.

Note that the effect of our high demand for raw material is a different matter from the desirability of the domestic control of mineral resources within the developing countries. Selling oil on the world market provides immense economic advantages to a developing country. Whether foreign interests should be represented in the extraction of raw materials is another question.

AUSTERITY, WASTE, AND NEED

E.C. Pasour, Jr

IN RECENT YEARS U.S. citizens have been encouraged to waste less and adopt a more austere life style. We have been chastized for our energy-intensive habits. We eat meat instead of cereals; we drive large cars, drive when we could walk, keep our homes warmer than needed, and engage in many other "wasteful" practices. More generally, U.S. citizens have been criticized for the use of luxury goods and for using a disproportionate amount of the world's resources.

There is a great deal of concern about the ability of reserves of fossil fuel and other resources to support projected levels of population. Extrapolation of current trends in resource use in the *Limits to Growth* and other "Doomsday Models" lead to the spectre of the world grinding to a halt within the next century or so as currently known resources are depleted. What should be our posture toward the use of energy in the production of agricultural and industrial commodities? What is the appropriate approach to take toward legislation mandating mileage standards for automobiles, mandating insulation standards for home and office buildings, banning the use of cereals for livestock feed, etc?

Conventional wisdom holds that we should eliminate waste in the use of gasoline, electricity, and other resources. The concepts "need" and "waste," however, are far more complicated than the widespread use of these terms suggest. The philosophy of austerity rests on an insecure basis and leads to measures that are harmful and contradictory. This essay explores the meaning of "waste" and "need" and compares the effects of market and non-market rationing procedures.

Waste and Need

During the energy crisis following the OPEC oil embargo, individuals were encouraged to reduce waste in the use of food, energy, and other raw materials. The same pleas were heard again in the winter of 1976-77 during the fuel crisis. If individual citizens are to respond intelligently, they must be able to identify waste.

The concept "waste," however, has little or no objective content and appears to be closely related to the term "need." Furthermore, the concept of "need," is meaningless as a guide in determining one's consumption pattern. It is impossible to define objectively the goods and services we "need." How much meat, housing, heating fuel or gasoline, for example, do we "need?" The concept "need" implies that, regardless of price, there is some minimum amount of meat, housing, heating fuel, or gasoline (or any other good or service) that is essential to our livelihood. We observe, however, that people reduce consumption when relative price rises for any good or service including gasoline, coffee, meat, and housing.

The amount "needed" of any good depends on the price of that good relative to the price of other goods. The "need" of Jones for coffee depends on his subjective valuation of coffee and coffee substitutes. It is an illusion of bureaucrats and consumerists that the U.S. Congress, the Federal Energy Administration, or Ralph Nader can determine our individual "needs." Even we as individuals can only determine our own needs in the sense that we decide how much we prefer at given prices. When the relative price of a good changes, however, our "needs" change in the sense that we prefer to purchase more or less of the good. There is no known example of any good where the quantity purchased remains the same when there is a large change in its price relative to the prices of other goods.

The concept "waste" implies that amounts of a product which exceed our "needs" level provide no additional satisfaction. If this were in fact the case, amounts of the product above this "needs" level could be removed at no loss of satisfaction to the consumer. When we investigate commonly cited examples of "waste," however, we find that the goods presumably wasted are, in fact, contributing to consumer welfare. If the goods alleged to be wasted were providing no satisfaction, coercion would not be required to reduce use.

The use of the family auto provides a good example of the point being made. It is frequently alleged that the typical motorist wastes gasoline by using his auto when it isn't "needed." We observe that people drive their car when they could walk, ride a bicycle, or form a car pool. This fact that people use their car when alternatives are available doesn't mean that the use of gasoline used in this way is "wasted."

There is no denying the fact that auto users could get by with much less use of the auto. The fact that motorists *could* reduce auto use however, does not imply that it is in their interest to do so. Each motorist will act in the way which is most beneficial to him given the costs and benefits of various alternatives as they are perceived by him. Consumers respond to relative transportation costs in deciding where to live, how to get to work, where to vacation, etc.

We observe that motorists economize in numerous ways when costs of auto travel increase. As gasoline prices increase, for example, there is substitution in favor of smaller autos, more bicycling, more car pooling, shorter vacation trips, etc. The fact that less gasoline is used when price increases does not mean that the larger amount used at the lower price was "wasted." One might, to the contrary, argue that at a lower price larger numbers of consumers are able to benefit from a particular good or service, and therefore a lower price is an efficient means of effecting a widespread distribution of some item of value. People economize in the use of scarce resources in responding to their own self interests. This does not mean that all people respond to a change in prices in the same way. We expect individuals to make different accommodations to any change in relative prices reflecting their own unique situation and preferences.

The preceding discussion implies that there is no objective basis by which an outside observer can determine the amount of gasoline any particular motorist "needs." If "needs" cannot be objectively determined, decisions made to limit use through administrative controls must be arbitrary and capricious.

Relative Prices vs. Need

It has become fashionable in recent years to suggest that people in the U.S. and other highly developed countries should base their consumption not on relative prices but on "need." The problem as indicated above is that need and waste concepts apart from relative prices provide no operational criterion by which the consumer can make choices. What do we "need," for example, in the way of clothing, electricity, gasoline or food? It has been shown that a nutritious minimum cost diet can be formulated for people as is done for livestock. Such a diet would enable U.S. families simultaneously to have a more nutritious diet and to reduce food expenditures to less than one-half their current level! Is this the food budget we "need?" Can all food expenditures in excess of this minimum cost nutritious diet properly be considered waste? Few of us will agree to sacrifice all palatability considerations and rely solely on cost considerations in choosing our diet.

The same problem arises in the use of gasoline, electricity, and in fact, all goods and services. How does one decide what size car to drive, amount of electricity to use, temperature to set the home thermostat, etc.? We, as consumers, respond to changes in relative prices in deciding what to eat, where to live, how to travel to work, heat our homes, etc. Each of us would do virtually everything we do in some other way if changes in relative costs were sufficiently large. Attempts to get people voluntarily to change consumption habits are unlikely to have much effect so long as relative prices remain unchanged. One the other hand, when relative prices change, no one has to encourage consumers to make adjustments in the mix of goods consumed—regardless of whether the good is coffee, beef, gasoline or heating oil. In recent years, as relative prices changed, we have observed pronounced changes in size of auto, in meat consumption, in coffee consumption, in use of electricity, etc. Market price will effectively ration goods when prices are not held down by administrative decree. There is no known example of a shortage persisting over time where price was relied upon as the rationing mechanism.

Problems of Non-Market Allocation Methods

Few people who suggest that consumers should make choices related to food, gasoline and other products on the basis of some criterion other than price recognize the problems to be overcome when the market mechanism is abandoned or market signals are ignored. Price in a market economy provides signals to both consumers and producers. The market is a highly useful mechanism for determining and transmitting information between all persons in the market. If the market is not permitted to coordinate the decisions of consumers and producers, central direction must be used. However, there is no way for the central planner to obtain all the information which enters into any real world market transaction. Knowledge about no good or resource exists in concentrated form or in a single mind. The economic problem, as Hayek has long stressed, is to secure the best use of our resources utilizing the knowledge of all members of society for ends whose relative importance only these individuals know.

In the absence of market signals, the regulator faces severe information problems. When the market is abandoned, the planner has no way to determine the information necessary to coordinate supply and demand. Consider the problem of determining the "appropriate" temperature setting, for example, where energy use is to be restricted not by price but by restricting fuel use. The regulator must not only predict the amounts which will be used at various settings, but also the amount which will be produced at the price level which is arbitrarily held below the market-clearing level.

Mandatory allocation or rationing schemes cannot be based on individual preferences and must be arbitrary and capricious. "Need" has no objective content as a guide to allocation, and in the absence of price signals there is no way for the central planner to make an allocation which reflects the subjective considerations of all persons in the market.

Problems are also created for the individual consumer who unilaterally adopts austerity measures and does not respond to market signals. Attempts by a single individual to reduce consumption of food, energy, and other raw materials beyond the level dictated by relative prices will serve little or no useful purpose. If an individual, acting alone, reduces the level of consumption, the effect on total consumption will be negligible. For example, consider the effect of a decision by one "socially aware" person to reduce gasoline use. The effect on total gasoline usage will be negligible but the inconvenience to that individual can be quite large. If the individual is in business such a decision will increase costs and under competitive conditions will seriously decrease profits or bankrupt the entrepreneur attempting to "do good."

What are the possibilities for group action? If large numbers of consumers reduce consumption below the level dictated by relative prices, producers receive incorrect signals. Consider, for example, the effect of consumer boycotts of beef, coffee and other goods. A reduction in demand will decrease price, given the level of supply. The decrease in price resulting from the boycott is likely to reduce future supply and cause future prices to be higher than they would otherwise be. Unless the product is effectively monopolized, there is no way for group action by consumers to reduce the long run price of the product.

What do these comments imply about moral suasion as a method of rationing and reducing resource use. Pleas to reduce use may be effective for a limited period of time when the situation is deemed to be urgent by a large part of the population. The problems enumerated above which arise when market signals are ignored are not eliminated by moral suasion. Moral suasion is also subject to another set of problems. First, there is the free rider problem common to all voluntary group activity. These activities confer benefits on people who cannot be made to pay for the benefits they receive. When the demand for gasoline is reduced by everyone except Jones, the price is decreased. Jones as a free rider will benefit through the lower price and use *more* gasoline. Moral suasion may be effective for a limited period of time but it cannot solve the free-rider problem associated with pleas to reduce gasoline use, lower thermostats, and otherwise reduce resource use below the level dictated by market prices.

Second, moral suasion to reduce resource use to one's "needs" or to reduce "waste" is subject to all the problems associated with identifying "waste" and "need" discussed above. There is no objective basis upon which the "socially concerned" citizen can determine the proper level of energy use. Consider, for example, the natural gas shortage during the winter of 1976-77 and the pleas to reduce "waste" in home heating. Upon what basis can the "socially concerned" homeowner decide whether to heat his home to 68° F, 65° F, 60° F, 55° F (or even lower)?

Third, even if moral suasion is effective and all citizens fully comply with, say, a request that thermostats be set at 62° F, the impact will differ greatly from person-to-person. Such a policy as in the case of mandatory restrictions assumes that everyone has similar circumstances. In reality, people differ greatly in their preferences for heat and in their preferred tradeoffs between heat and other sources of expenditure including size of house. The differential impact of any restriction in use of any good or resource is greatly magnified if an attempt is made to indicate the appropriate *amount* to use, *e.g.,* 10 gallons of gas per car per week or 1000 KWH of electricity per month. The circumstances of time and place vary greatly from person-to-person.

Conservation and Rationing

A great deal of effort has recently been devoted to reduce "waste" of natural resources. The effects of such efforts, however, may be inconsistent with other closely related goals. First, consider action to reduce waste of *renewable* resources. Environmentalists and consumerists have been active in promoting recycling of paper in recent years as a way of conserving trees and protecting forests. Moves to reduce "waste" by reducing the use of paper and other wood products will be counterproductive for people who desire large amounts of forests for their esthetic value since such action will *reduce* the number of trees being grown. Tree producers respond to economic incentives just as other producers do. Thus, the more wood products used, the more trees will be produced. The more trees produced, the more land required for tree production and the larger the forest acreage. People who like to see trees growing are working against their own self-interest when they discourage the use of forest products.

Another simplistic solution, bans on the use of cereals in feeding livestock or forced reductions in meat consumption, will likewise not accomplish the intended goal. Jean Mayer, the famed nutritionist, has suggested, for example, that America could release enough grain to feed 60 million people by reducing meat consumption by 10 percent. Measures to reduce grain fed to livestock or to reduce meat consumption, however, are unlikely to be effective in providing food to the world's hungry people. The hamburger not eaten in the United States will not miraculously appear in the hands of a hungry person in another land.

How about the use of non-renewable resources including oil, coal, etc.? It is in the owner's interest at any point in time to exploit these resources in such a way as to maximize the wealth, or value of these resources. As non-renewable resources are used, increasing scarcity will be reflected in two ways. First, the price will be bid up as the resources become progressively scarcer. When this happens consumers are induced to economize on the use of the resources. The market provides an effective system of rationing scarce resources both at a given time and over time.

When prices are arbitrarily held down by government as in the case of natural gas, a shortage is created or exacerbated. The natural gas crunch in January 1977 can be traced directly to current and past government price controls. These price controls have kept prices artificially low have given incorrect signals to gas producers and consumers. Low consumer prices have encouraged the "wasteful" use of gas by discouraging the use of home insulation, alternative fuel sources, lower thermostat settings, etc. Low producer prices, at the same time, have reduced the incentives of suppliers of natural gas and served to decrease production.

In addition to involving less government intervention and red tape, price rationing provides a far more predictable method of restricting energy resource use when contrasted with administrative decrees such as mandatory standards for home insulation and auto mileage. When price increases, people will economize in different ways depending upon their own subjective evaluations. If the price of home heating fuel increases, for example, some people will reduce the temperature in all rooms, some will apply more insulation, some will close off rooms, etc.

When mandatory conservation standards are imposed, on the other hand, there is little latitude left for individual ingenuity. Everyone is forced to meet the same standard even though people having different tastes and preferences would economize in different ways if left free to do so; or alternatively, individual ingenuity is now channeled to the circumvention of the mandatory standards rather than to the solution of the problem for which those standards were allegedly imposed. There is no way mandatory standards can cater to the diversity of individual tastes or take into account the differences existing in literally millions of different circumstances. The fundamental inequity of treating people in unequal circumstances the same way is ignored. Why, for example, should the homeowner with children now away from home be forced to insulate his entire house through heating and using only half the house? In this and numerous other examples it is easy to see how the individual homeowner can make an accommodation much easier to higher prices than to mandatory standards which cannot reflect different tastes and circumstances.

As non-renewable resources are exhausted, increasing scarcity is reflected in a second way. As price is bid up, the development of substitute resources is encouraged. The cotton price support program instituted in the 1930's, for example, encouraged the development of nylon and other substitutes none of which were predictable when cotton prices were increased. Similarly, increases in prices of fossil fuel will serve to increase the development of new energy sources. Higher fossil fuel prices serve both to make alternative fuel sources currently available more profitable and to encourage the development of energy sources not currently available.

Finally, the market minimizes conflicts when compared with non-market rationing methods. The market is based on voluntary exchange so that all parties gain when a market transaction takes place. There is little basis for concern or antagonism when everyone can purchase all of a product he desires at a specified price.

The situation is much different when rationing is performed by non-market methods. The creation of antagonism and conflict is inherent in non-market rationing procedures since more of a product is desired than is available at the price arbitrarily held below the market clearing level. In such a situation an individual can legitimately feel that he is in competition with other consumers for the product whose price is controlled. The equity problems endemic in non-market rationing procedures were discussed above.

Allegations of antisocial conduct frequently arise where nonmarket allocation procedures are used. Each person has a vested interest in reductions in consumption by other people when there is a shortage. Individuals consuming more than they "need" as perceived by the outside observer are alleged to be wasteful. Since "waste" is in the eye of the beholder, efforts to reduce waste must be authoritarian in nature. Such efforts must be based on the values as perceived by the state and not on the values of individual decision makers.

Nonmarket allocation procedures also give rise to will-of-the-wisp attempts to determine whether producers are holding back production and whether costs of production are excessive. There will always be a perceived conflict between producers and consumers when production is subject to price controls. Consider the action by the Secretary of the Interior in February 1977 to determine whether producers of natural gas were "holding back production." The allegation was made that producers might be acting against the "public interest" by holding back on production under the expectation that future prices might be higher. Thus, a producer who reduces the amount of natural gas available for immediate consumption is "holding back" and is thereby "anti-social," but a consumer who reduces immediate consumption practices "conservation" which is "socially desirable." If a producer were restricting production anticipating higher future prices, would such conduct be antisocial? If producers do not follow market signals, they have no way to make production decisions including how much to produce or when to produce.

There are only two ways to allocate goods and resources—the market and central direction. The market permits people to choose on the basis of relative prices. Since each party gains under voluntary exchange, conflicts are minimized. When economic goods are rationed by nonmarket methods, conflicts are inevitable. Since more is desired than is available at the controlled price, measures must be taken to reduce consumption. Austerity measures with pleas to eliminate "waste" and reduce consumption to the "needs" level are endemic in nonmarket allocation procedures. Mandatory "conservation" measures mean that consumers are made poorer by being forced to do without.

Conclusions

How then ought choices be made in a world of "finite resources?" Conflicts will be minimized when the rules of the game are such that the market is mainly relied upon to ration goods and the individual citizen bases his decision on his preferences and on relative prices. Admonitions to satisfy essential needs, forego waste, and live in austerity are unlikely to have the effect of feeding the hungry or sheltering the homeless throughout the world. Forced austerity works against people's willingness to work. If people are prohibited from buying the goods they desire, they will work less and take more of their real income in the form of leisure.

What does this mean about the level of living for us as individuals? The attitude that consumers should not be prohibited from acquiring the "luxury" goods they desire does not mean that we as individuals should follow a pattern of conspicuous consumption. The question of what goods and services each of us consumes, is a matter which must be answered by each of us as individuals. One person cannot identify "waste" in consumption by another individual except by imposing his own standard of values. There is little question that many of us could benefit from a more austere life style. Yet, moves to impose life styles upon us are at variance with the tenets of a free society.

VOLUNTARY SIMPLICITY AND THE DUTY TO LIMIT CONSUMPTION

K.S. Shrader-Frechette

BECAUSE THEY REALIZE that resources are finite, that high patterns of consumption and production are environmentally destructive, and that most humans on the planet are malnourished, a growing number of persons is becoming dissatisfied with levels of US consumption. The emotive accusation, leveled by one contemporary Jeremiah, is that *"The American way of life is a criminal act* ... . Any man who owns more than the necessities of life is an *ecological criminal."*[1] Against such charges, many persons have argued that high patterns of consumption in the US have enabled this country to develop an expanding economy. This, they say, has helped to reduce domestic poverty, to increase employment, and to provide a broader fiscal base for foreign aid to underdeveloped nations. Who is right, the moral idealists demanding that we change our ways, or the financial realists who argue that even social justice takes money, and that only healthy consumption and production patterns can provide it?

According to both Harris and Roper polls taken in 1975, 1976, and 1977, a majority of Americans believes that the nation "must cut back" on production and consumption in order to conserve resources. 79 percent of Americans indicated that they want to place greater emphasis on "teaching people how to live with ... essentials," rather than on "reaching higher standards of living." 76 percent of all US citizens said that they prefer "learning to get ... pleasure out of nonmaterial experiences" rather than on "satisfying ... needs for more goods and services." In order to reduce consumption of goods, 92 percent of Americans said they were willing to eliminate annual model changes in automobiles. 91 percent said they were willing to forego meat one day per week, 90 percent indicated that they would be glad to do away with annual fashion changes in clothing, and 73 percent said that they would be willing to wear old clothes until they wore out. Moreover, 57 percent of Americans would like to see a national policy that would make it much cheaper to live in multiple-unit apartments than in single family homes. The results of these polls, said Harris, suggest that "a quiet revolution may be taking place in our national values and aspirations."[2]

What Is Voluntary Simplicity?

Both environmental and personal reasons are responsible for this new emphasis on conserving resources and consuming fewer goods.[3] People are beginning to wonder whether growing levels of production and consumption are in their best interests. Some of them are saying that, in order to remain healthy, happy, and freer of pollution, we should embrace a life of voluntary simplicity, of voluntary limits on consumption. The lifestyle many envision focuses on following principles formulated by the Quakers. According to these criteria, one ought to own, buy, or consume something only if doing so will promote his activity, self-reliance, and involvement, rather than his passivity or dependence. A second norm is that one's patterns of consumption ought to be based on serving real needs and authentic satisfaction, rather than on meeting mere wants or the expectations of others. Finally, the Quakers suggest that, in a life of voluntary simplicity, one's patterns of consumption ought in part to be dictated by their estimated impacts both on other people and on the well-being of the earth.[4]

Although attending to the Quaker criteria is likely to lead one to reduce personal consumption of goods, proponents of voluntary simplicity or poverty point out that it is not the same thing as enduring destitution. Without a modicum of goods, they say, one would have neither the leisure nor the psychic ability to investigate his patterns of owning and buying and their effects on the environment and on his freedom and growth. They warn that destitution could be the opposite of living in voluntary poverty or simplicity, since it might make life a struggle, rather than an opportunity to live in awareness of the value of the earth and the needs of others.

Just as voluntary simplicity is not the same as living in destitution, so also it is not identical with a lifestyle based on the back-to-nature movement. Urban existence is in no way imcompatible with living in simplicity, since both rural and urban dwellers are able to base their consumption on needs rather than on wants or on the expectations of others.

Living in voluntary simplicity also should not be confused with living under government rationing of goods and resources. The question of politically imposed limits to consumption raises a whole host of troubling, social, ethical, and legal issues not central to the question of whether an individual ought to live in voluntary simplicity. What I will be advocating in this section is that it might be desirable for one to live in voluntary simplicity; I will not address the much broader and more difficult question of whether government ought to use non-market means of allocating goods and resources. Resolving this latter problem would involve various considerations (e.g., whether rationing is at variance with the tenets of a free society) which are better left to a more lengthy treatment.[5]

In arguing that it might be desirable for one to adopt a lifestyle of voluntary simplicity, first I will examine some of the considerations that support this belief. Many people have maintained that voluntary limits on consumption will help to avoid planetary environmental crisis, to achieve equity among peoples, and to enable persons to develop psychologically and spiritually. I will review these lines of reasoning and then analyze the major arguments that have been used against voluntary simplicity. These include the beliefs that lowered consumption is not necessary to avoid global crisis and that increased consumption in both developed and underdeveloped countries is necessary to help the poor. Finally, I will close with several practical suggestions for those who might want to live in voluntary simplicity.

Arguments in Favor of Voluntary Simplicity

From the point of view of environmentalists, the most compelling reason for living in voluntary simplicity is that patterns of consumption in most developed countries are causing crises of resource depletion and pollution. On this basis, one can make at least two arguments in favor of voluntary limits on consumption. First, if massive pollution and resource depletion are not slowed, then the future will bring growth, followed by collapse into a dismal, depleted existence. Secondly, the high consumption patterns characteristic of wealthy persons and nations are responsible, in part, both for underdeveloped countries' receiving less than their fair share of the planet's resources and for a crisis of personal freedom in developed countries. Let us look at each of these arguments.

Limiting Consumption in Order to Avoid Environmental Crisis

The belief that current patterns of consumption are likely to lead to the collapse of the planet and its population, and therefore that it is desirable to limit consumption, has been expressed forcefully by a team of MIT scientists in their report, *The Limits to Growth.*[6] According to the authors of this study, four of the five parameters contributing most significantly to world collapse are heightened by high levels of consumption. These are industrialization, pollution, food production, and resource depletion. (The fifth factor is population growth.) They argue that, unless the people and the nations of the world adopt a consumption pattern leading to global equilibrium regarding industrialization, pollution, food production, and resource depletion, then catastrophic limits to growth will be suddenly forced upon the planet within the next one hundred years. There will be a "sudden and uncontrollable decline in both population and industrial capacity" because we will have used most nonrenewable resources and will have exceeded the limits of pollution above which life and society cannot be sustained.[7]

The authors of the MIT report point out that the world usage rate of every natural resource, including land, is growing exponentially. World population, average consumption per person, and every pollutant (that has been measured as a function of time) are also increasing exponentially.[8] This means that, although the upper limit for resource depletion, industrialization, and pollution is not known precisely, it is clear that, because the earth is finite, there is a limit to its ability to withstand all these effects of growth. In fact, the limit has already been surpassed in many local environments.

To lead society to accommodation to the limits to growth in a finite world, the MIT authors say that two things must be done. We must stop exponential increases in the population and exponential growth in average consumption per person. This, in turn, could inhibit acceleration of pollution and resource depletion.[9]

If such restraint is not forthcoming, they say, then we will have a "tragedy of the commons," a catastrophe in which misguided self-interest leads each of us to enlarge the size of his possessions and his rate of consumption. The result of this action will be that everyone loses, because the effects of our patterns of consumption will have exceeded the carrying capacity of "the commons." This means that the rational strategy of each individual, to get as much as possible from the finite wealth of the earth, is socially irresponsible because it leads to disaster for all. "The tragedy is that," as Schedler puts it, "even though the outcome is perfectly predictable, no one acts to avert the disaster."[10] When gas was rationed recently in California, many people stepped up their buying, so as to have a personal reserve. As a result, the shortage worsened.

Compelling as is the explanation of the "tragedy of the commons" and the MIT discussion, *The Limits to Growth,* they have been attacked widely, both inside and outside the community of economists. One of the better of these analyses, *Thinking about the Future: A Critique of "The Limits to Growth,"* by K.L. Pavitt, C. Freeman, and M. Jahoda, was published in 1973. Pavitt and colleagues, and other writers, correctly point out that it is impossible to term the MIT conclusions correct, because they rest on many unknown parameters whose values have been extrapolated or merely guessed at because there is no way to determine them accurately.

While critics of *The Limits to Growth* have made a number of valid points, in perspective there are several reasons for believing that the MIT authors are close to the truth. For one thing, apart from the rate at which resources are being used, or whether the limit of pollution will be reached in one hundred years, it is clear that many resources are finite and nonrenewable. Hence, it seems reasonable to suppose that eventually the MIT predictions will come true, if world consumption continues on its present course. If this is correct, then it appears logical to stave off resource depletion and pollution as long as is possible. One way of doing so is to limit consumption and production. Secondly, while it is true that new technologies and discoveries of additional bodies of resources will be made in the future, this fact is offset by several difficulties. No one knows for sure whether technology will be able to rescue us from resource depletion and pollution; to assume that it will do so seems unwarranted, especially because of the finite nature of the many goods that might be necessary to employ future technologies. Moreover, even if the MIT authors are incorrect in asserting that present patterns of consumption and production will lead to environmental crisis within one hundred years, there is still the problem of whether current levels of consumption and production will lead to political crisis.

The beginning of political catastrophe could come when a number of underdeveloped countries have the capacity to build and use nuclear warheads. As wealthy nations use more and more resources, the less-privileged nations could become more hostile toward the countries which are better off. Many persons have argued that these nations' aggressiveness will very likely result in worldwide political insecurity.[11] (More about the social and political effects of disproportionate allocation of resources will be mentioned shortly.)

If it is correct that increasing levels of consumption will eventually lead to both political and environmental crisis, then it may well be desirable to avoid high levels of consumption. Further, "if survival is a minimum objective of social effort" and if "every society strives to improve its ability to sustain life,"[12] then whatever threatens survival ought to be avoided. One of the least objectionable ways of doing so might be for many persons to practice a lifestyle of voluntarily chosen austerity.

Admittedly, however, voluntary simplicity might not be effective in curbing consumption and therefore in enabling us to avoid catastrophe. If it is not, then limitations to growth might come in other, less desirable, ways. If the numerous predictions about ecological crises are correct, then these constraints will come about either by physical means (resource depletion, famine, pollution-induced death) or by political ones (governmental controls, e.g., rationing). Viewed in this light, it seems desirable to practice a lifestyle of voluntary simplicity rather than to wait for nature or the government to force its own type of austerity on us. As N.O. Schedler puts it: "Limitations will come. Either we choose those who will limit us or we will be limited by those we did not choose."[13]

Limiting Consumption in Order To Achieve Equity

Avoiding environmental crises and limitations on freedom, however, are not the only reasons why one might adopt a lifestyle of voluntary simplicity. Many persons have also argued that equity demands that all persons have access to the goods of the earth, rather than that wealthy nations use the lion's share of nonrenewable resources. They maintain that equitable distribution of scarce goods and opportunities forms part of the total requirements of distributive justice, and that these bind one, not merely to equity in the present, but also to achieving equity in the future.[14] As one moral philosopher put it, "a form of proprietorship which permits owners to violate the duty of preservation they owe to future generations would be defective."[15]

The most famous current ethical framework, providing justification for principles of distributive justice, is that of John Rawls. According to him, persons are obligated to follow "the principle of fairness."[16] Rawls defends "the principle of fairness," in part, by the following explanation. If one were ignorant of his natural ability, place in society, and other factors that might influence his share of society's goods, says Rawls, then he would be able to assent to principles of justice that were the consequence of fair agreement or bargain, rather than the result of his own hope for advantage. In such a situation, he says, no rational person would agree to principles of utility that would sanction "an enduring loss for himself in order to bring about a greater net balance of satisfaction." Instead, people would choose principles of "equality in the assignment of basic rights and duties." They would sanction inequalities in wealth, use of resources, and authority only to the extent that "they result in compensating benefits for everyone, and in particular for the least advantaged members of society."[17]

For Rawls and for other ethicians, justification of principles of distributive justice regarding world resources is a plea, not only for egalitarianism, but also "for rational priorities." Although developed societies claim to be rational in their social organizations, they often do not follow the dictates of "rationality on a world scale," says one expert.[18]

If we followed such a rational theory of equal justice, equal opportunity, and equal protection on a planetary scale, then distributive justice might dictate a minimal list of essential rights and duties. According to Denis Goulet, these would include the following tenets:

> 1. that every person in every society is entitled to the resources needed to sustain a minimum standard of life for all its members;
> 2. that the people of poor nations have prior claims on the resources within their boundaries;
> 3. that the people of nations with more resources are under obligation to grant some access to nations with fewer resources;
> 4. and that the rights mentioned in items (1), (2), and (3) are operative among individuals within a nation.[19]

In fact, all the rights mentioned in this minimal list can be defended not only on the rational grounds that they are required by distributive justice, but also on the legal and political basis that they already have been recognized as rights by the United Nations. In its 1969 "Declaration on Social Progress and Development," for example, the UN reaffirmed the right of persons in all nations to have "proper nutrition"; it also sanctioned the right of persons in all countries to have "equality of opportunity" to use all the resources of earth, regardless of whether they were citizens of poor or wealthy nations.[20] In justifying its position, the UN cited requirements of equity, moral advancement, and protection of the environment.[21]

If one follows the reasoning outlined by the UN, then it can be argued that Americans ought not to consume, for example, 42 percent of the world's aluminum, 44 percent of the world's coal, 33 percent of the world's copper, 40 percent of the world's molybdenum, 63 percent of the world's natural gas, and 38 percent of the world's nickel. All of these are nonrenewable natural resources and the bulk of them is being used by a country with only 6 percent of the world's population.[22]

The problem with such criticisms of US consumption, however, is that they ignore two facts. Often US patterns of buying materials from underdeveloped countries aid these nations economically, especially since they sometimes lack the industrialization to utilize all their resources. Also, a more valid perspective on US consumption could be given by noting what percentage of manufactured goods in the world are supplied by the US and not merely by citing what percentage of raw materials are used. Once these two facts are included in the analysis, then one might have a more accurate picture of the effects of US patterns of use and waste.

Within an accurate assessment, several important conclusions would emerge. These suggest that, despite their benefits, high levels of production and consumption are often built upon questionable suppositions. One such assumption is that one can ignore the fact that our world is a closed, not an open system, and that many resources are finite. Essentially this comes down to ignoring the second law of thermodynamics, which stipulates that the universe is continually "running down." If resources are finite, then current levels of energy consumption in some developed nations might mean not only that some persons are using more than their fair share of the goods of the earth, but also that this disparity will have negative consequences thousands of years from now.

A second reason why US patterns of consumption might be questionable is they imply that there is no need to develop a system of priority needs in resource use. According to John Kenneth Galbraith, an important requirement of modern development planning is a theory of consumption, a theory of what production is ultimately "for." Yet, in the way Americans use and misuse resources, often no clear theory of consumption or of priority needs is apparent. Contrary to Galbraith's advice, many nations have not attended to feeding the poor before using their resources for their own luxuries. Instead, he says, there "will be time to worry about leisure, contemplation, the appreciation of beauty and the other higher purposes of life when everyone has had a decent meal."[23]

One theory of priority needs, that might structure our patterns of consumption in a more equitable way, is built around three classes of goods. These are needs of first necessity, enhancement needs, and luxury needs.[24] Items in the first category include food, clothing and shelter. Goods in the second class, enhancement needs, are somewhat more difficult to define. Goulet maintains that they enable humans to attain transcendence, to create, to express themselves, and to choose their futures. These are the needs, says Goulet, which are "most intimately human." They go beyond mere survival.[25] Luxury goods, on the other hand, also go beyond mere survival, but are not essential to our actualization, as are enhancement needs. They include what some have termed "waste," since they are not essential to human life or well-being. Although the fulfillment of luxury needs does not necessarily corrupt people, they are often seen to be as noble or as debased as is the use to which they are put or the spirit in which they are enjoyed.[26]

To the degree that one can distinguish priorities (such as the three just mentioned) among needs, then to that extent, it might be rational for both individuals and societies to attempt to meet first-order needs first. Admittedly, however, this is a difficult injunction to follow. For one thing, if Paul Ehrlich and others are correct, it may be impossible, even with determined efforts, to meet the minimal needs for food, shelter, medical care, and clothing for all inhabitants of the earth. For another, a number of political constraints currently prevent equitable distribution of resources so that all persons have access to a minimum standard of living.

In spite of the fact that it may not be possible to follow a different mode of resource allocation, E.F. Schumacher has warned that current patterns of resource consumption undergird a way of life which can "have no permanence." If developed nations continue to deplete nonrenewable resources, fail to distribute them justly, and hence do not seek lower levels of consumption, then they are headed for destruction, he says.[27] They are "living parasitically." Moreover, says Schumacher, such behavior is "an act of violence against nature which must almost inevitably lead to violence between men."[28]

Many people believe that high-consumption patterns of life constitute violence against nature because high consumers sometimes have used unethical means to obtain and keep the lion's share of the planet's resources. This, however, might not be the only reason for questioning the failure to limit consumption. As Goulet and others explain it, even when a nation or person has used ethical means to secure an advantageous and prosperous position in the world, he is responsible nevertheless to help others obtain their fair share. This responsibility exists, they say, neither by virtue of guilt nor because of the necessity to provide reparation for injustice. It exists simply because rich persons with high levels of consumption have (perhaps only *indirectly*) impeded the economic development of others.[29] Hence, they argue, their lowered consumption is necessary to help promote equality of opportunity for all. As Gandhi put it, "Whenever I live in a situation where others are in need ... whether or not I am responsible for it, I have become a thief."[30]

If Gandhi's claim is correct, then virtually everyone in the world is a thief. This in turn suggests, however, that his warning is probably not meant to be taken completely literally. If not, apart from whether Gandhi and Goulet are correct in their political and economic assumptions, then they have provided us with a moral *ideal,* to meet the needs of all persons, even though they have not established the existence of a rationally defensible *duty.*

Limiting Consumption in Order to Enhance Personal Development

Even if we do not believe that we have a duty, however, to be our brothers' keepers, there are a number of reasons why it might be in one's self-interest to limit consumption. Voluntarily chosen simplicity can sometimes prepare one for greater self-awareness, enhance the quality of one's life, and provide for more personal freedom. Voluntary austerity might also be an effective way to control the prospensity toward greed and egotism.

For centuries, voluntary simplicity has been embraced by persons seeking religious goals. Their motivation, on the whole, has not been to condemn material goods but to detach themselves from them as a preparation for greater self-awareness. For the Zen Buddhist, for example, coveting material goods is a form of mental slavery; one can attain Buddahood only when he is purified and freed from material things.[31] Poverty, in fact, has been said to be "the most fundamental" of all Buddhist virtues and to correspond ontologically to emptiness and psychologically to selflessness or innocence.[32] Hence practicing voluntary simplicity might be one way of returning to an original state of selflessness or innocence, one way of clearing away external clutter so as to be freer to explore the inner life. Likewise in Vedānta philosophy, for example, the Self (Ātman) "is obtainable by" austerity. *"Brahman* is austerity."[33] As the *Bhagavad-Gita* puts it: "United with Brahman,/Cut free from the fruit of the act, / A man finds peace / In the work of the spirit. Without Brahman, / Man is a prisoner, / Enslaved by action, / Dragged onward by desire."[34]

Although Christian, Buddhist, Vedānta, and other traditions differ in their conception of the spiritual life, they and nearly all religious philosophies are founded on the principle that the aim of life ought not to be pursuit of material wealth. Rather, they all agree that one's main goal ought to be spiritual. As Schumacher put it, the materialist is interested in more goods, while the Buddhist or Christian, the practitioner of voluntary simplicity, "is mainly interested in liberation" and peace of mind.[35]

Voluntary austerity is said by its practioners to bring liberation because it frees one from the illusory happiness provided by consumption of material things. These objects alone, they say, cannot provide us with happiness; they arouse desires they cannot alleviate. Perhaps one reason, why humans are often not satisfied with increasing levels of consumption of material goods, is that what is "enough" has no objective meaning.[36]

Although excessive poverty clearly stands in the way of human happiness and fulfillment, it is impossible to say what types and numbers of material goods are "enough" for a given person. But if what is "enough" is different for each person, then one cannot say that increasing consumption is, *a priori,* a good thing or something that will lead to human happiness and fulfillment.

Goulet, who has lived and worked with numerous Third-World peoples, is convinced that increasing the levels of consumption of persons in underdeveloped countries may prevent them from attaining authentic happiness. He points out that, "even in the best of circumstances, development exacts a high toll in value destruction."[37] At times, he says, the damage is so great that it negates economic and educational benefits that come from a lifestyle of higher production and consumption.

The value crises generated by technological expansion and increased consumption often threaten basic relations among family members, workers, and citizens. Goulet, for example, witnessed the deterioration of family relations among the Bedouins of the Algerian Sahara when they began to adopt a lifestyle of increased production and consumption and access to Western technology. Once traditional value systems are shattered in this manner, says Goulet, the affected societies are left with two options, both unpleasant. First, they can maintain their belief in the old system, even when it has become incompatible with norms governing day-to-day behavior; this option, he says, causes painful cultural fragmentation. Or, says Goulet, people can fashion a new nexus between the values which give meaning to their lives and those which provide rules for action. This second option is not easily achievable, argues Goulet, since groups undergoing their first experience with modern technology are unlikely to develop a synthesis of meaning. Even "advanced" countries, with several centuries of experience of technology-induced advancement of production and consumption, have failed, he maintains, to provide a framework of goals against which to measure the accomplishments of industry and science. Hence, says

Goulet, Third-World countries are "doomed to social disruption" by most development programs focusing on expanding production and consumption.[38]

The problem with Goulet's reasoning, however, is that it is built on the presupposition that social disruption ought to be avoided at all costs. Clearly this is not the case. Many needed social improvements (e.g., implementation of integration in the US, have resulted in "social disruption"). The key issue, therefore, is not the presence or absence of upheaval, but whether it is worth the gains to be won through it. This question, and especially the problem of balancing short-term disruption with long-term progress, are not addressed by Goulet.

Apart from whether Western-style consumption and development are desirable for Third-World nations, there do seem to be some instances in which voluntary austerity may actually improve the quality of life. Learning to do without some of the unnecessary trappings of society might enable one not to work so hard. Just as Thoreau found that he could live on six weeks labor per year,[39] so also might contemporary persons find that a lifestyle of voluntary simplicity would enable them to live less laboriously and less hurriedly. Voluntary austerity might also allow one to escape a definition of "reality" in terms of money and competition (i.e., "the prevailing rat race for economic superiority and oversufficiency").[40] Instead, one might be able to define reality in terms of learning, appreciating, and loving what is around oneself. This lifestyle might be far more enjoyable, challenging, and ethical than one built around accumulating and using more material goods. As Daniel Kozlovsky puts it, "In our attempts to manufacture the good life ... we have deprived ourselves of fundamentally important aspects of living"; these include "clean air and clean water, open natural spaces, quiet in the night, darkness, serenity, individual purpose and opportunity to perceive beauty."[41]

While our quest for consumption has endangered goods, such as clean air and water, it has not always given us much in return. Some persons believe that since the middle of the 1960s, the correlation between increased consumption (measured via GNP) and improved quality of life in developed countries has "broken down"; they argue, for example, that our heightened levels of consumption have not brought us an expanded lifespan for males.[42] Our increasing rates of consumption and our growing GNP have also not helped us to achieve equity among persons.

In the famous MIT report, *The Limits to Growth,* it was shown that present patterns of growth in consumption and industrialization "are actually increasing the gap between the rich and the poor on a worldwide basis."[43] And in his famous work, *Energy and Equity,* Ivan Illich argues that "only a ceiling on energy use can lead to social relations that are characterized by high levels of equity." Above a certain threshold, says Illich, "energy grows at the expense of equity," and further energy affluence then means decreased distribution of social control of goods and opportunities.[44]

To illustrate his belief that increasing one's level of consumption does not necessarily increase the quality of one's life, Illich uses a classic example: the automobile. According to him, the US uses an inordinate amount of energy to keep cars on the road.[45] The typical American male, he says, devotes more than 1,600 hours a year to his car; he sits in it, searches for it, and earns money to pay for it. Why, says Illich, should the model American put in 1,600 hours to get 7,500 miles? This is less than five miles per hour. In countries where most citizens are deprived of the ability to consume cars and gas, notes Illich, people also manage to travel at the rate of five miles per hour. Moreover, he says, they allocate only 3 to 8 percent of their society's "time budget" to traffic, instead of 28 percent for that of the average American.[46] Hence, reasons Illich, why should persons in countries dependent upon high consumption of gasoline and automobiles be said to have a higher quality of life than persons whose lifestyles are not so characterized?

Although Illich's example is an instructive one, the correctness of his conclusion is undercut by several facts. First, in many areas of the country, motor vehicles are necessary just to obtain the basics of life; mass transit and the use of bicycles, for example, are often not feasible. Secondly, although America's use of the automobile may detract from the quality of life, it also enhances it in ways not mentioned by Illich. Cars enable one, for example, to obtain speedy medical care when it is needed; they afford one a broader range of job opportunities, since long-range commuting is possible. The merit of Illich's insight, I think, is that he has pointed out how allegedly "freeing devices," such as cars, often enslave us.

Perhaps this insight is valid, even if it is not true in all cases. Perhaps the failure to practice voluntary austerity does place certain limits on the freedom of many persons. After all, in many developed countries, "abundance" has become "the opiate of the people." For many persons, "to be" someone means "to have" things.[47]

Compulsive consumption has often become the order of the day, and many persons have allowed themselves to be manipulated into wanting more and more. As a result, both advertising and technology, rather than buyers themselves, sometimes control the levels of consumption. It might even be the case that the more one consumes, the more likely it is that one is not free to choose whether or not to consume.[48]

A high-consumption lifestyle might encourage limits on individual freedom in ways other than societal manipulation of one's desire mechanisms. It could also lead to a personal situation in which one is unable to face want or abundance with equanimity. If "the good life" is defined in terms of greater accumulation of goods, and if one's identity is tied up with this definition, then he might not be free to enjoy situations of lesser abundance or of lowered consumption. Provided that his basic needs are met, the truly free person might be able to find peace and happiness in situations where he has very few material goods. As the Chinese philosopher, Shao Yung expressed it: "When the mind retains its unity and is not divided, it can respond to all things. Thus the mind of the superior man in vacuous (absolutely pure and peaceful) and is not disturbed."[49]

The superior freedom of the person who can endure both limited want and abundance equally well has often been a theme in classical philosophical writings. In Plato's *Symposium,* for example, Socrates was recognized as a wise, good, and free person because he was equally happy, whether he had wine and shoes, or whether he did not. For Socrates, voluntary simplicity enhanced freedom because it promoted self-reliance. Often, however, this mode of freedom is not recognized. People frequently view increased production and consumption of goods as an enchancement of freedom because more goods, *externally,* are available for choice. The freedom of voluntary simplicity, on the other hand, is one in which *internal* constraints are often lifted. It aims, not to provide more goods for choice, but instead to nurture a person better able to choose wisely, apart from the number of alternatives present before him. If a person has many options externally, but internally is unable to free himself from superfluous desires or manipulation of his wants, then that individual may well be less free than the person with fewer options but who also has freedom from want manipulation and from the need to engage in compulsive consumption.[50]

If one does not define himself in terms of the goods he is able to possess or consume, and if one practices voluntary austerity, then it might also be likely that he or she will be less subject to the pangs of greed than otherwise might have been the case. Perhaps, if one is freed from compulsive consumption, he will be better able to recognize the needs of those less fortunate, and to help forge a brotherhood and sisterhood of universal human solidarity.[51]

Arguments Against Voluntary Simplicity

In spite of the environmental, ethical, and personal reasons that often make a lifestyle of voluntary simplicity desirable, there are several reasons why persons have argued against it. Most basically, they have claimed that current patterns of consumption and production will not bring environmental crisis, as has been alleged, and that lowered consumption will neither help the poor nor serve equity. Finally, scholars have shown that there are several conceptual problems with the notion of voluntary reduction in consumption. Let us look at each of these arguments, in turn, and determine whether they constitute grounds for the failure to limit consumption.

Is Voluntary Simplicity Necessary To Avert Environmental Crisis?

Perhaps the major argument against the desirability of practicing voluntary simplicity is that it is not needed. Many proponents of this viewpoint have maintained that the imminence of ecological catastrophe has been exaggerated. In such exaggerations, they note, people forget about "the discovery of new sources" of resources and about development of "new techniques."[52] They do not remember, for example, that "techniques for exploration for and extraction of metals seem to have kept ahead of scarcity."[53] Moreover, goes the argument, even if technological developments do not enable us to find substitutes for nonrenewable goods, environmental crisis and resource depletion will not occur because price will act as a means of rationing scarce materials. Hence, they continue, it is economically naive to believe that price will not protect reserves of endangered resources and that consumption will not decrease as a result of increased costs to the consumer.[54]

Because arguments in favor of limiting consumption ignore future technological advances and the effect of price on demand, say opponents of voluntary simplicity, they exaggerate the likelihood of environmental crisis. They are based on "emotive phrases," and their proponents play "fast and loose with statistics."[55] Because the alleged environmental crisis is exaggerated, they maintain, there is no justification for affirming the desirability of voluntary limits on consumption.[56]

Compelling as this argument is (against the desirability of voluntary simplicity), there are several ethical assumptions which render it open to question. The most serious of these presuppositions is that it is ethically acceptable to continue to employ nonrenewable resources at an exponential rate, since future technological developments will protect posterity from the consequences of current patterns of use and waste.

While future technological advances are likely, this probability alone is not enough to justify present levels of consumption. This is because the presupposition contains the assumption that it is ethical to contract a debt even when one does not know with certainty that he will be able to repay it. The debt contracted is that to future generations, who deserve compensation for our use of their share of nonrenewable resources. Equity demands that obligations be met and that, as a corollary, debts not be contracted unless one is certain that they can be paid.

If one assumes that a future "technological fix" will compensate for the pollution and resource depletion caused by current high levels of consumption in certain developed nations, then one is risking violating the rights of future generations to equal justice and equal opportunity. By the National Environmental Policy Act (NEPA) of 1969, US citizens are required to make environmental choices which will result in future generations' bearing no risks greater than those borne by present persons.[57] If one does not know whether and when technology will protect us from the disastrous effects of pollution and resource depletion, then it is also doubtful whether sanctioning current levels of consumption is consistent with meeting the mandates of NEPA.

Another problem associated with assuming that technology will save us from the disastrous consequences of present patterns of consumption is that this presupposition ignores the fact that the planet and its goods are finite. Hence, even if substitutes for nonrenewable resources are found in the next century, by the century after next, for example, substitutes for the first substitutes may be needed. Eventually we must come to the end of the line, so long as we are dealing with a finite world. A third problem with this assumption is that even if mankind does succeed in finding a number of technological "fixes" to avoid environmental catastrophe and to maintain current styles of life, and even if these patterns somehow prove to be equitable to future generations, one still needs to address the question of whether they are equitable to all present persons, especially those in underdeveloped countries.

As was shown in the subsection on limiting consumption in order to enhance personal development, even if environmental crisis and the necessity to guarantee equity were not valid reasons for living simply, there might still be a number of other reasons to do so. These include enhancing one's personal development, especially freedom, and enjoying a higher quality of life. Hence any complete argument against voluntary austerity ought to address not only the social and environmental reasons for curbing consumption but also the personal desirability of doing so.

A final problem with the thesis that current levels of consumption will not lead to environmental crisis has to do with the notion that prices will curb resource depletion. While it is true that economic incentives will prevent rapid use or waste of scarce (and therefore costly) goods, this "solution" to resource shortages raises questions of its own. One such issue is whether it is equitable to distribute scarce goods purely on the basis of the ability of the purchaser to pay for them. If my earlier arguments are correct, then such a means of allocation would violate principles of distributive justice. If it is true that all people ought to have equal rights to the wealth of the earth, then it is not equitable for economic happenstance to determine who has the opportunity to exercise those rights. Considerations about the effect of price on demand address only the problem of resource depletion and not the question of the equity of distribution. For this reason they might not be sufficient grounds for arguing against the desirability of voluntary simplicity.

Is Voluntary Simplicity Necessary To Help the Poor?

A second, often-used reason for arguing that limits on consumption are not desirable is that they will not necessarily help either the poor or future generations. As one author put it, there is a "mathematical relationship between prosperity and the consumption of nonrenewable resources."[58] Slowing consumption might result, for example, in great increases in unemployment.[59]

Another, somewhat opposed, argument is that practicing voluntary austerity would not help the poor because current difficulties with equity do not involve overconsumption by the rich, but instead are problems of redistribution and resource transfer. One writer maintains that "the hamburger not eaten in the United States will not miraculously appear in the hands of a hungry person in another land."[60]

A fourth argument against the thesis that voluntary austerity is desirable in order to help the poor is that limiting consumption is unfair, since the poor have the right to produce and consume as much as the rich have. In other words, if a member of a developed nation argues in favor of limiting consumption, it is somewhat like a person on board a life-raft arguing that the ladder should be brought up after him, in spite of the fact that there are other survivors waiting to get aboard. As one critic of "ecological asceticism" put it, "The affluent eco-enthusiast who sees a connection between his giving up his electric can-opener and poor people giving up their revolutionary aspirations for prosperity is the victim of very sick thinking."[61] Let us examine, in turn, each of these four arguments against voluntary simplicity and determine whether any of them provides convincing reasons not to limit consumption.

The first argument, viz., one ought not to limit consumption since there is a direct relationship between prosperity and utilization of nonrenewable resources, is a familiar one. It pits the interests of environmentalists against the welfare of the poor. Yet, if economists such as Ezra Mishan and E.F. Schumacher are correct, there is no authentic relationship between prosperity and consumption of material goods. According to Mishan, there are several reasons for this lack of correlation. For one thing, even though real national income may rise substantially over a period, the poor generally remain in the same position relative to the rest of the population. Hence even if consumption caused prosperity, the "hard core" proverty group would not share directly in the growing real income of the community. Moreover, says Mishan, even if economic growth and increased consumption, as well as reduction in poverty all occurred over time, this would be the result of a "purely political decision" to "transfer expenditures away from ... other goods in order to give more to the poor" and not the result of "economic necessity."[62] Because a large proportion of the indigent are unemployable, "only direct government action, using the method of selective benefits, can make an immediate impact on the problem of hard core proverty."[63]

Just as economic growth and increased consumption do not always result in benefits to the poor in one's own country, so also, says Mishan, they do not always enable us to help the destitute in underdeveloped countries. He believes that the scale of foreign aid is unimpressive and suggestive of conscience money. Even the help of the largest donor, the US, is well under one percent of its GNP.[64] Hence, claims Mishan, unless foreign aid patterns change radically, increased consumption in developed countries, with consequent increased production and augmented GNP, will not result in significant betterment of the plight of the poor of the world. Moreover, even if more money were spent in foreign aid, and even if the indigent shared in growing real wealth, rising GNPs and rising rates of consumption still might not always be said to have brought prosperity to the poor. This assertion is doubtful because it is based on the assumption that the GNP is in fact an authethic measure of prosperity, a presupposition challenged by many, if not most economists.[65] Also, as I argued earlier in this essay, heightened consumption may well bring environmental crisis rather than affluence and well-being. For all these reasons, it is conceivable that there is no direct relationship between consumption and prosperity.

What of the second argument, that voluntary austerity is not desirable because it might heighten unemployment? This is a critical issue, but several facts help to put it in perspective. First, while voluntary limits on consumption would mean "substantial changes of direction," as Backstrand and Ingelstam point out,"much of the shift might well be inside the same field (e.g., from the manufacture of automobiles to the maintenance of a public transport system)."[66] Secondly, if more unemployment did occur as a result of voluntarily limiting consumption, this effect should be weighed against the possible results of not practicing voluntary austerity. Some of the consequences of the failure to constrain consumption might include global political and environmental crisis. Another negative result might be the inability to achieve equity in the distribution of the goods of the earth.

It is not clear, however, that increases in consumption are proportional to increases in employment. In the period 1977-1979, for example, total real unemployment, along with a succession of economic difficulties, have taken place "while national energy use has been at an all-time high and increasing."[67] What explains this phenomenon?

Part of the explanation is that, for the last 40 years, total employment in the manufacturing sector has declined; goods-producing industries have sought to use fewer workers and to increase output per worker. The net result has been that "the productivity index is really an automation index," a key to the degree to which energy has been substituted for jobs.[68] Aluminum industries provide an excellent example of manufacturing interests that have a high-energy, low-job ratio. Likewise increased consumption of energy, fertilizers, chemicals and equipment in agriculture has led to a steady decline in the number of persons employed there.[69] What new jobs have become available, especially in the last 30 years, have not necessarily been the result of higher consumption of goods and products, but instead the consequence of an expansion of the service sector of the economy. Between 1947 and 1977, for example, employment in the service areas increased 95 percent, more than in any other sector.[70] What all this suggests is that there may be no firm basis for saying that because consumption of goods (e.g., energy) is growing, therefore jobs are increasing. In fact, just the opposite might be the case.

What about the argument that voluntary austerity will not help the poor, because the problem is one of redistribution? Is it true that goods "given up" in developed country A will not necessarily help underdeveloped country B, or that within a country, the austerity practiced by the rich will not, in itself, alleviate the needs of the poor?

To some extent, this is a valid point. On the national scale, the welfare of the hard-core poor is almost always ultimately dependent upon political decisions. For voluntary simplicity to be an effective means of helping the destitute, at least two conditions would have to be met. First, all or nearly all large industrial countries would have to make the transition to lowered consumption in order for the resource savings to be substantial. Secondly, there would have to be "effective mechanisms for the transfer of resources."[71] In other words, decreasing consumption in developed countries is a necessary, but not a sufficient, condition for redistribution of wealth to benefit the poor. However, if it is correct that reapportionment is a good to be achieved, then it is not rational to fail to meet the necessary criterion (limiting consumption) for it, simply because that condition is not also sufficient.

If one has a moral obligation to do something (e.g., redistribute resources) then that obligation is not removed simply because conditions necessary for redistribution are not also sufficient ones. Obviously, for example, only the proper behavior of a majority of jury members is sufficient to insure that a jury has served justice. This, however, does not mean that every individual is free from duty because one jurist alone cannot guarantee justice. Therefore it is not acceptable to argue against voluntary austerity solely on the grounds that it is not sufficient to achieve a desired end, viz., reapportionment of resources.

A final argument, that practicing simplicity would not help the poor, is based on considerations of equity. According to proponents, if self-chosen limitation of consumption were said to be desirable for all persons, then the poor might never have the same opportunity to enjoy the luxury and affluence already appreciated by wealthy individuals or nations. Although, under *voluntary* simplicity, the poor could still choose to consume inordinately when given the opportunities and resources, the point of this argument is that it is wrong to say that limitation of consumption is desirable for them.[72]

While this reasoning is appealing because of its emphasis on equity and fair play, it fails because it implicitly begs the question. For one to argue that since A (e.g., a wealthy nation) has had the chance to do C (e.g., to avoid limits on consumption), therefore it is morally desirable for B (e.g., a poor nation) also to have the chance to do C, is to assume what one is trying to prove (viz., that C is morally desirable). Obviously, if C were not morally desirable, then one ought not argue that because A has been able to do C, therefore B also ought to be able to do it. For example, if C were "to engage in unjustifiable murder," then clearly one would not argue that because A has been able to do C, therefore B also ought to be able to do it. But if this is the case, then any argument, that B ought to be able to do C, simply because A has done C previously, clearly presupposes the moral desirability of C. Because of this presupposition, the reasoning assumes the very proposition it attempts to prove and is therefore logically fallacious. Because of this logical fallacy and the arguments given earlier in the presentation, there is reason to believe that it might not be morally desirable to engage in unlimited consumption.

Methodological Problems with Voluntary Austerity

Besides the foregoing considerations against voluntary simplicity, a number of writers have mentioned that the notion of limiting consumption is beset with several conceptual problems. Two of these difficulties are "the free rider problem" and the impossibility of defining the term 'enough'. Let us examine each of these issues.

According to E.C. Pasour, one problem with the notion of voluntary limitation of consumption is that it might "confer benefits on people who cannot be made to pay for them."[73] For example, if all persons except X decreased their use or waste of gas and therefore caused the price to go down, X would be a "free rider" who would benefit from the lower price, even though he had not himself "paid" for this benefit.

While the free rider problem is an unfortunate consequence of all voluntary group activity, there are several reasons for believing that it does not negate the moral desirability of practicing austerity. First, if this result were the sole basis of arguing against freely chosen group behavior, then many otherwise worthwhile activities would have to be said to be improper. One such example might be giving alms to the poor; this is clearly a good thing to do, even though it confers benefits on people who cannot be made to pay for them. Secondly, even though the "free rider" problem might be an unwanted consequence of the practice of voluntary simplicity, it would not be sufficient grounds for denying that self-chosen limits on consumption are desirable. Rather, the rational procedure to follow is to weigh all the positive and negative effects that are likely to follow from the practice of voluntary simplicity against all those that are probable results of the failure to engage in it. If my earlier arguments are correct, then the negative consequences (e.g., planetary crisis and continued violations of equal justice) of neglecting to limit consumption are probably graver than the undesirable results (e.g., the "free rider" problem) of self-chosen restraints on the use and waste of goods.

A second conceptual problem with the notion of practicing simplicity is that it is impossible to define precisely terms such as 'waste', 'need', and 'enough'. Therefore, reason some writers, it is impossible to argue clearly that voluntary limits on consumption are desirable.[74]

Although the search for clarity in definition of terms is admirable, the basic problem with this argument is that it wrongly presupposes that there is no subjective component in legitimate value judgments.[75] In other words, it presupposes that because one cannot define objectively terms such as 'waste', 'need', and 'enough', therefore one can make no authentic ethical choices. To say, "one ought not to *waste* goods but instead ought to use only what is *enough* to satisfy his basic *needs,*" for example, is to make a value judgment. Because each of these three terms might be defined somewhat differently for each individual, however, does not necessarily mean that the given statement is incorrect or completely incoherent. To see why this is so, consider the following value judgment: "one ought to love his neighbor." Even though the term 'love' might be defined differently for various persons, this does not mean that the proposition itself is false or completely incoherent. Likewise, the value judgment, "do good and avoid evil," is in all probability a correct and understandable injunction, even though there is no clear and agreed-upon definition for the term 'good'. Some might define 'good' as "pleasure," while others might define it as "happiness" or as "utility." Regardless of these variations, the legitimacy of the original value judgment is probably accepted by most people.

As Denis Goulet points out, even though 'enough' is a relative term, we can make sense of it. We know, for example, "as a minimum, to have 'enough' must mean that men's basic biological needs are met sufficiently so that they can devote part of their energies to affairs beyond subsistence."[76] The objection, that the terms 'enough', 'need', and 'waste' set no operational limits to the kinds and amounts of goods that one ought to consume, is a reasonable one. One can meet it, however, "by making judgments about the suitability of different goods" for different people. Such decisions involve, in part, the observation that some persons have little, while others have much. In other words, the meaning of terms like 'enough' is "determined by men's conscious perceptions at different times, not by minimum standards set by experts in objective terms."[77] Obviously, for example, if there are a greater number of goods or resources available to all, 'enough' will be defined according to much higher standards than if there were only a limited amount of goods for everyone.

Despite the fact that there are no sufficient conditions for defining what is meant by 'enough', it is clear that in certain specific circumstances, such as extreme poverty, most people are able to make value judgments about whether humans have what they need. As Goulet puts it, "it is trivial to claim that the notion of 'enough goods' is a relative one."[78] Even though it is true, it misses the more relevant fact that millions of persons lack "enough" food, medicine, clothing and shelter to live. If this is true, and if one is bound both to practice distributive justice and to avoid environmental catastrophe, then it might be desirable to limit consumption (i.e., to practice voluntary simplicity).

Practical Applications: How To Live in Voluntary Simplicity

There are, however, numerous difficulties inherent in the notion of voluntarily limiting consumption. As has just been seen, it is sometimes difficult to know what is "enough." Several writers who have dealt with the questions of how to practice self-chosen austerity have made suggestions in this regard, and it would be well to consider some of their ideas.

Goran Backstrand and Lars Ingelstam, both on the staff of the Swedish government's Secretariat for Future Studies, have given five proposals directed at enabling the Swedish people to consume less. They have suggested that persons ought to (1) set maximum consumption levels on meat, so as to "assure a more equitable distribution" of it; (2) place a ceiling on oil consumption; and (3) use buildings more efficiently. To accomplish item (3), Backstrand and Ingelstam maintain that government regulations could be enacted to favor resource-saving techniques in the construction and utilization of facilities. Such regulations could provide for penalizing owners of under-utilized dwellings and rewarding those of multi-purpose buildings housing different activities/groups at various times of the day and week.[79]

The final two proposals of the Swedish experts are (4) to increase the durability of consumer goods and (5) to rent rather than own a car. This (4) could be accomplished, they say, by establishing a legal "average life" for key products and by making manufacturers conform to these standards. Other means for achieving item (4) would be to use government intervention to sell certain basic commodities (e.g., work clothes and shoes) on a nonprofit basis. In order to implement proposal (5), the authors maintain that government could offset the need for cars by forbidding private auto traffic in city centers and by improving public transport. For longer-range travel, they say, municipally owned rental companies could provide cars at rates corresponding with actual cost. These measures, together with speed limits of 90 kilometers per hour (approximately 55 miles per hour), argue Backstrand and Ingelstam, would have at least five desirable consequences. The number of automobiles in Sweden would fall to 60 to 70 percent of today's level; bus, rail, and air transport would improve; the need for gasoline and oil would decrease by 0.1 to 0.2 tons per capita; there would be a shift toward smaller cars; and finally, there would be a sharp decline in traffic fatalities.[80]

The MIT authors of the famous report, *The Limits to Growth,* also have some ideas for reducing consumption. Although theirs are generally broader than the suggestions of the two Swedish theorists, several of the points are the same. D.L. Meadows and his associates call for (1) new methods of waste collection and recycling, so as to reduce resource depletion; (2) better product design to increase product lifetime and easy repair; (3) implementation of solar energy and natural forms of pest control; and (4) development of contraceptive techniques that would facilitate the equalization of the birth and death rates.[81]

Tom Bender, one of the leading theoreticians of *Rain* magazine, an ecology publication edited in Portland, Oregon, also has several practical suggestions for limiting consumption. In his article, "Why We Need To Get Poor Quick," Bender argues that we should stop using the GNP as a measure of prosperity and should instead use a more accurate and ecologically sound formula.[82] Another proposal of his is to "stop tourism" and instead to develop the communities in which we live.[83] For him, however, voluntary austerity is tied to a conversion in values, more than to a change in techniques, habits, or government regulations. He suggests following some "watchwords" for the transition to lifestyles of voluntary simplicity, and urges people to emphasize "stewardship not progress," "austerity not affluence," "permanence not profit," "responsibility not rights," "betterment not biggerment," "equitization not urbanization," "work not leisure," and "tools not machines."

Apart from whether one agrees with the suggestions of Backstrand and Ingelstam, Meadows and his MIT associates, or Bender and the editors of *Rain,* each of them provides some challenging ways to think about what is really meant by "the good life" and how we might come closer to attaining it. Perhaps, paradoxically, some of us might become more humane only by consuming less. Perhaps, too, if voluntary simplicity frees us from compulsive consumption, we will discover that as our wants decrease our freedom and our joy increase.

Notes

1. Daniel Kozlovsky, in *An Ecological and Evolutionary Ethic,* Englewood Cliffs, New Jersey, Prentice-Hall, 1974, p. 102; hereafter cited as: *Ethic.*

2. Results of these polls are cited in D.S. Elgin and A. Mitchell, "Voluntary Simplicity," *The Futurist,* XI (August 1977): 208-209; hereafter cited as: Simplicity.

3. Kozlovsky, *Ethic,* p. 101, also believes that the two reasons for choosing a lifestyle of voluntary poverty are "ecological responsibility" and "personal liberty and development."

4. The Quaker criteria are cited in Elgin and Mitchell, Simplicity, p. 203; see also pp. 202, 257.

5. Some of the problems raised by central (governmental) direction of the allocation of goods and resources, such as the one mentioned here, are found in E.C. Pasour, "Austerity, Waste, and Need," *The Intercollegiate Review,* 13 (Winter-Spring 1978): 79-85; hereafter cited as: Austerity.

6. D.H. Meadows, et al., *The Limits to Growth,* New York, New American Library, 1974; hereafter cited as: *Limits.*

7. Meadows, *Limits,* pp. 29.

8. Meadows, *Limits,* pp. 40, 60, 69, 81.

9. Meadows, *Limits,* pp. 93-94. This same point is made by Goran Backstrand and Lars Ingelstam, "Should We Put Limits on Consumption?" *The Futurist,* XI (June 1977): 157-162; hereafter cited as: Consumption.

10. N.O. Schedler, "Our Destruction of Tomorrow," in W.R. Durland and W.H. Bruening (eds.), *Ethical Issues,* Palo Alto, Mayfield, 1975, pp. 262-63; hereafter cited as: Destruction. See also Denis Goulet, *The Cruel Choice,* New York, Atheneum, 1971, pp. 273-94; hereafter cited as: *Choice.*

11. Goulet, *Choice,* p. 134, makes this point, as do Backstrand and Ingelstam, Consumption, p. 162; Elgin and Mitchell, Simplicity, p. 256; and Tom Bender, "Why We Need To Get Poor Quick," *The Futurist,* XI (August 1977): 210; hereafter cited as: Need.

12. D.A. Goulet, "An Ethical Model for the Study of Values," *Harvard Educational Review,* 41 (May 1971): 213; hereafter cited as: Values.

13. Schedler, Destruction, p. 262. See also Elgin and Mitchell, Simplicity, p. 256.

14. This point is made by R.M. Green, "Intergenerational Distributive Justice and Environmental Responsibility," *BioScience,* 27 (April 1977): 261; hereafter cited as: Justice.

15. Alan Donagan, *The Theory of Morality,* Chicago, University Press, 1977, p. 99; hereafter cited as: *Morality.* See also Charles Fried, *Right and Wrong,* Cambridge, Harvard University Press, 1978, p. 147; hereafter cited as: *Right.*

16. John Rawls, *A Theory of Justice,* Cambridge, Harvard University Press, 1971, pp. 108-117; hereafter cited as: *Theory.*

17. Rawls, *Theory,* pp. 14-15, 342-50.

18. Goulet, *Choice,* p. 246.

19. Goulet, *Choice,* pp. 118-19.

20. *Human Rights: A Compilation of International Instruments of the United Nations,* New York, United Nations, 1973, p. 99; see especially pp. 97-102. Hereafter cited as: *Human Rights.*

21. *Human Rights,* pp. 99 (Article 13) and 97.

22. Bender, Need, p. 210; see also Backstrand and Ingelstam, Consumption, p. 161; Meadows, *Limits,* pp. 64-67; and D.E. Engel, "Elements in a Theology of Environment," *Zygon: Journal of Religion and Science,* 5 (September 1970): 227.

23. J.K. Galbraith, *Economic Development in Perspective,* Cambridge, Harvard University Press, 1962, p. 43, and J.K. Galbraith, *The New Industrial State,* Boston, Houghton Mifflin, 1967, p. 407.

24. Goulet, *Choice,* p. 240.

25. Goulet, *Choice,* p. 242.

26. Goulet, *Choice,* p. 245.

27. E.F. Schumacher, *Small Is Beautiful,* New York, Harper and Row, 1973; hereafter cited as: *Small.*

28. Schumacher, *Small,* p. 57.

29. Goulet, *Choice,* p. 135.

30. Cited in Goulet, *Choice,* p. 133.

31. See, for example, Lucien Stryk (ed.), *World of the Buddha,* Garden City, New York, Anchor, 1968, pp. 364-65, 370.

32. Thomas Merton, *Zen and the Birds of Appetite,* New York, New Directions, 1968, p. 115.

33. S. Radhakrishnan and C. Moore (ed.), *A Source Book in Indian Philosophy,* Princeton, University Press, 1957, pp. 54, 63; see also pp. 60-65, 90, 95, 173, 183, 260.

34. S. Prabhavananda and C. Isherwood (tr.), *The Song of God: Bhagavad-Gita,* New York, New American Library, 1951, p. 58.

35. Schumacher, *Small,* p. 54.

36. Goulet, *Choice,* p. 130, makes this same point.

37. Goulet, "An Ethical Model for the Study of Values," *Harvard Educational Review,* 41 (May 1971): 225; hereafter cited as: Model.

38. Goulet, Model, p. 212; see also pp. 211-212; 225-27.

39. Cited by Kozlovsky, *Ethic,* p. 102.

40. Kozlovsky, *Ethic,* p. 102.

41. Kozlovsky, *Ethic,* p. 101.

42. Backstrand and Ingelstam, Consumption, pp. 157-58. See also Bender, Need, p. 212.

43. Meadows, *Limits,* p. 183.

44. Ivan Illich, *Energy and Equity,* New York, Harper and Row, 1974, p. 5; hereafter cited as: *Equity.*

45. Illich, *Equity,* p. 11.

46. Illich, *Equity,* pp. 18-19.

47. See Goulet, *Choice,* p. 128.

48. See Goulet, Model, p. 217.

49. Wing-Tsit Chan (tr.), *A Source Book in Chinese Philosophy,* Princeton, University Press, 1963, p. 494.

50. See Illich, *Equity,* p. 6. See also Goulet, Model, p. 217.

51. See Goulet, *Choice,* p. 143; see also Elgin and Mitchell, Simplicity, p. 208.

52. Ansley J. Coale, "An Economist's View of Resource Exhaustion," from *Science,* 170 (October 9, 1970): 132-33; hereafter cited as: Exhaustion.

53. John Maddox, *The Doomsday Syndrome,* London, Macmillan, 1972, p. 88; hereafter cited as: *Syndrome.*

54. For arguments to this effect, see E.C. Pasour, "Austerity, Waste, and Need," *The Intercollegiate Review,* 13 (Winter-Spring 1978): 3-4; hereafter cited as: Waste. See also Coale, Exhaustion, pp. 3-4.

55. Maddox, *Syndrome,* p. 86, and Richard Neuhaus, *In Defense of People,* New York, Macmillan, 1971, p. 258; hereafter cited as: *Defense.*

56. See Neuhaus, *Defense,* pp. 260, 262, who discusses this claim.

57. See Environmental Protection Agency, "Criteria for Radioactive Wastes," *Federal Register,* 43 (Wednesday, November 15, 1978): 53262-53263.

58. Maddox, *Syndrome,* p. 87.

59. Backstrand and Ingelstam, Consumption, p. 161.

60. Pasour, Waste, p. 83.

61. Neuhaus, *Defense,* p. 252.

62. E.J. Mishan, *21 Popular Economic Fallacies,* New York, Praeger, 1969, pp. 232-33; hereafter cited as: *Fallacies.*

63. Mishan, *Fallacies,* p. 237.

64. Mishan, *Fallacies,* pp. 237-38.

65. See, for example, K.E. Boulding, "Environment and Economics," in W.W. Murdoch (ed.), Environment: Resources, Pollution, and Society, Stamford, Connecticut, Sinauer, 1971, pp. 359-367; E.J. Mishan, *Technology and Growth: The Price We Pay,* New York, Praeger, 1969; K.E. Boulding, "Fun and Games with the Gross National Product," in R.T. Roelofs, J.N. Crowley, and D.L. Hardesty (eds.), *Environment and Society,* Englewood Cliffs, Prentice-Hall, 1974, pp. 126-136.

66. Backstrand and Ingelstam, Consumption, p. 161.

67. R. Grossman and G. Daneker, *Jobs and Energy,* Washington, D.C., Environmentalists for Full Employment, 1977, p. 1; hereafter cited as: *Jobs.*

68. Grossman and Daneker, *Jobs,* p. 2.

69. This example is used by Denis Hayes, *Energy: The Case for Conservation,* Washington, D.C.: Worldwatch Institute, 1976.

70. Grossman and Daneker, *Jobs,* p. 2.

71. Backstrand and Ingelstam, Consumption, p. 160, also make these two points.

72. See note 61.

73. Pasour, Waste, p. 82.

74. See, for example, Pasour, Waste, p. 82, who uses this argument.

75. Goulet, Model, p. 220 also makes the point that there is an essential subjective component in all value judgments.

76. Goulet, *Choice,* p. 130.

77. Goulet, *Choice,* p. 131.

78. Goulet, *Choice,* p. 132.

79. Backstrand and Ingelstam, Consumption, pp. 159-60.

80. Backstrand and Ingelstam, Consumption, p. 160.

81. Meadows, *Limits,* p. 182.

82. Bender, Need, p. 212.

83. Bender, Need, p. 212; see also Elgin and Mitchell, Simplicity, pp. 254-55.

Further Reading

Backstrand, G., and Ingelstam, L. 1977. Should We Put Limits on Consumption? *The Futurist* 11 (June): 157-162.

Bender, T. 1977. Why We Need To Get Poor Quick. *The Futurist* 11 (August): 210-215.

Chan, W.T., trans. 1963. *A Source Book in Chinese Philosophy,* esp. pp. 494; 713-714; 740-741. Princeton, New Jersey: Princeton University Press.

Cheney, E.S. 1974. U.S. Energy Resources: Limits and Future Outlook. *American Scientist* 62 (January-February): 14-22.

Clark, W. 1974. *Energy for Survival.* New York: Anchor.

Cloud, P.E. 1968. Realities of Mineral Distribution. *Texas Quarterly* 11: 103-126.

Commoner, B. 1974. *The Closing Circle,* New York: Bantam.

Dasmann, R.F. 1968. *Environmental Conservation.* New York: Wiley.

______. 1975. *The Conservation Alternative.* New York: Wiley.

Donagan, A. 1977. *The Theory of Morality,* esp. pp. 94-100. Chicago: University of Chicago Press.

Dumas, L.J. 1976. *The Conservation Response.* Lexington, Massachusetts: Lexington Books.

The editors of *The Ecologist,* 1973. Muddling Our Way to Extinction. In *Priorities for Survival,* ed. W.P. Lineberry, pp. 10-23. New York: H.W. Wilson.

Elder, F. 1970. *Crisis in Eden.* New York: Abingdon Press.

Elgin, D.S., and Mitchell, A. 1977. Voluntary Simplicity. *The Futurist* 11 (August): 200-209, 254-261.

Energy Policy Project. 1974. *Exploring Energy Choices—A Preliminary Report.* Washington, D.C.: Ford Foundation Energy Policy Project.

Engel, D.E. 1970. Elements in a Theology of Environment. *Zygon* 5 (September): 216-227.

Fried, C. 1978. *Right and Wrong,* esp. pp. 147-150, 170. Cambridge, Massachusetts: Harvard University Press.

Gaedeke, R.M. and Etcheson, W.W., eds. 1972. *Consumerism: Viewpoints from Business, Government, and the Public Interest.* San Francisco: Canfield.

Goulet, D. 1971a. *The Cruel Choice: A New Concept in the Theory of Development,* esp. pp. 121-122, 128-143, 206-207; 236-249, 271-294. New York: Atheneum.

______. 1971b. An Ethical Model for the Study of Values. *Harvard Educational Review* 41 (May): 205-27.

______. 1973. Seminar Evaluation. *IDOC—North America* 50 (January-February): 42-46.

Grayson, M.J., and Shepard, T.R. 1973 *The Disaster Lobby: Prophets of Ecological Doom and Other Absurdities,* pp. 174-201. Chicago: Follet.

Green, R.M. 1977. Intergenerational Distributive Justice and Environmental Responsibility. *BioScience* 27 (April): 260-265.

Grossman, R., and Daneker, G. 1977. *Jobs and Energy.* Washington, D.C.: Environmentalists for Full Employment.

Hardin, G. 1978. *Exploring New Ethics for Survival,* esp. pp. 119-132. New York: Viking.

Holdren, J., and Herrera, P. 1971. *Energy.* San Francisco: Sierra Club.

Illich, I. 1974. *Energy and Equity,* esp. pp. 5-12. New York: Harper and Row.

Kozlovsky, D.G. 1974. The Pleasures of Poverty. In *An Ecological and Evolutionary Ethic,* pp. 101-103. Englewood Cliffs, New Jersey: Prentice-Hall.

Maddox, J. 1972. The End of the Lode. In *The Doomsday Syndrome,* pp. 82-93. London: MacMillan.

Mayda, J. 1970. Conservation, "New Conservation" and Ecomanagement. *Environment Law Review—1970,* pp. 21-33. New York: Clark Boardman.

McKelvey, V.E. 1972. Mineral Resource Estimates and Public Policy. *American Scientist:* 60: 32-40.

Meadows, D.L., et al. 1974. *The Limits to Growth.* New York: New American Library.

Merton, T. 1968. *Zen and the Birds of Appetite,* esp. pp. 93-138. New York: New Directions.

Mesarovic, M., and Pestel, E. 1974. *Mankind at the Turning Point,* esp. Tug of War for Scarce Resources, pp. 83-100, and Epilogue, pp. 143-157. New York: New American Library.

Miller, G.T. 1975. *Living in the Environment,* pp. 7-21, 201-245. Belmont, California: Wadsworth.

Murdoch, W.W., ed. 1971. *Environment: Resources, Pollution, and Society.* Stamford, Connecticut: Sinauer.

National Research Council, National Academy of Sciences, 1969. *Resources and Man.* San Francisco: W.H. Freeman.

Nelson, W.N. 1974. Special Rights, General Rights and Social Justice. *Philosophy and Public Affairs* 3 (Summer): 410-430.
Neuhaus, R. 1971. *In Defense of People,* pp. 248-265. New York: Macmillan.
Novick, D. 1977. Facing Up to a World of Scarcities. *The Futurist* 11 (August): 217-223.
Nozick, R. 1973. Distributive Justice. *Philosophy and Public Affairs* 3 (Fall): 45-126.
Pavitt, K.L., Freeman, C., and Jahoda, M., eds. 1973. *Thinking About the Future: A Critique of "The Limits to Growth."* London: Sussex University Press.
Prabhavananda, The Swami, trans. 1951. *The Song of God: Bhagavad-Gita,* esp. pp. 49-62; 119-130. New York: New American Library.
Radhakrishnan, S., and Moore, C.A., ed. 1957. *A Sourcebook in Indian Philosophy,*esp. pp. 54, 60-65, 90, 95, 173, 183, 260. Princeton, New Jersey: Princeton University Press.
Rawls, J. 1958. Justice as Fairness. *The Philosophical Review* 67 (April): 164-194.
______. 1971. *A Theory of Justice,* esp. pp. 266-270, 336. Cambridge, Massachusetts: Harvard University Press.
Rescher, N. 1966. *Distributive Justice.* Indianapolis: Bobbs-Merrill.
Roosevelt, N. 1970. *Conservation: Now or Never.* New York: Dodd, Mead.
Rose, D.J. 1974. Energy Policy in the U.S. *Scientific American* 230: 20-29.
Schedler, N.O. 1975. Our Destruction of Tomorrow. In *Ethical Issues,* ed. W.R. Durland and W.H. Bruening, pp. 247-269. Palo Alto, California: Mayfield.
Schumacher, E.F. 1973. Buddhist Economics. In *Small Is Beautiful: Economics As If People Mattered,* pp. 50-58. New York: Harper and Row.
Skinner, B.J. 1969. *Earth Resources.* Englewood Cliffs, New Jersey: Prentice-Hall.
Stern, P.C., and Kirkpatrick. E.M. 1977. Energy Behavior: Conservation without Coercion. *Environment* 10 (December): 10-15.
Stryk, L., ed. 1969. *World of the Buddha.* Garden City, New York: Anchor.
Weinberg, A.M., and Hammond, R.P. 1972. Global Effects of Increased Use of Energy. *Bulletin of the Atomic Scientists* 28 (March): 5-8.
Woodwell, G.M. 1974. Success, Succession and Adam Smith. *BioScience* 24: 81-87.
Zurhorst, C. 1970. *The Conservation Fraud.* New York: Cowles.

SECTION THREE:

Social Justice and the Environment

8

THE ENVIRONMENT VERSUS THE ECONOMY

Introduction

THE SAHARA CLUB is a group of Southern-California-based miners, ranchers, off-road-vehicle owners, and loggers who have declared war on environmentalists. Members of the Sahara Club claim that "tree huggers and wimps" have interfered with the profits of ranchers, miners, and loggers. The largest members of the group (over 220 pounds) call themselves the "Sahara Clubbers." They carry baseball bats to "take care" of environmentalists who don't mind their own business. Similar confrontations occur between persons all over the world who think that economic and environmental interests are opposed. Coal miners, and even some Presidents of groups like the United Mine Workers, see themselves as endangered species, victims of an environmental movement that will destroy their jobs and other economic foundations of their communities.

In one sense, environmental and economic interests are sometimes opposed. It is often cheaper, over the short term, to mine, log, and ranch in ways that destroy the environment. Over the longer term, however, environmental losses hurt everyone, including the miners, loggers, and ranchers. If industry is not environmentally sustainable, then over the long term, it will not be economically sustainable either. Moreover, environmental economists like Wicke, Eden, Potier, Freeman, Haveman, and Kneese, have repeatedly shown that reasonable environmental measures reduce more environmental damages than they cost. For every dollar spent to clean up the air, for example, we receive from $3 to $15 in avoided environmental damage. Also, if we do not act so as to preserve air, water, species, and habitats, our failure to do so will cause economic losses. Annual environmental damage in western countries is typically 6 percent of the GNP, according to Wicke.

Apart from environmental protection being economical, clean and healthy environments also provide quantifiable benefits for recreation, aesthetic enjoyment, and study. Americans spend more than $14 billion annually on nonconsumptive wildlife-related recreation. To do so, they need a healthy environment, one that provides watershed protection, regulates climate, and keeps options open for the future. Healthy environments and communities of species also contribute to photosynthesis; breaking down and absorbing pollutants; cycling nutrients; binding soil; degrading organic waste; controlling gasses in the air, radiation, and climate; and fixing solar energy.

Environmental protection likewise provides easily measured economic benefits. Resale dollars lost on USA homes near polluted rivers and lakes are $5 billion annually. Moreover, when a habitat is damaged, society incurs numerous costs, e.g., providing waste-water treatment facilities, once wetlands are destroyed. Duplicating nature's water treatment facilities would cost more than $205,000 per hectare per year. In West Germany, the annual increase in the cost of drinking water, because of pollution, is DM 6 billion. Likewise, for example, air-pollution damage, from the San Bernadino Freeway alone, has been estimated at $27 million per year. The damage occurs in part because of the freeway's interfering with plants' ability to absorb other pollutants, bind the soil, and control radiation. More generally, roads and canals are typically seen as economical, yet when one tallies all the costs and benefits (excluding ecological and psychological factors), the market-based costs are typically twice as great as the benefits. These costs are the reason one author was able to show, for example, that it would save Lexington, Massachusetts taxpayers money if they would buy up 2,000 acres of vacant land. The cost of schools, fire and police protection, sewage treatment, drainage, and welfare costs for new residents on the 2,000 acres would add up to far more than the annual cost to retire the acquisition bond for the property.

Given all the economic benefits of environmentally protective actions, why is there apparent conflict between economic and environmental interests? One reason, already mentioned, is that persons often do not distinguish between short-term and long-term economic goals; the latter are always consistent with environmentally sound actions, but the former may not be.

A second reason for the apparent controversy between economic and environmental interests is that those who use or damage the environment—loggers, miners, and so on—often do not pay the full cost of their actions or damages. They do not pay for them, because classical market mechanisms count air, water, soil, and environmental amenities as "free" goods becaue they are typically not traded on a market. However, they ought not be counted as free goods if their destruction causes economic losses to others. Clean air, into which polluters freely dump emissions, imposes health and environmental costs on others. Just because the polluters do not pay for these costs does not mean that the costs are zero. To counteract the tendency to treat environmental well being as a free good, economists need to cost the value of environmental destruction and preservation. They need to internalize "externalities"—factors like clean air or polluted water that are allegedly external to the market—and include them in their benefit-cost analyses.

In the readings that follow, the authors disagree as to whether environmental protection is economical. If one includes the cost of externalities in one's calculations, then one obtains a different answer from those who, like Drucker and classical economists, do not include externalities in their calculations.

Drucker defends the market system and an expanding economy, while Larkin argues for a no-growth system and attacks classical market institutions. Drucker maintains that only an expanding economy is easily able to provide jobs to low-skilled workers and to cover the cost of cleaning up the environment. Without the funds generated by industrial expansion, additional monies will have to be taken, says Drucker, from budget sectors such as health care and education. He claims, moreover, that pollution is not due to capitalism or the profit motive but is an unavoidable consequence of the "desire to bring a decent living at the lowest possible cost to the largest possible number of people.." For Drucker, pollution is a necessary evil tolerated to obtain a greater economic good.

A quite different view of the moral permissibility of pollution generated by a growing economy is found in the second selection. Here Andrew Larkin, of the Economics Department of the University of Minnesota, argues that industrial growth in developed countries is ethically reprehensible. Larkin, who holds degrees in mathematics, philosophy, and economics, builds much of his case on the MIT report *The Limits to Growth.* This study shows that since an expanding economy is able to avoid pollution only at a high cost, exponential increases of environmental damages are an eventual consequence of economic growth. Hence growth inevitably leads, according to Larkin and the authors of the MIT report, to a collapse of the world system and to death for many people. After tracing the cultural factors responsible for catastrophic industrial expansion, Larkin answers the arguments for economic growth presented by Drucker and others. He shows why it does not provide an improved standard of living and why it is not a cure for the problems created by past growth. In the fourth section of his essay, Larkin argues against economic expansion. He attacks pseudogrowth, internally generated wants, conspicuous consumption, the passion for haste, scarcity, selfishness, and false ideas of freedom, all of which are allegedly produced by the growthmanship ethic.

Larkin ends his analysis with some suggestions for resolving the environmental problems aggravated by following classical economic theories. His solution depends heavily on the work of E.F. Shumacher, the top economist for the British Coal Board and (like Larkin) a close student of Gandhi, nonviolence, and ecology. With Shumacher, Larkin stresses the fact that a sound economy can be built around a new ethic regarding growth, but only if humankind is willing to relinguish greed and to pay the price of redistribution of wealth. He is convinced that Gandhi was right, that "earth provides enough to satisfy every man's need, but not for every man's greed."

Readings

1. Drucker, P.F. 1972. Saving the Crusade. *Harper's Magazine* 244 (January): 66-77. Copyright ©1971 by *Harper's Magazine*. Reprinted by permission of Peter F. Drucker and *Harper's Magazine*.
2. Larkin, A. 1979. The Ethical Problem of Economic Growth Versus Environmental Degradation. Written expecially for inclusion in this volume.

SAVING THE CRUSADE

Peter Drucker

EVERYBODY today is "for the environment." Laws and agencies designed to protect it multiply at all levels of government. Big corporations take full-color ads to explain how they're cleaning up, or at least trying to. Even you as a private citizen probably make some conscientious effort to curb pollution. At the same time, we have learned enough about the problem to make some progress toward restoring a balance between man and nature. The environmental crusade may well become the great cause of the seventies—and not one moment too soon.

Yet the crusade is in real danger of running off the tracks, much like its immediate predecessor, the so-called war on poverty. Paradoxically, the most fervent environmentalists may be among the chief wreckers. Many are confused about the cause of our crisis and the ways in which we might resolve it. They ignore the difficult decisions that must be made; they splinter the resources available for attacking environmental problems. Indeed, some of our leading crusaders seem almost perversely determined to sabotage their cause—and our future.

Consider, for example, the widespread illusion that a clean environment can be obtained by reducing or even abolishing our dependence on technology. The growing pollution crisis does indeed raise fundamental questions about technology—its direction, uses, and future. But the relationship between technology and the environment is hardly as simple as much antitechnological rhetoric would have us believe. The invention that has probably had the greatest environmental impact in the past twenty-five years, for instance, is that seemingly insignificant gadget, the wire-screen window. The wire screen, rather than DDT or antibiotics, detonated the "population explosion" in underdeveloped countries, where only a few decades ago as many as four out of five children died of such insect-borne diseases as "summer diarrhea" or malaria before their fifth birthday. Would even the most ardent environmentalist outlaw the screen window and expose those babies again to the flies?

The truth is that most environmental problems require technological solutions—and dozens of them. To control our biggest water pollutant, human wastes, we will have to draw on all sciences and technologies from biochemistry to thermodynamics. Similarly, we need the most advanced technology for adequate treatment of the effluents that mining and manufacturing spew into the world's waters. It will take even more new technology to repair the damage caused by the third major source of water pollution in this country—the activities of farmers and loggers.

Even the hope of genuine disarmament—and the arms race may be our worst and most dangerous pollutant—rests largely on complex technologies of remote inspection and surveillance. Environmental control, in other words, requires technology at a level at least as high as the technology whose misuse it is designed to correct. The sewage-treatment plants that are urgently needed all over the world will be designed, built, and kept running not by purity of heart, ballads, or Earth Days but by crew-cut engineers working in very large organizations, whether businesses, research labs, or government agencies.

Who Will Pay?

The second and equally dangerous delusion abroad today is the common belief that the cost of cleaning the environment can be paid for out of "business profits." After taxes, the profits of all American businesses in a good year come to $60 billion or $70 billion. And mining and manufacturing—the most polluting industries—account for less than half of this. But at the lowest estimate, the clean-up bill, even for just the most urgent jobs, will be three or four times as large as all business profits.

Consider the most efficient and most profitable electric-power company in the country (and probably in the world): the American Power Company, which operates a number of large power systems in the Midwest and upper South. It has always been far more ecology-minded than most other power companies, including the Government's own TVA. Yet cleaning up American Power's plants to the point where they no longer befoul air and water will require, for many years to come, an annual outlay close to, if not exceeding, the company's present annual profit of $100 million. The added expense caused by giving up strip mining of coal or by reclaiming strip-mined land might double the company's fuel bill, its single largest operating cost. No one can even guess what it would cost—if and when it can be done technologically—to put power transmission lines underground. It might well be a good deal more than power companies have ever earned.

We face an environmental crisis because for too long we have disregarded genuine costs. Now we must raise the costs, in a hurry, to where they should have been all along. The expense must be borne, eventually, by the great mass of the people as consumers and producers. The only choice we have is which of the costs will be borne by the consumer in the form of higher prices, and which by the taxpayer in the form of higher taxes.

It may be possible to convert part of this economic burden into economic opportunity, though not without hard work and, again, new technology. Many industrial or human wastes might be transformed into valuable products. The heat produced in generating electricity might be used in greenhouses and fish farming, or to punch "heat holes" into the layer of cold air over such places as Los Angeles, creating an updraft to draw off the smog. But these are long-range projects. The increased costs are here and now.

Closely related to the fallacy that "profit" can pay the environmental bill is the belief that we can solve the environmental crisis by reducing industrial output. In the highly developed affluent countries of the world, it is true that we may be about to de-emphasize the production-orientation of the past few hundred years. Indeed, the growth sectors of the developed economies are increasingly education, leisure activities, or health care rather than goods. But paradoxical as it may sound, the environmental crisis will force us to return to an emphasis on both growth and industrial output—at least for the next decade.

Overlooked Facts of Life

There are three reasons for this, each adequate in itself.

1. Practically every environmental task demands huge amounts of electrical energy, way beyond anything now available. Sewage treatment is just one example; the difference between the traditional and wholly inadequate methods and a modern treatment plant that gets rid of human and industrial wastes and produces reasonably clear water is primarily electric power, and vast supplies of it. This poses a difficult dilemma. Power plants are themselves polluters. And one of their major pollution hazards, thermal pollution, is something we do not yet know how to handle.

Had we better postpone any serious attack on other environmental tasks until we have solved the pollution problems of electric-power generation? It would be a quixotic decision, but at least it would be a deliberate one. What is simply dishonest is the present hypocrisy that maintains we are serious about these other problems—industrial wastes, for instance; or sewage or pesticides—while we refuse to build the power plants we need to resolve them. I happen to be a member in good standing of the Sierra Club, and I share its concern for the environment. But the Sierra Club's opposition to any new power plants today—and the opposition of other groups to new power plants in other parts of the country (e.g., New York City)—has, in the first place, ensured that other ecological tasks cannot be done effectively for the next five or ten years. Secondly, it has made certain that the internal-combustion engine is going to remain our mainstay in transportation for a long time to come. An electrical automobile or electrified mass transportation—the only feasible alternatives—would require an even more rapid increase in electrical power than any now projected. And thirdly it may well, a few years hence, cause power shortages along the Atlantic Coast, which would mean unheated homes in winter, as well as widespread industrial shutdowns and unemployment. This would almost certainly start a "backlash" against the whole environmental crusade.

2. No matter how desirable a de-emphasis on production might be, the next decade is the wrong time for it in all the developed countries and especially in the United States. The next decade will bring a surge in employment-seekers and in the formation of young families—both the inevitable result of the baby boom of the late forties and early fifties. Young adults need jobs; and unless there is a rapid expansion of jobs in production there will be massive unemployment, especially of low-skilled blacks and other minority group members. In addition to jobs, young families need goods—from housing and furniture to shoes for the baby. Even if the individual family's standard of consumption goes down quite a bit, total demand—barring only a severe depression—will go up sharply. If this is resisted in the name of ecology, environment will become a dirty word in the political vocabulary.

3. If there is no expansion of output equal to the additional cost of cleaning up the environment, the cost burden will—indeed, must— be met by cutting the funds available for education, health care, or the inner city, thus depriving the poor. It would be nice if the resources we need could come out of defense spending. But of the 6 or 7 percent of our national income that now goes for defense, a large part is cost of past wars, that is, veterans' pensions and disability benefits (which, incidentally, most other countries do not include in their defense budgets—a fact critics of "American militarism" often ignore). Even if we could—or should—cut defense spending, the "peace dividend" is going to be 1 or 2 percent of national income, at best.

But the total national outlay for education (7 to 8 percent), and the inner city and other poverty areas (almost 5 percent) comes to a fifth of total national income today. Unless we raise output and productivity fast enough to offset the added environmental cost, the voters will look to this sector for money. Indeed, in their rejection of school budgets across the nation and in their desperate attempts to cut welfare costs, voters have already begun to do so. That the shift of resources is likely to be accomplished in large part through inflation—essentially at the expense of the lower-income groups—will hardly make the environmental cause more popular with the poor.

The only way to avoid these evils is to expand the economy, probably at a rate of growth on the order of 4 percent a year for the next decade, a higher rate than we have been able to sustain in this country in the postwar years. This undoubtedly entails very great environmental risks. But the alternative is likely to mean no environmental action at all, and a rapid public turn—by no means confined to the "hard hats"—against all environmental concern whatever.

Making Virtue Pay

The final delusion is that the proper way to bring about a clean environment is through punitive legislation. We do need prohibitions and laws forbidding actions that endanger and degrade the environment. But more than that, we need incentives to preserve and improve it.

Punitive laws succeed only if the malefactors are few and the unlawful act is comparatively rare. Whenever the law attempts to prevent or control something everybody is doing, it degenerates into a huge but futile machine of informers, spies, bribe givers, and bribe takers. Today every one of us—in the underdeveloped countries almost as much as in the developed ones—is a polluter. Punitive laws and regulations can force automobile manufacturers to put emission controls into new cars, but they will never be able to force 100 million motorists to maintain this equipment. Yet this is going to be the central task if we are to stop automotive pollution.

What we should do is make it to everyone's advantage to reach environmental goals. And since the roots of the environmental crisis are so largely in economic activity, the incentives will have to be largely economic ones as well. Automobile owners who voluntarily maintain in working order the emission controls of their cars might, for instance, pay a much lower automobile registration fee, while those whose cars fall below accepted standards might pay a much higher fee. And if they were offered a sizable tax incentive, the automobile companies would put all their best energies to work to produce safer and emission-free cars, rather than fight delaying actions against punitive legislation.

Despite all the rhetoric on the campuses, we know by now that capitalism has nothing to do with the ecological crisis, which is fully as severe in the Communist countries. The bathing beaches for fifty miles around Stockholm have become completely unusable, not because of the wicked Swedish capitalists but because of the raw, untreated sewage from Communist Leningrad that drifts across the narrow Baltic. Moscow, even though it still has few automobiles, has as bad an air-pollution problem as Los Angeles—and has done less about it so far.

We should also know that greed has little to do with the environmental crisis. The two main causes are population pressures, especially the pressures of large metropolitan peopulations, and the desire—a highly commendable one—to bring a decent living at the lowest possible cost to the largest possible number of people.

The environmental crisis is the result of success—success in cutting down the mortality of infants (which has given us the population explosion), success in raising farm output sufficiently to prevent mass famine (which has given us contamination by insecticides, pesticides, and chemical fertilizers), success in getting people out of the noisome tenements of the nineteenth century city and into the greenery and privacy of the single-family home in the suburbs (which has given us urban sprawl and traffic jams). The environmental crisis, in other words, is very largely the result of doing too much of the right sort of thing.

To overcome the problems success always creates, one has to build on it. The first step entails a willingness to take the risks involved in making decisions about complicated and perilous dilemmas:

- What is the best "trade-off" between a cleaner environment and unemployment?
- How can we prevent the environmental crusade from becoming a war of the rich against the poor, a new and particularly vicious "white racist imperialism?"
- What can we do to harmonize the worldwide needs of the environment with the political and economic needs of other countries, and to keep American leadership from becoming American aggression?
- How can we strike the least agonizing balance of risks between environmental damage and mass starvation of poor children, or between environmental damage and large-scale epidemics?

An Environmental Crime?

More than twenty years ago, three young chemical engineers came to seek my advice. They were working for one of the big chemical companies, and its managers had told them to figure out what kind of new plants to put into West Virginia, where poverty was rampant. The three young men had drawn up a long-range plan for systematic job creation, but it included one project about which their top management was very dubious—a ferroalloy plant to be located in the very poorest area where almost everybody was unemployed. It would create 1,500 jobs in a dying small town of 12,000 people and another 800 jobs for unemployed coal miners—clean, healthy, safe jobs, since the new diggings would be strip mines.

But the plant would have to use an already obsolete high-cost process, the only one for which raw materials were locally available. It would therefore be marginal in both costs and product quality. Also the process was a singularly dirty one, and putting in the best available pollution controls would make it even less economical. Yet it was the only plant that could possibly be put in the neediest area. What did I think?

I said, "forget it"—which was, of course, not what the three young men wanted to hear and not the advice they followed.

This, as some readers have undoubtedly recognized, is the prehistory of what has become a notorious "environmental crime," the Union Carbide plant in Marietta, Ohio. When first opened in 1951 the plant was an "environmental pioneer." Its scrubbers captured three quarters of the particles spewed out by the smelting furnaces; the standard at the time was half of that or less. Its smokestacks suppressed more fly ash than those of any other power plant then built, and so on.

But within ten years the plant had become an unbearable polluter to Vienna West Virginia, the small town across the river whose unemployment it was built to relieve. And for the last five years the town and Union Carbide fought like wildcats. In the end Union Carbide lost. But while finally accepting Federal and state orders to clean up an extremely dirty process, it also announced that it would have to lay off half the 1,500 men now working in the plant—and that's half the people employed in Vienna. The switch to cleaner coal (not to mention the abandonment of strip mining) would also put an end to the 800 or so coal-mining jobs in the poverty hollows of the back country.

There are scores of Viennas around the nation, where marginal plants are kept running precisely because they are the main or only employer in a depressed or decaying area. Should an uneconomical plant shut down, dumping its workers on the welfare rolls? Should the plant be subsidized (which would clearly open the way for everybody to put his hand in the public till)? Should environmental standards be disregarded or their application postponed in "hardship" cases?

If concern for the environment comes to be seen as an attack on the livelihood of workers, public sympathy and political support for it is likely to vanish. It is not too fanciful to anticipate, only a few years hence, the New (if aging) Left, the concerned kids on the campus, and the ministers in a protest march against "ecology" and in support of "the victims of bourgeois environmentalism."

Third World Ecology

In the poor, developing countries where men must struggle to make even a little progress in their fight against misery, any industry bears a heavy burden of high costs and low productivity. Burdening it further with the cost of environmental control might destroy it. Moreover, development in these countries—re gardless of their political creed or social organization, in Mao's as well as in Chiang Kai-shek's China and in North as well as in South Vietnam—cannot occur without the four biggest ecological villains: a rapid increase in electric power, chemical fertilizers and pesticides, the automobile, and the large steel mill.

That poor countries regard those villains as economic saviors confronts us with hard political choices. Should we help such countries get what they want (industrialization), or what we think the world needs (less pollution)? How do we avoid the charge, in either case, that our help is "imperialistic?" To complicate matters, there is a looming conflict between environmental concern and national sovereignty. The environment knows no national boundaries. Just as the smog of England befouls the air of Norway, so the chemical wastes of the French potash mines in Alsace destroy the fish of the lower Rhine in Belgium and Holland.

No matter what the statistics bandied about today, the United States is not the world's foremost polluter. Japan holds this dubious honor by a good margin. No American city can truly compete in air pollution with Tokyo, Milan, Budapest, Moscow, or Dusseldorf. No American river is as much of an open sewer as the lower Rhine, the Seine, or the rivers of the industrial Ukraine such as the lower Dnieper. And we are sheer amateurs in littering highways compared to the Italians, Danes, Germans, French, Swedes, Swiss, and Austrians—although the Japanese, especially in littering mountainsides and camp grounds, are clearly even more "advanced."

If not the worst polluter, however, the United States is clearly the largest one. More important, as the most affluent, most advanced, and biggest of the industrial countries, it is expected to set an example. If we do not launch the environmental crusade, no one else will.

We shall have to make sure, however, that other nations join with us. In the absence of international treaties and regulations, some countries—especially those with protectionist traditions, such as Japan, France, and even the United States—may be tempted to impose ecological standards on imports more severe than those they demand of their own producers. On the other hand, countries heavily dependent on exports, especially in Africa and Latin America, may try to gain a competitive advantage by lax enforcement of environmental standards

One solution might be action by the United Nations to fix uniform rules obliging all its members to protect the environment; and such action is, in fact, now under official study. The United States might help by changing its import regulations to keep out goods produced by flagrant polluters—allowing ample time for countries with severe poverty and unemployment problems to get the cleanup under way. We have good precedent for such an approach in our own history. Forty years ago we halted the evils of child labor by forbidding the transportation in interstate commerce of goods produced by children.

Such a course, however, will demand extraordinary judgment. Unless we persuade other nations to join with us—and set an example ourselves—we may well be accused of trying again to police the world.

Choosing the Lesser Evils

The hardest decisions ahead are even more unprecedented than those we have been discussing. What risks can we afford to take with the environment, and what risks can we *not* afford to take? What are the feasible trade-offs between man's various needs for survival?

Today, for example, no safe pesticides exist, nor are any in sight. We may ban DDT, but all the substitutes so far developed have highly undesirable properties. Yet if we try to do without pesticides altogether, we shall invite massive hazards of disease and starvation the world over. In Ceylon, where malaria was once endemic, it was almost wiped out by large-scale use of DDT; but in only a few years since spraying was halted, the country has suffered an almost explosive resurgence of the disease. In other tropical countries, warns the UN Food and Agricultural Organization, children are threatened with famine, because of insect and blight damage to crops resulting from restrictions on spraying. Similarly, anyone who has lately traveled the New England turnpike will have noticed whole forests defoliated by the gypsy moth, now that we have stopped aerial spraying.

What is the right trade-off between the health hazard to some women taking the pill and the risk of death to others from abortions? How do we balance the thermal and radiation dangers of nuclear power plants against the need for more electricity to fight other kinds of pollution? How should we choose between growing more food for the world's fast-multiplying millions and the banning of fertilizers that pollute streams, lakes, and oceans?

Such decisions should not be demanded of human beings. None of the great religions offers guidance. Neither do the modern "isms," from Maoism to the anarchism popular with the young. The ecological crisis forces man to play God. Despite the fact that we are unequal to the task, we can't avoid it: the risks inherent in refusing to tackle these problems are the greatest of all. We have to try, somehow, to choose some combination of lesser evils; doing nothing invites even greater catastrophe.

Where to Start

Cleaning up the environment requires determined, sustained effort with clear targets and deadlines. It requires, above all, concentration of effort. Up to now we have had almost complete diffusion. We have tried to do a little bit of everything—and tried to do it in the headlines—when what we ought to do first is draw up a list of priorities in their proper order.

First on such a list belong a few small but clearly definable and highly visible tasks that can be done fairly fast without tying up important resources. Removing the hazard of lead poisoning in old slum tenements might be such an action priority. What to do is well known: burn off the old paint. A substantial number of underemployed black adolescents could be easily recruited to do it.

Once visible successes have been achieved, the real task of priority-setting begins. Then one asks: (1) what are the biggest problems that we know how to solve; and (2) what are the really big ones that we don't know how to solve yet? Clean air should probably head the first list. It's a worldwide problem, and getting worse. We don't know all the answers, but we do have the technological competence to handle most of the problems of foul air today. Within ten years we should have real results to show for our efforts.

Within ten years, too, we should get major results in cleaning up the water around big industrial cities and we should have slowed (if not stopped) the massive pollution of the oceans, especially in the waters near our coastal cities.

As for research priorities, I suggest that the first is to develop birth-control methods that are cheaper, more effective, and more acceptable to people of all cultures than anything we now have. Secondly, we need to learn how to produce electric energy without thermal pollution. A third priority is to devise ways of raising crops for a rapidly growing world population without at the same time doing irreversible ecological damage through pesticides, herbicides, and chemical fertilizers.

Until we get the answers, I think we had better keep on building power plants and growing food with the help of fertilizers and such insect-controlling chemicals as we now have. The risks are now well known, thanks to the environmentalists. If they had not created a widespread public awareness of the ecological crisis, we wouldn't stand a chance. But such awareness by itself is not enough. Flaming manifestoes and prophecies of doom are no longer much help, and a search for scapegoats can only make matters worse.

What we now need is a coherent, long-range program of action, and education of the public and our lawmakers about the steps necessary to carry it out. We must recognize—and we need the help of environmentalists in this task—that we can't do everything at once; that painful choices have to be made, as soon as possible, about what we should tackle first; and that every decision is going to involve high risks and costs, in money and in human lives. Often these will have to be decisions of conscience as well as economics. Is it better, for example, to risk famine or to risk global pollution of earth and water? Any course we adopt will involve a good deal of experimentation—and that means there will be some failures. Any course also will demand sacrifices, often from those least able to bear them: the poor, the unskilled, and the underdeveloped countries. To succeed, the environmental crusade needs support from all major groups in our society, and the mobilization of all our resources, material and intellectutal, for years of hard, slow, and often discouraging effort. Otherwise it will not only fail; it will, in the process, splinter domestic and international societies into warring factions.

Now that they have succeeded in awakening us to our ecological peril, I hope the environmentalists will turn their energies to the second and harder task: educating the public to accept the choices we must face, and to sustain a worldwide effort to carry through on the resulting decisions. The time for sensations and manifestos is about over; now we need rigorous analysis, united effort, and very hard work.

THE ETHICAL PROBLEM OF ECONOMIC GROWTH VS. ENVIRONMENTAL DEGRADATION

Andrew Larkin

The Catastrophic Situation: What Price Growth?

In his novel *The Centaur,* John Updike includes a little story about a bacterium, the leptothrix, that was responsible for removing pure iron from ferric salt and thus creating the huge iron deposits, like the Mesabi Range, which have long been mined by human beings. Updike's character makes the observation that "to win World War Two, we gouged out these battleships and tanks and jeeps and coke machines out of it and left the poor old Mesabi Range like an old carcass the jackals had chewed. ..." (1963)

The story also fits other resources which have existed in abundance. Three-hundred million years ago teeming vegetation died, and without oxidizing, formed massive seams of coal and left the oxygen, produced by the plants, in the atmosphere to be used in the development of animal and human life. But present generations of humans are wastefully and carelessly burning nearly all available coal, and also oil and natural gas.

In a great cycle, the sun continually evaporates water from the oceans and lifts it into the atmosphere, whence it will return to earth, some to be used by humans for power, for cooling, for drinking, for irrigation, and for flushing sewage; some will seep into the earth to be used later; some will form pools and streams, sources of food and recreation. But even here world-wide pollution of oceans, lakes, and streams could interrupt access to one of the most important resources—water.

"The impairment of our environment and the damages and costs resulting therefrom constitute one of the most fundamental, dangerous and long-run issues which mankind has ever faced." (Kapp, 1970). In this discussion I will analyze the continuing destruction of our physical environment that is occurring at a rapidly increasing pace in our industrial age. We need to determine the ultimate sources of our problems of interrelation with the environment in order to redirect our efforts into less harmful channels, and an important part of any analysis of problems stemming from industrialization is the question of economic growth. Current growth patterns have culminated in widespread environmental degradation that has already imposed burdens on contemporary society, such as shortages of resources and health hazards to humans. Both common sense and evidence analyzed by mathematical methods show that even worse burdens will be imposed on future societies. Thus, present economic activity is a moral problem with respect to the welfare of future generations, and it is this problem with which I want to deal. In my conclusion, after presenting and analyzing arguments for and against growth, I will propose several methods of alleviating the environmental problems due to growth; in particular, I will advocate more effective democratic social control of the economy. But first, let us look sketchily at some of the problems.

Resource Depletion. Humans are wasteful; we use too much; we gouge out, dig out, pump out, and deplete the resource deposits not only for interspecies killing but also for consumption—of both necessities and non-necessities. We mine most resources at rates many times faster than they are or were deposited. We are using oxygen faster than plants can replenish it, and even then, gradually reducing the amount of plant life on the earth's surface; we are replacing the oxygen with poisons and carcinogens. And we are depleting our underground aquifers and turning surface channels and reservoirs into sewers.

Chemical Contamination. Human beings, the other animals, as well as plants have been evolving within a natural environment which consists of certain elements, chemicals, compounds, and the like. We are adapted to these in that we can use some of them for nourishment and metabolism; we are indifferent to others, and only a few are dangerous poisons. However, within the past century, employing the manufacturing processes of the Industrial Revolution, we have introduced thousands of new chemicals into our biosphere, compounds to which we are not adapted, and we have enormously increased the concentration of others. In our attempts to control our environment and to make it more productive for our growing population and our growing wants and needs, we have deliberately introduced dangerous chemicals to kill "pests" and "weeds"—aldrin, dieldrin, DDT, dioxin, 2,4-D, 2,4,5-T, and many others. These chemicals succeed by the very fact that many current life forms are not adapted to live with them, but they spread to life forms other than the pests themselves. As a consequence, birds, and even people, are harmed. Because plants and animals often cannot cope with these chemicals, they remain poisonous in their bodies, which are eaten by other animals and passed through the food chains. The chemicals eventually show up as poisons in human beings.

Pesticides have not even proven to be a solution to the pest problem. Because insects have rapid rates of reproduction, those few that are immune to a particular chemical poison can thrive and procreate (while the susceptible insects die), and consequently generate whole populations of immune insects. This results in a new insect problem which is "solved" by developing a new chemical pesticide and introducing it into the biosphere.

An Environmental Protection Agency (EPA) study conducted in 1975 concluded that "99 percent of the samples [of mother's milk] taken from more than 1400 nursing mothers contained dangerous pesticide compounds. Among these compounds were suspected carcinogens like dieldrin and DDT." (*Mother Jones,* Sept./Oct., 1977). Furthermore, the poisons have already been concentrated to the extent that "according to American biologists, most mother's milk in the United States contains so much DDT that it would be declared illegal in interstate commerce if it were sold. ..." (Kapp, 1970).

Air Pollution. Perhaps even more serious than our deliberate poisonings of our environment are our inadvertent ones, and among the worst offenders are automobiles, power plants, and industrial plants because of their use of fossil fuels. Incomplete oxidation of fossil fuels produces carbon monoxide, the asphyxiating agent of automobile exhausts; even small, wind-swept prairie cities like Lincoln, Nebraska, report carbon monoxide levels above EPA standards. Because fossil fuels contain elements other than carbon and hydrogen, even complete oxidation results in the emission of oxides of nitrogen and sulfur. Combined with water in the atmosphere, these oxides form acids which precipitate and injure biota, destroy buildings, ruin soil, and change the pH of lakes. "Today, in more than 50 percent of all Adirondack lakes above 2000 feet in elevation ... all the fish have died. ..." (MacRobert, 1977). This has been called "acid rain."

Radiation. Bob Dylan warned us of the "hard rain," radiation, which is also beginning to fall. "One manufacturer [of microwave ovens] had to recall 300,000 units recently because of radiation leakage." (*Mother Jones,* July, 1977). Among the forms of pollution, radiation is most insidious because so little is known about its effects, especially over long periods, and because what is known is that radiation can be quite harmful. (It destroys living tissue, and seems to be a major cause of unpredictable genetic mutations.) Radiation will certainly leave many problems for future generations because of the long half-life of many radioactive by-products. Already the present generation has found it necessary to deal with problems created in the second decade of this century, when radium was processed in Denver. In early 1979 EPA researchers discovered potentially harmful amounts of low-level radiation being emitted by 60-year-old tailings left as waste products from the earlier operations. Now Denver or Colorado or US citizens must spend money to clean up the mess.

The threatened explosion of the Three Mile Island nuclear plant awoke millions of people to the hazards of radiation and to the problems of energy use in industrial societies. Political and business leaders have said that nuclear energy is the only alternative to coal power in the face of declining petroleum production. But are air pollution from coal or radiation from atomic reactors our only choices?

Cancer. The level of carcinogens and other harmful agents in the human environment has increased geometrically since the end of World War II. On an average day over 1000 people die from cancer in the United States—that is 400,000 people annually (Bureau of the Census, 1977; Epstein, 1978). Cancer is believed to be induced by foul air, foul water, cigarettes, drugs, chemicals, and even our food. "One out of four of us will get a serious cancer. About 17 percent of deaths can be laid at its door. ... We know with certainty that environmental factors are the causes in 40 percent of all human cancers." (Drummond, 1977). Barry Commoner (1976), a prominent ecologist, estimates that "about 75-80 percent of cancer incidence is due to environmental disease."

The MIT-Club of Rome Report—Total Control of World Economy vs. Complete Destruction

By now most people are aware of at least a few of the environmental problems caused by pollution; many of them, however, do not yet consider resources to be finite and depletable. In order to understand the interactions of all forms of pollution and resource depletion, their impact on the environment, and, in turn, the impact on human society, we need an overall model which will capture all of this information at once. One such model is offered by the MIT-Club of Rome group in their world famous report, *The Limits to Growth.* (Meadows 1972; Forrester, 1971).

In this model the MIT-Club of Rome group considers the earth as one ecosystem composed of any number of subsystems (e.g., national economic systems). Using the consensus of originally thirty, and later as many as seventy, members from around the world, the Club chose "five basic factors that determine, and therefore, ultimately limit growth on this planet—population, agricultural production, natural resources, industrial production, and pollution." (Meadows, 1972). These factors interact in their growth paths, as, for example, the depletion of natural resources affects the growth of industrial production, which in turn affects the rise in pollution. This model includes, for example, feedback loops like the birth rate and the death rate and their effects on population growth. Also included are other effects of such factors as pollution and agricultural production on the birth and death rates.

Using computers to handle the mass of data and the numerous equations, the Club arrived at three broad conclusions: If present trends continue, (1) the next 100 years will see a termination of growth of all five factors, and the decline in population will be sudden and uncontrollable; but (2) we can alter the growth trends of these factors in such a way that they will become stable yet permit each person on earth an access to necessities and the opportunity to live a truly human life; and (3) the sooner we reject catastrophic growth and accept ecological and economic stability, the more likely it will be that we can then achieve the human alternative of equal access and opportunity (Meadows, 1972).

In order to draw these conclusions, the Club actually ran several models. The "standard" model is based on actual data gathered between 1900 and 1970; it forecasts a very rapid decline in resources in the early twenty-first century, simultaneous with a peak and decline of two other parameters: food per capita and industrial output per capita. These events are soon followed by a peak in the rate of pollution, and then a bit later in the twenty-first century, there is a peak in the population growth curve.

Other models developed by the Club alter the input variables of one or more of the five factors; that is, they depart from the historical data in order to project different possible outcomes. For example, the model based on double the known amount of resource reserves in 1970 shows a tremendous increase in the growth of pollution and again a peak in the population growth curve in mid twenty-first century. The Club also altered variables in order to project more stable models. In these models they assumed that, beginning in 1975, there would be: (1) a birth rate equal to the death rate, (2) a 75 percent decrease in resource use, (3) a decrease in the production of material goods and an increase in services, (4) a 75 percent reduction of pollution, (5) diversion of capital to food production, (6) preservation and enrichment of soil, and (7) increased durability of industrial capital. The Club concluded the necessity of making these changes soon because one model shows that if they are not made until the year 2000, stability will not be attainable until after 2100.

This is a condensation of one planetary view of the human situation with respect to our environment. Its virtue is that it is not limited in either space or time, but shows what is likely to happen throughout the world if we continue our present paths of population increase, pollution, resource depletion, and economic growth: we will exhaust the ability of the ecosystem to sustain human life.

The Arguments for Growth

Economic growth has not just happened; it has been chosen by society for its supposed benefits, especially in recent decades. To understand some of the problems experienced both by the economy and by the environment we need to grapple with arguments presented in favor of growth; we need to see whether they are all that valid and whether there really are some good reasons for continuing the present growth of economies at the expense of the environment.

1. Growth as the Economic Panacea. The major argument for growth is that it provides an improved standard of living; a growing economy would make more material goods and more services available to everyone, especially to the poor, as the argument goes; ideally we could even be seeing an improvement in the quality of goods available. However, we don't really have a meaningful measure of the standard of living. The GNP is a measure of all final goods and services, and therefore it includes trivial knickknacks and trinkets, which we see more and more, and yet it neglects recycled products. The GNP is not able to distinguish "enjoyment-producing" or "happiness-producing" goods from mere junk, and it is just as incapable of measuring increased pollution, ugliness, and congestion (Mishan, 1967). For this reason, we cannot determine whether growth has been beneficial just by using GNP data.

Although we cannot accurately measure the standard of living, we can measure relative access to our increased output by consulting income statistics. In 1947, just after World War II, the poorest 20 percent of American people received 4 percent of the national income and the wealthiest 20 percent received 42 percent; by 1976, thirty years later, the figures had hardly changed—the lowest 20 percent were receiving 5 percent while the wealthiest were dividing 41 percent (and the other 3 quintiles remained just as constant) (Bureau of the Census, 1977). Thus, despite the tremendous growth of the American economy in the post-war period, despite the tremendous damage perpetrated upon the physical environment as well as on the social environment, the poorer people have not improved their relative share of this increased output.

Apparently this argument for growth has not considered all of the aspects of the original unequal wealth distribution; perhaps because the wealthy are more powerful both economically and politically, they tend to take more and more of a growing national income. The problem is that more money available to everyone has increased the capacity to consume, thereby aggravating pollution and depletion problems. Also, wealthier groups can afford to invest in the environmentally destructive industry of the present period—while increased numbers of poor people around the world have less access even to necessities.

These criticisms of distribution apply to the production of necessities because it has been argued that growing economies would produce more necessities, and that these necessities would become available to poorer people around the world. However, necessities sent to less developed countries (LDCs) have been appropriated by the rich and powerful, and then, if possible, sold to the poor. These problems with the distribution of necessities apply to the production of necessities within the LDCs as well (Lappe, 1977; George, 1977).

A corollary to the income distribution argument is that a growing economy would provide more employment of human beings. There is some truth to this argument, but in order to employ more people, our type of economy also has to employ more natural resources with which these people may work, and it has to generate more waste products. Thus, it would be just as well for our society to provide incomes without requiring people to work and produce material goods.

2. Growth for the Sake of Growth. Increased economic growth has also been advocated as a cure for the problems created by past growth; it has been argued that the economy will need the tools provided by increased output in order to clean the air and the water and to get rid of the junk, waste, pollution, and congestion we've accumulated. However, there are several problems with this contention. Growth cannot reclaim the availability of spent fuels such as coal and petroleum. Furthermore, growth certainly cannot re-create species which are being eradicated by poisons like DDT (Murdoch and Connell, 1971).

Using growth to solve the problems of growth violates the second law of thermodynamics, which states that the *available* energy of a closed system is being constantly depleted. This is the reason it is impossible to reclaim oxidized fossil fuels; the energy stored in fossil fuels is converted to low-level unavailable heat when it is used to do work in the production process.

If we consider the earth to be an open system with respect to the sun, thus increasing our potential source of energy enormously, we are still within the closed planetary system which can absorb only a finite amount of wastes, the other problem of unlimited growth. Systems theorists like Jay Forrester (1971) argue that, within a social system, attempts to solve one problem tend to aggravate other problems (1971); attempts to solve pollution problems by using energy ultimately result in an increase in thermal pollution representing the work exerted to clean the system. It would also result in radioactive pollution if nuclear energy were used as a resource for "cleaning up the environment" (Daly, 1971).

It might be concluded that this tinkering with the environment, which is supposed to be cleaning it up but is really "a shift of the disposal of residuals from one environmental medium to another" (Kapp, 1977), is nothing more than a pseudo-event which tricks us into thinking something is being done when, in fact, not much at all is being done (Murdoch and Connell, 1971).

3. Freedom. It might be argued that economic growth contributes to personal freedom. The naive conception of freedom is to do as one pleases without regard for others, including those of future generations. Economic growth seems to have fostered this conception of egoistic freedom, since we use up as many resources as possible for our own sake, leaving a radioactive junkpile. A more complex notion of freedom is the opportunity to develop oneself as a human being in more noble aspects. This requires and enables altruistic interrelation with other human beings, the social environment, including, of course, people of other generations, as well as with the physical environment. Material consumption has not proven to be a method of developing one's humanity, just as it cannot improve one's happiness, especially when that consumption is at the expense of others. Materialistic consumption is thus an important argument against growth, yet this is an insight which our generation is all too ready to ignore.

Some Arguments Against Growth

1. *Pseudo-growth.* What has been taken to be growth has often been a "pseudo-growth"; society has not always received more or better goods or services. Not much credit can be given to a growing economy which relies on the "annual model" for a large portion of its output; these offer consumers no genuine increase in their range of choices. Nor does the practice of built-in obsolescence, which is also quite wasteful, contribute to the satisfaction of consumers. A third major contributor to economic growth, which offers consumers no direct satisfaction, is the production of armaments; in terms of environmental deterioration, weapons and means of destruction are to be condemned for their needless depletion, disruption, and waste.

2. Internally Generated Wants. The satisfaction of wants that have actually been created by the economic system itself, through advertising, cannot all be claimed as contributions to society by the growing economy. These internally generated wants are symptomatic of the difficulties of keeping the economy expanding. Even the satisfaction of any fashions which are not producer dictated, but which arise more from the consumption psychology itself, is not a genuine contribution to society. Examples include the demand for a second or third car because the Joneses have a car for every member of the family above driving age, or for things a person might not really want but feels he "must" have, such as a trip to Europe.

Wilden (1978) decries "the well-known tendency of capitalism to produce more and more essentially useless goods under a system of planned obsolescence, and to rely on advertising and other means to make such goods 'needed'." Several times per hour, tens of thousands of times per year, our population is bombarded with the message that to consume is to be happy. Both within America and internationally, from the more developed to the less developed countries, advertising is a tremendous stimulus to further growth; its main purpose is to encourage people to increase their consumption in order to enlarge the incomes of producers. The ethic of advertising ignores quality or value, not only of the products advertised, but also with regard to the products' impact on the physical, biological, and social systems (unless, of course, environmental concern can be used to help sell a product).

3. Conspicuous Consumption. In a society where classes are determined by income levels, bizarre events are bound to occur, such as the practice of demonstrating one's income level or status by consuming. Professional athletes and movie stars buy Rolls Royces. College teachers, who have received as little as a 10 percent increase in their salaries, have been known to change friends and to move to more expensive houses. For many, a larger income results in a larger car or power boat. Houses are no longer homes, where several generations can spend their entire lives, but they are investments; individual dwelling units are of higher status than shared dwelling units, as well as being more wasteful of the earth's resources.

4. The Consumption of Haste and Separation. Economic growth in America has resulted in a much higher consumption of non-necessities which do not fulfill actual human needs. (Consider plastic imitation-flowers.) Growth has resulted in increased haste, which in turn has aggravated physiological and psychological illnesses. Consider the hair dryer, which does nothing that wouldn't have happened anyway, but does it faster, with the major result being that it is now socially gauche to be seen with wet hair (not to mention that it has been found that many hair dryers emit asbestos fibers, thus increasing the incidence of lung cancer). The major culprit behind our haste is the automobile, which does not seem at all to have increased our available useful time, but more importantly has depleted our iron and petroleum resources and has contributed most to the pollution of the atmosphere. The automobile is also the major cause of the overwhelming congestion and constant noise of contemporary cities.

Since the dominant commodity is not steel or food or housing but a mere means of transportation, the automobile, we devote vast amounts of resources just to move around; whatever we do after we get there is economically secondary. The automobile has increased the ability of people to live farther away from each other—from family, friends, shops, and jobs—in separate houses which require separate sets of household appliances and consumer durables. One of these appliances, the television, is hardly more than a medium for advertising even more consumption (Mander, 1978).

Thus we have a consumption-advertising vicious circle which worsens as households become smaller and smaller, requiring more housing and appliances, as it becomes more and more common for each person to have separate facilities, such as television sets in the children's bedrooms. Now the automobile is indispensable for consuming other goods, even necessities, because shopping areas are located too far from most homes for access by walking and are too dangerously surrounded by congested highways to be accessible by bicycle. The autombile has demanded the shopping center, which now demands the automobile (Mishan, 1967, 1978).

5. Maintaining High-Level GNPs. Economic growth is most insidious in the industrialized countries whose people consume more per capita; these economies are already so large that they require ever greater exploitation of the environment in order to maintain the high amounts of growth desired by these countries. For example, the world produces some 51 million barrels of oil per day, but 20 million barrels, or 40 percent, are consumed in the United States. With just over 220 million people in a world population of over 4 billion, the US population amounts to about 5 percent of the world population. Therefore, the average American consumes 8 times as much petroleum as the average human being. Similar statistics apply to other resources. The problem is that the amount of resources needed to sustain American and other large economies is enormous.

It is becoming increasingly difficult to get more out of industrial economies, according to economists like Kenneth Boulding. Boulding (1970) observes that more and more of the GNP goes to maintaining the present level in such ways as commuting expenses and government defense budgets. The British economist, E.J. Mishan (1978), writes that in the wealthier countries, economic growth

> can be sustained only by the unremitting efforts of industry to create dissatisfaction with existing possessions and to promote unbridled convetousness accompanied and aggravated by a degenerated environment and by ... increasingly congested and polluted metropolises and their suburbs.

Suggestions for Ameliorating Ecological Degradation Due to Growth

Can anything be done about the growing environmental problems of pollution and resource depletion? Perhaps it is already too late. Perhaps there is a process, operating on its own, carrying the human race along as an unwitting or unwilling tool. But most of us think that we can do something to change the path that society and its economy seems to be on. If so, what might we do?

1. More Services, Fewer Goods.

Certainly we must at least slow the growth of GNPs, perhaps halt them altogether, and perhaps even decrease the larger GNPs to much lower levels (Hardesty, 1971). This seems especially necessary for material goods output, so a good suggestion is to convert our economies more towards the production of services such an entertainment, education, health care, social work, and so on, while attempting to provide physical necessities to all members of the human race instead of concentrating on luxuries for a few. Environmental degradation could also be lowered by increased re-cycling of wastes and by-products.

2. Leisure. We must adopt a more approving attitude toward leisure. Currently we praise only those who work, even though the work be the tedious production of wasteful junk, and then condemn those who don't work and consume little. As has been pointed out many times before in human history, material consumption cannot by itself make anyone into a better human being, but less consumption means less production and less employment. So we have to reverse our negative attitude toward unemployment and leisure and encourage more adult education, scientific exploration, and the like in hopes of better, rather than more, output. "Work that wounds the environment or makes the world ugly is wrong and wasteful" (Roszak, 1978).

3. Intermediate Technology. The present economic system is criticized from various viewpoints for its exploitation of human labor. Workers are employed with the same attitude that accompanies the use of machinery, but people have many more facets which require fulfillment in their daily lives. Any changes in the socioeconomic system should attempt to satisfy more of these human needs. Work could be made less wasteful, less polluting, and also less demeaning if it were done with less dominating technology. E.F. Schumacher (1973) has introduced the concept of intermediate technology, which still means tools, but it also means people—people who contribute more than just a tiny part to the production process, and thus can achieve some satisfaction from their work. Intermediate technology might mean, for example, farming with small machinery so that people are closer to their soil and their plants and animals, with less chemicals so that the soil stays alive and absorbs moisture rather than eroding and polluting waterways. Such practices bring humans closer to the ecosystem they live in.

4. Quality Production. There is certainly room to increase the emphasis on quality rather than quantity, not only because large quantity production and consumption are destroying our ecosystem, but also because good quality is less alienating and more humane, more in harmony with healthy human life. Quality production, like useful work, goes hand in hand with respect for and enjoyment of the human ecosystem.

5. An Evolutionary View. Human beings need to adopt an evolutionary attitude, (1) that living beings develop under certain conditions, and changing those conditions might change the life forms, and (2) that the species is of the same importance as the individual. We need to take more of a long-run attitude rather than emphasize the maximization of short-run consumption (Wilden, 1978; Forrester, 1971).

6. An Ecological View. We also need to adopt an ecological attitude, (1) that every act has repercussions, many of which are unpredictable, (Forrester, 1971), (2) that some massive actions will cause irreversible changes, (3) that a greater population means greater problems, and (4) that there is a symbiotic relationship among all living creatures (McHarg, 1971). Some have suggested that we introduce a measure of the destruction of the human environment. much as the GNP measures economic output, but others have suggested that the GNP itself might be a useful measure of environmental destruction since any product is simultaneously a cost. As Boulding has pointed out, there is a growing need to see the economy of the world as a closed system; material resources, including energy, are not limitlessly obtainable from the environment (1971).

One of Wilden's points (1978) is that "what survives in nature is not simply the fittest individual but, more significantly, the fittest ecosystem." No individual lives alone; no species lives alone; multiple species live together. Humans are, and can only be, members of an ecosystem; we are not separate from our environment except in conception. Hence, any harm to the environment is harm to human beings.

7. An Aggregate View. Another source of trouble in our socioeconomic system is our tendency to look only at small aspects of problems. For many people a shortage of gasoline is a problem to be solved by simply providing more gasoline. But we all know that if refineries produce more gasoline from given quantities of crude oil, they must produce less of other products such as diesel fuel or heating oil. Thus, solving individual problems tends to create or aggravate other difficulties (Forrester, 1971).

In order to become effective in dealing with the problems of the socioeconomic system, we must look at the system, as a whole. Orthodox economists argue that as a resource becomes scarce, other resources will be substituted for it, but this argument fails to deal with the simultaneous depletion of all resources of the system. Similarly, pollutants aggregate their effects, and "waste products of various kinds may not only react upon each other, but also upon other elements in the environment and in this way give rise to additional toxological effects ... with delayed cumulative consequences on human health" (Kapp, 1970).

Developing an aggregate view will be difficult because "evolutionary processes have not given us the mental skill needed to properly interpret the dynamic behavior of the systems of which we have now become a part" (Forrester, 1971). Fortunately we have developed the technology of the computer sufficiently to aid us in thinking about these problems; we are able, for example, to employ models like those of Forrester and the Club of Rome.

8. Social Decision-Making Teams. There is a special need for the cooperation of teams of social scientists and social philosophers with political leaders and other specialists in the decision-making procedures that affect the physical as well as the social environment. For example, the physical environment and potential changes in the human ecosystem must be thoroughly examined by social experts before society makes decisions such as whether to use nuclear or coal-powered energy or whether even to produce additional energy. After the social experts have made thorough studies, they must propose their solutions to the public through Congress and Parliaments, using the communications media The burden of these ethical decisions will then fall partially on the team of social experts which will be morally obligated to choose the interests of society over the interests of individual egoism.

One of the major problems of scientific thought and research is its dichotimization into smaller and smaller specialties, which means that scientists know less and less about the subject matter of other fields. As a social consequence, decisions are made concerning problems of small scope, thus introducing the ramification of problems within the socioeconomic system. For example, monetary economists make decisions concerning the rate of inflation without being able to foresee the effects on mental illness or crime. Kapp (1977) argues that "it is necessary that social and natural scientists concerned with environmental problems and policies engage in transdisciplinary research." Teams of scientists must come together so that they can pool their expertise, but just as important, more members of society must become more broadly educated so that we have people who can bring together the advances of the numerous scientific disciplines and then make less detrimental decisions.

Education should include the thesis that human society is not separate from its environment.

9. Democratic Social Controls. It has also become necessary to form national political programs and movements to put into effect these suggestions, and the many others neglected in this paper, keeping in mind one of the warnings of Forrester and the Club of Rome, that half measures will not work. In order to prevent global ecological catastrophe, we must engage in worldwide democratic decision making in order to gain political control over our dangerous and endangered national economies, especially those of the more industrialized countries. This means that we must also change our mental attitude against controlling and planning the industrialization and growth of our economies; after all, society doesn't permit other forms of death and destruction. At the same time, as much decision making as possible should be decentralized in order to take advantage of the manageability of problems at the local level.

References (Cited by Larkin)

Boulding, K. 1970. *Economics as a Science.* New York, McGraw-Hill.

______. 1971. The Economics of Coming Spaceship Earth. In W.A. Johnson and J. Hardesty, eds., *Economic Growth Vs. the Environment.* Belmont, California, Wadsworth Publishing Company.

Bureau of the Census. 1977. *Statistical Abstract of the United States.*

Commoner, B. 1976. *The Poverty of Power.* New York, Alfred A. Knopf.

Daly, H.E. 1971. Toward a New Economics—Questioning Growth. In W.A. Johnson and J. Hardesty, eds., *Economic Growth Vs. the Environment.* Belmont, California, Wadsworth Publishing Company.

Drummond, H. 1977. Dr. Drummond on the Big Casino. *Mother Jones,* II:10 (December, 1977).

Epstein, S.S. 1978. *The Politics of Cancer.* San Francisco, Sierra Club Books.

Forrester, J.W. 1971. Counterintuitive Behavior of Social Systems. *Theory and Decision,* II:2, pages 109-114.

George, S. 1977. *How the Other Half Dies; the Real Reasons for World Hunger.* Montclair, New Jersey, Allanheld, Osmun, & Co.

Hardesty, J. et al. 1971. The Political Economy of Environmental Destruction. In W.A. Johnson and J. Hardesty, eds. *Economic Growth Vs. the Environment.* Belmont, California, Wadsworth Publishing Company.

Kapp, K.W. 1970. Environmental Disruption and Social Costs: A Challenge to Economics. *Kyklos,* XXIII:4, pages 833-847.

______. 1977. Environment and Technology: New Frontiers for the Social and Natural Sciences. *Journal of Economic Issues,* XI:3, pages 527-540.

Lappe, F.M. and Collins, J. 1977. *Food First; Beyond the Myth of Scarcity.* Boston, Houghton Mifflin Co.

MacRobert, A. 1977. Look What They've Done to the Rain. *Mother Jones,* II:10 (December, 1977).

McHarg, I.L. 1971. Values, Process, and Form. In W.A. Johnson and J. Hardesty, eds. *Economic Growth Vs. the Environment.* Belmont, California, Wadsworth Publishing Company.

Mander, J. 1978. *Four Arguments for the Elimination of Television.* New York, William Morrow and Company.

Meadows, D.H., et al. 1972. *The Limits to Growth.* New York, Universe Books.

Mishan, E.J. 1967. *The Costs of Economic Growth.* New York, F.A. Praeger.

______. 1978. Whatever Happened to Progress? *Journal of Economic Issues,* XII:2, pages 405-425.

Mother Jones. July 1977. Set Ahead for Rising Sun.

______. September/October 1977. How Sweet It Is.

Murdoch, W., and Connell, J. 1971. All About Ecology. In W.A. Johnson and J. Hardesty, eds. *Economic Growth Vs. the Environment.* Belmont, California, Wadsworth Publishing Company.

Roszak, T. 1978. *Person/Planet.* Garden City, New York, Doubleday.

Schumacher, E.F. 1973. *Small is Beautiful,* New York, Harper & Row.
Updike, J. 1963. *The Centaur,* New York, Alfred A. Knopf.
Wilden, A. 1978. Ecosystems and Economic Systems. In M. Maruyama and A.M. Harkins, eds. *Cultures of the Future.* The Hague, Moulton Publishers, pages 101-124.

Further Reading

Ayres, R.V., and Kneese, A.V. 1975. Economic and Ecological Effects of a Stationary Economy. *Annual Review of Ecology and Systematics* 2: 1-22.
Bain, J.S. 1973. *Environmental Decay: Economic Causes and Remedies.* Boston: Little, Brown.
Barrett, S. 1988. "Economic Guidelines for the Conservation of Biological Diversity." Paper Presented at Workshop on Economics. Costa Rica: IUCN General Assembly.Boulding, Boulding, K.E. 1971. Environment and Economics. In *Environment: Resources, Pollution, and Society,* ed. W.W. Murdoch, pp. 359-367. Stamford, Connecticut: Sinauer.
Brown, G.M., and Goldstein, J.H. 1984. A Model for Valuing Endangered Species. *J. Environmental Ecoonomics and Management* 11: 303-309,
Brunner, D., Miller, W., Stockholm, N., eds. 1981. *Corporations and the Environment.* Stanford: Stanford University.
Buchanan, A. 1985. *Ethics, Efficiency, and the Market.* Totowa, New Jersey: Rowman and Allanheld.
Chandler, W. 1986. *The Changing Role of the Market in National Economies.* Washington, D.C.: Worldwatch Institute.
Commoner, B. 1974. The Economic Meaning of Ecology. In *The Closing Circle,* pp. 249-291. New York: Bantam.
Crocker, T.D. 1971. *Environmental Economics.* Hinsdale, Illinois: Dryden Press.
Duda, M. 1987. *Floridians and Wildlife.* Technical Report No. 2, Tallahassee, Florida: Game and Freshwater Fish Commission, Nongame Wildlife Program.
Eden, E. 1990. Ecology Guides Young Environmentalists. *Focus* 12, No. 2 (Spring): 2.
Edwards, S. 1987. In Defense of Environmental Economics. *Environmental Ethics* 9, No. 2 (Spring): 73-86.
Enthoven, A.C., and Freeman, A.M. 1973. *Problems of the Modern Economy: Pollution, Resources, and the Environment.* New York: W.W. Norton.
Fox, I.K. 1973. Environment, Economics, and Institutions. In *Interpretating Environmental Issues,* ed. C. Schoenfeld, pp. 61-66. Madison, Wisconsin: Dembar Educational Research Services.
Freeman, C., Haveman, R., and Kneese, A. 1973. *The Economics of Environmental Policy.* New York: Wiley.
Galbraith, J.K. 1967. *The New Industrial State.* Boston: Houghton Mifflin.
Garvey, G. 1972. *Energy, Ecology, Economy.* New York: W.W. Norton.
Goddard, H.C. 1972-73. Environmental Policy as Economic Policy. *Environmental Affairs* 2: 630-637.
Healy, R.G. 1979. Land Use and the States. In *Land in America,* ed. R. Andrews, pp. 7-23. Lexington, Massachusetts: Lexington Books.
Henderson, H. 1981. *The Politics of the Solar Age.* Garden City, New York: Doubleday.
Herfindahl, O.C., and Kneese, A.V. 1965. *Quality of the Environment: An Economic Approach to Some Problems in Using Land, Water, and Air.* Baltimore: Johns Hopkins Press.
Hunt, E.K., and d'Arge, R.C. 1971-72a. Economic Orthodoxy and Externalities Revisited. *Environmental Affairs* 1: 845-853.
______., 1971-72b. Environmental Pollution, Externalities, and Conventional Economic Wisdom. *Environmental Affairs* 1: 266-286.
Johansson, P-O. 1987. *The Economic Theory and Measurement of Environmental Benefits.* London: Cambridge University Press.
Johnson, W.A., and Hardesty, J., eds., 1971. *Economic Growth Vs. the Environment.* Belmont, California: Wadsworth.
Kapp, K.W. 1970. Environmental Disruption and Social Costs: A Challenge to Economics. *Kyklos,* 23: 837.
Kelman, S. 1981. *What Price Incentives?* Boston: Auburn House.

Kneese, A.V., Ayres, R.U., and d'Arge, R.C. 1973. Economics and the Environment. In *Problems of the Modern Economy: Pollution, Resources, and the Environment,* ed. A.C. Enthoven and A.M. Freeman, pp. 25-36. New York: W.W. Norton.

Krutilla, J.V., and Fisher, A.C. 1975. *The Economics of Natural Environments: Studies in the Valuation of Commodity and Amenities Resources.* Baltimore, Maryland: Resources for the Future. Johns Hopkins University Press.

Lecomber, R. 1975. *Economic Growth Versus the Environment.* New York: Wiley.

Maler, K.G. 1974. *Environmental Economics,* Baltimore: Johns Hopkins Press.

McCaull, J. 1976. Energy and Jobs. *Environment* 18 (January/February): 18-20.

McNeeley, J.A., et al. 1990. *Conserving the World's Biodiversity.* Washington, D.C.: IUCN.

Meadows, D., et al. 1974. Pollution and the Limits to Economic Growth. In *The Limits to Growth,* pp. 78-94; 137-151. New York: New American Library.

Miller, G.T. 1975. *Living in the Environment,* pp. 343 ff. Belmont, California: Wadsworth.

Mishan, E.J. 1969. *Technology and Growth: The Price We Pay.* New York: Praeger.

Murray, B.G. What the Ecologists Can Teach the Economists. In *Earthkeeping,* ed. C. Juzek and S. Mehrtens, pp. 17-25. Pacific Grove, California: Boxwood Press.

North, D.C. 1977. Political Economy and Environmental Policies. *Environmental Law* 7 (Spring): 449-462.

North, D.C. 1977. Political Economy and Environmental Policies. *Environmental Law* 7 (Spring): 449-462.

Pearse, W.D. 1976. *Environmental Economics.* London: Longmans.

Rauber, P. 1990. The Stockbroker's Smile. *Sierra* 75, No. 4 (July/August): 18-21.

Rauber, P. 1991. Ugly Guys with Baseball Bats. *Sierra* 76, No. 3 (May/June): 73.

Ray, M.E. 1974. *The Environmental Crisis and Corporate Debt Policy.* Lexington, Massachusetts: Lexington Books.

Sagoff, M. 1985. *Risk-Benefit Analysis.* Dubuque: Kendall/Hunt.

Sagoff, M. 1988. Some Problems with Environmental Economics. *Environmental Economics* 10, No. 1 (Spring): 55-74.

Sanchez-Cardona, V., et al. 1975. The Struggle for Puerto Rico. *Environment* 17 (June): 34-40.

Sauer, C.O. 1971. Theme of Plant and Animal Destruction in Economic History. In *Environ/mental,* ed. P. Shepard and D. McKinley, pp. 52-60. New York: Houghton Mifflin.

Schumacher, E.F. 1973. The Economics of Permanence. In *Small Is Beautiful: Economics As If People Mattered,* pp. 21-31; 35-37. New York: Harper and Row.

Schurr, S.H., ed. 1972. *Energy, Economic Growth, and the Environment.* Baltimore: Johns Hopkins Press.

Seneca, J.J. 1974. *Environmental Economics.* Englewood Cliffs, New Jersey: Prentice-Hall.

Shrader-Frechette, K.S. 1985a. *Science Policy, Ethics, and Economic Methodology.* Boston, Massachusetts: Reidel.

Shrader-Frechette, K.S. 1985b. *Risk Analysis and Scientific Method.* Boston, Massachusetts: Reidel.

Shrader-Frechette, K.S. 1988. Agriculture, Ethics, and Restrictions on Property Rights. *Journal of Agricultural Ethics* 1, No. 1: 21 ff.

Shrader-Frechette, K.S. 1991. *Risk Analysis and Rationality. Berkeley: University of California Press.*

Sinclair, T.C. 1973. Environmentalism and Pollution. In *Thinking about the Future: A Critique of "The Limits of Growth,"* ed. C. Freeman, M. Jahoda, and K.L. Pavitt, pp. 175-184, 188-191. London: Sussex University Press.

Stillman, P. 1984. Morality, Economics, and Environmental Policy. *Environmental Ethics* 6, No. 4 (Spring): 95-96.

Sullivan, J.J., and Arias, H.F. 1972-73. Environmental Economics. *Environmental Affairs.* 2: 597-613.

Theobald, R. 1970. *The Economics of Abundance.* New York: Pitman. Tybout, R.A. 1975. Two Market Biases. In *Environmental Quality and Society,* ed. R.A. Tybout, pp. 239-259. Columbus, Ohio: Ohio State University Press.

US Congress. 1976. *Technology Assessment Activities in the Industrial, Academic, and Governmental Communities,* Hearings Before the Technology Assessment Board of the Office of Technology Assessment, 94th Congress, Second Session, June 8-10, 14, 1976. Washington, D.C.,: US Government Printing Office.

Weisskopf, W. 1971. *Alienation and Economics.* New York: E.P. Dutton.

Westman, W.E. 1977. How Much Are Nature's Services Worth? *Science* 197: 960-964.

Wicke, L. 1990. Environmental Damage Balance Sheets. In *Maintaining a Satisfactory Environment,* ed. N. Akerman, pp. 34-53. San Francisco: Westview Press.

9

THE ENVIRONMENT VERSUS THE POOR

Introduction

TROPICAL rain forests cover approximately only 7 percent of the earth's land surface, yet they contain more than half of all the species in the entire world biota. Harvard zoologist E. O. Wilson claims that the rainforests are being destroyed so rapidly that they will be mostly gone within the next century.

Environmentalists and conservationists in both developed and developing nations are trying to stop the rainforest destruction. Native groups like the Kayapo Indians of the Brazilian rainforest have stopped World Bank development in the 1980s and 1990s. Foreign and multinational organizations like the World Wildlife Fund (WWF) also have protected the rainforest by trading foreign debts for nature preservation. In August 1989, for example, the WWF spent $950,000 to buy $ 2.1 million of Malagasy (Africa) debt; they paid only 45 cents on the dollar. Banks holding such debts sometimes sell them at a discount to environmental groups and then donate the proceeds back to the organization, enabling the bank to take a tax deduction both on the loss of the debt and on the donation.

Protecting rainforests is neither simple nor controversial, however, because tropical nations often see rainforest development as the basis for economic prosperity. They also accuse environmentalists and developed nations of opposing this development because they wish to keep poor nations poor. Hence, rainforest development vs. preservation has been couched as an issue oif helping poor nations versus helping the environment. In March 1989, Brazil and 7 other South American nations that form the "Amazon Pact" denounced "foreign meddling" on the issue of preserving the rain forest. They accused environmentalists of trying to keep Brazil from becoming a world power.

Conflicts between development of poor regions versus their environmental protection do not occur merely in tropical rainforests. They also take place in highly developed nations. Throughout the 1980s and the early 1990s loggers and environmentalists have been pitted against each other in the Pacific Northwest of the USA over whether to try to preserve the spotted owl. This endangered species needs old-growth forest to survive, yet the poor people of the region need the jobs in the logging industry.

Conflicts between development of poor regions versus their environmental protection do not occur merely in tropical rainforests. They also take place in highly developed nations. Throughout the 1980s and the early 1990s loggers and environmentalists have been pitted against each other in the Pacific Northwest of the USA over whether to try to preserve the spotted owl. This endangered species needs old-growth forest to survive, yet the poor people of the region need the jobs in the logging industry.

Conflicts like those over rainforest development or logging of the old-growth spotted-owl habitat raise a number of important issues in environmental ethics. Are those who actively support adoption of new environmental ethics wrong in diverting public interest from more important concerns such as poverty and social justice? Or would adoption of policies protective of our surroundings benefit rich and poor alike? Blacks and blue-collar workers have argued that environmental protection is often not in their interests because opposition to heavily polluting industries, such as chemical companies and power plants, destroys opportunities for jobs.

Whether they occur in developed or developing nations, the alleged conflicts between environmental interests and the welfare of the poor raise two main questions for environmental ethics. (1) How should the resources of the earth be distributed, according to egalitarian, utilitarian, libertarian, or some other principles? (2) When environmental welfare appears to conflict with principles of social justice and human rights, how does one decide which priority ought to come first?

Equalitarians like John Rawls of Harvard University would argue that the resources of the earth ought to be distributed so as to benefit the persons who are the least well off. Hence, an environmentalist following Rawls' egalitarian position would likely argue that developed countries must compensate developing nations if they wish them to forego development of the rainforest, for example. Rather than relying on egalitarian principles, utilitarians like Smart would ask whether rainforest destruction or preservation serves the greater long-term good. Libertarians like Robert Nozick of Harvard University, however, would claim that the issue of environmental versus social-justice interests ought to be decided purely on the grounds of who is entitled to property rights. If developing nations have property rights to the rainforest, for example, and if they have acquired the rights justly, then there are no egalitarian or utilitarian grounds for taking the rainforest from them or for interfering with their use of it.

Deciding the issue of whether human or environmental priorities ought to come first, especially in cases involving poverty and deprivation, is in part a question of whether one ought to followan anthropocentric or a biocentric environmental ethics. We raised this question earlier in Chapter 1. In order to determine whether specific types of environmental protection lead to the greatest good and are ethical, especially as regards distribution of benefits to rich and poor, one also needs to question the extent to which failure to prevent pollution results in costs which are borne equitably by all classes. On the one hand, many Blacks argue that neighborhood pollution is of little concern to them if they are struggling to exist, and that only members of elite classes can afford to worry about the quality of the environment. On the other hand, militants. Leftists also charge that this same environment-motivated exploitation continues beyond the borders of urban ghettos in the USA. Pressure for domestic conservation and for stiff regulation of pollution has encouraged American industrialists to exploit both the world's natural resources and the laborers of the world. (See the introduction to Chapter 11.)

In the first selection, Sam Love, editor of *Earth Tool Kit,* coeditor of *Ecotage,* and former editor of *Environmental Action,* takes issue with the view that environmental concern competes with interest in the poor. As he explains, both minority and ecological problems have a common cause, profit.

Nathan Hare, editor of *The Black Scholar* and author of *The Black Anglo-Saxons,* is the author of the second essay. Unlike Love, Hare believes that the interests of the Black community and those of the environmental movement "stand in contradiction to each other." Also unlike Love, Hare believes that the causes of ecological problems are "fundamentally different in the suburbs and ghetto." Hare believes that environmentalist morality is wrongly directed toward reform of the status quo, while "Black environmental correction" is rightly aimed at social and political revolution.

What both readings in this chapter suggest is that everyone ought to bear equitably both the costs and the benefits of pollution control. A factual difficulty is whether Blacks receive greater costs and fewer benefits than other members of society. An important ethical issue is how to rank environmentalism within a system of social priorities, and whether ecological concerns ought to be addressed before other basic human needs and rights have been satisfied. Whether one sympathizes with Blacks, such as Hare, ecologists such as Love, or whether one sees no conflict between them, the readings in this chapter invite examination of a number of important moral issues. One such question is whether a social system may be said to be ethical when it forces persons to fight for jobs that contribute to a greater health risk for themselves and their children. Problems such as this make it clear that environmental ethics must be understood in the context of a larger framework of social and political philosophy.

Readings

1. Love, S. 1972. Ecology and Social Justice: Is there a conflict? *Environmental Action* 4 (August 5): 3-6. Copyright 1972 by *Environmental Action,* Reprinted by permission of *Environmental Action,* 1346 Connecticut Avenue, N.W., Washington, D.C. 20036. A one-year subscription is $15.
2. Hare, N. 1970. Black Ecology. *The Black Scholar* 2 (April): 2-8. Reprinted by permission of *The Black Scholar* and Nathan Hare.

ECOLOGY AND SOCIAL JUSTICE

Sam Love

DURING its Denver hearing on the 1972 Democratic Party platform, the platform committee heard Colorado state Senator George Brown, one of the Senate's black members, unleash an all-too-familiar attack on the environmental movement. After listening to testimony from environmental advocates about what planks should be in the platform, Brown blasted the environmentalists as diverting attention from the real issues. He then proceeded to tick off a list of black demands which ran the gamut from community control of police to guaranteed annual income.

The distrust of the environmental movement which surfaced in Denver runs deep in the minority communities. The basis for the mistrust stems from a long history of affluent whites spending millions to preserve natural areas while children suffer the pangs of hunger. Put simply, minorities are concerned by what they see as a series of distorted priorities being pursued by environmentalists. Furthermore, they are frightened by the genocidal tinge to some of the more common population control arguments. This legacy of our movement cannot be ignored; yet, as an understanding evolves that many minority and ecological problems have common causes, the gap between the social justice and ecology movements is closing.

West Virginia Congressman Ken Hechler articulated the interrelationship in his pronouncements on strip mining during his 1972 reelection campaign. "I told them I was against the exploitation of land because it was also the exploitation of people," Hechler said. His analysis is really not new. In *Capital,* Karl Marx said that labor and land (by which he also meant resources) are the sources of wealth and that the capitalist would abuse both in the pursuit of profit.

Many conditions have changed since the 1860s from which Marx drew his analysis, but exploitation of resources and humans still occurs because our social system places a higher priority on profit than on human or natural welfare. The blatancy of the human exploitation which occurred in the factories and fields has been curbed by an organized labor force, but more sophisticated forms of exploitation abound. Because today's workers wear white collars and labor in air-conditioned offices does not mean that exploitation disappeared with the sweatshop. It may be harder to spot, but it still exists and the struggle to create a just society is still underway by civil rights, minority, labor, and women's groups.

The government and industry's handling of the supersonic transport contract vividly illustrates the nature of the modern profiteering. The SST contract which Boeing and the subcontractors secured from the government provided for full repayment of the research and development investment of the private corporations in the event that the government decided to terminate the project. After a fight which pitted environmentalists and budget cutters against the aerospace industry and their unions, Congress did vote to stop the SST and the contractors recovered from Uncle Sam their $85-million investment in the project. Because the SST contractors had been writing off each year for tax purposes their share of the research and development costs, the refund was merely added to the corporations' normal profits. Thus, in the Boeing situation alone, the company's executives laid off almost 7,000 workers in the same fiscal quarter that they beefed up their reserve fund by declaring a 91-cents-per-share extra credit on their stock.

Exploitation doesn't end inside the corporation. The corporation's resource acquisition activities (such as mining, drilling for oil or cutting down forests), production (factory air or water pollution), and market manipulation (conspiring to keep prices high or advertising to stimulate sales) have an impact on the ecology and the society—not to mention the problems stemming from consumer use of a company's products. And this impact is not always equitably distributed.

Until now the ecology movement has operated on the assumption that everyone breathes the same air and drinks the same water; therefore, it is in everyone's interest to take care of the air and water. The harsh reality that our movement must accept is that the rich have a cleaner environment than the poor and the middle class. Let's take a look at some of the day-to-day differences in the rich man and the poor man's lives.

The average man does not breathe the same air that the rich man breathes. The Council on Environmental Quality's 1971 report to the Congress breaks down air pollution exposure by income. Utilizing data from industrialized areas, such as Chicago, and non-industrial areas, such as Washington, D.C., the report reveals that the lowest income neighborhoods are in the areas of highest air pollution concentrations. The report's exposure charts document the fact that lower-income people are exposed to higher amounts of air pollutants. These results can be explained by the fact that residential real estate values are lower near polluting factories and in the central city; therefore, people with lower incomes, forced into houses and apartments with cheaper rents, are exposed to more pollutants. So if air pollution continues to get worse, the rich will produce the gas masks but they will not be the first to have to buy them.

After air, the Council considered water. The report does not contain much information on who gets the dirtiest water, but it does point out that residents in older buildings can get large doses of lead from the cementing compounds which seal their water pipes. This type of compound is no longer used in construction; consequently, those who live in newer houses will get lower doses of lead in their water.

Residents in lower income areas are also more likely to be uprooted by freeway construction. The Council's report says that federally aided highway projects have displaced an average of 60,000 people from their homes and businesses each year. Of those "dislocated," three-fourths were urban dwellers, most of whom lived in middle- or low-income housing units that cost less than $15,000 or rented for less than $110 per month. Those who are displaced may be lucky in comparison to the lot of those who are left to live next to a freeway and continuously face high levels of noise and air pollution.

Even though bombardment by noise and air pollution may make the average man's home approach the intolerable, it is a monument to pristine purity in comparison to his work environment. A 1968 U.S. Public Health Service study among one million workers in the Chicago area found that 46 percent were exposed to "serious and urgent" health hazards. This situation exists because the government has consistently allowed workers to be exposed to higher levels of dangerous pollutants than citizens in the communities surrounding the industrial facilities. For example, workers are allowed to breathe 50 parts per million (ppm) of carbon monoxide (it was 100 ppm until recently) though the Environmental Protection Agency's general carbon monoxide standard for a 24-hour period is 8 ppm (it is allowed to rise to 35 ppm for a single day in any year). Workers, then, are permitted to take in five times more carbon monoxide every day on the job than the citizen walking down the streets.

The impact of such lax policies on the production worker's health is undeterminable because the corporations and the government have not conducted adequate tests or maintained proper health records. In one of the few studies which have been conducted, Dr. Irving J. Selikoff of the Mount Sinai School of Medicine found that among 230 textile factory workers exposed to asbestos between 1941 and 1946 the death rate from lung cancer is five times normal.

Equally startling statistics are coming to light with coal miners who have contracted black lung from breathing coal dust, textile workers who have developed brown lung from inhaling cotton fibers, and uranium miners who have gotten cancer from exposure to radiation. The recent emergence of this health and mortality data helped persuade Congress to pass the nation's first Occupational Health and Safety Law in 1970. Although the law marks a breakthrough in the concept of worker protection, most experts consider it inadequate to do the job, and labor leaders are fond of pointing out that we have more fish and game wardens than occupational health specialists who can identify workplace environmental problems.

With a socioeconomic system which blesses the rich with profits and the poor with pollution, we should not be surprised at minority and poor communities' skepticism of our Earth Day-type admonitions to all link arms and fight pollution. Our stress should be on social justice, because the fact that the poor live in a worse environment means that they have more to gain from a clean-up.

We must be concerned with much more than just stopping pollution. We must take a hard look at clean-up costs and at who will pay them. Corporations can only meet the costs of pollution by raising prices, getting a government subsidy, or taking a cut in profits. Corporate history is not rich with examples of corporations cutting profits for altruistic reasons, so it may be safe to assume that without public pressure the money for pollution control will either come from raising prices or securing a government subsidy.

Generally, government policies have placed responsibility for preserving and cleaning up the environment on the corporations which degrade it; therefore, the government is sanctioning a general policy of product pricing which will reflect the clean-up costs. This allows the corporations to pass pollution control costs on to consumers by raising the price. The Council on Environmental Quality's 1971 annual report states that higher prices which result from pollution control will force the costs to be "borne disproportionately by those with lower incomes because they spend a larger percentage of their incomes on such products." "Borne disproportionately" is an economic euphemism for giving someone the shaft.

The CEQ's economists estimate that meeting the standards of the Clean Air Act and making reasonable progress against water pollution will cost private industry over $35 billion between 1969 and 1975. "Most of that will be passed on to consumers in the form of higher prices," the report states. The costs of putting air pollution control devices on each car will be $350 according to a study by the government's Office of Science and Technology. Auto industry estimates put the cost per vehicle at over $500. Thus, if we allow the auto industry to tack the cost boost on each car, the average man's $3000 sedan will get a disproportionately higher cost increase than the same increase on the $5000 luxury car.

Raising prices can also serve the function of discouraging customers in most cases. In Denver, development pressures are so great that a $500 fee is required to hook up to the city's water and sewage system. A real need exists to limit consumption of many products and steeper prices can be an effective curb on sales. Unfortunately, the maldistribution of wealth which exists in America and the world creates a situation in which higher costs are used to exclude the have-nots. And for this reason our movement must not blanketly embrace present policies which finance pollution controls by upping the price. A reduction in profits offers a more just answer to the question of who will pay the clean-up costs.

Our efforts to shift the burden of clean-up costs from the consumer to the stockholder must be matched by a movement to redistribute wealth. A society with a reasonably fair distribution of wealth could use cost increases as a fair limiter of consumption. Redistribution also offers an added benefit to us in our fight to reduce consumption levels and create an equilibrium state. It can help reduce the lure of wealth which tempts all of us to want more from a planet that can increasingly produce less.

In 1848 John Stuart Mill described a "stationary state" economy in which people would not be addicted to wanting more. He wrote: "Most fitting, indeed, is it, that while riches are power, and to grow as rich as possible the universal object of ambition, the path to its attainment should be open to all, without favour or partiality. But the best state for human nature is that in which, while no one is poor, no one desires to be richer, nor has any reason to fear being thrust back, by the efforts of others to push themselves forward."

Mill believed that the creation of such a state required redistribution of wealth and a "restraint of population." To redistribute wealth he suggested that people be allowed to keep what was earned "with the final fruits" of their industry, but that there should be a "limitation of the sum which any one person may acquire by gift or inheritance, to the amount sufficient to constitute a moderate independence."

Mill's work is reverberating through history. With the publication of works such as *Limits to Growth*, a renewed interest is evolving in the stationary state and equilibrium economics. Critics, such as Vice President Spiro Agnew, may denounce it as "anti-progress," but Mill anticipated Agnew by over a hundred years. In *Principles of Political Economy* he wrote:

> It is scarcely necessary to remark that a stationary condition of capital and population implies no stationary state of human improvement. There would be as much scope as ever for all kinds of mental culture, and moral and social progress; as much room for improving the Art of Living, and much more likelihood of its being improved, when minds cease to be engrossed by the art of getting on. Even the industrial arts might be as earnestly and as successfully cultivated, with this sole difference, that instead of serving no purpose but the increase of wealth, industrial improvements would produce their legitimate effect, that of abridging labour.

The achievement of a stationary state or even a move in that direction threatens today's economic status quo. Indian Prime Minister Indira Ghandi touched upon the radical critique that our movement offers when she addressed the U.N. Environment Conference. "All of the 'isms' of the modern age—even those which in theory disown the private profit principle—assume that man's cardinal interest is acquisition. The profit motive, individual or collective, seems to overshadow all else. This overriding concern with Self and Today is the basic cause of the ecological crisis."

As we challenge the basic tenet of capitalism that happiness and success can be found through the acquisition of more material goods we will promote the ultimate act of subversion—non-consumption. And it is important for us to remember that the abuses associated with consumption occur at the top rungs of the economic ladder, not the bottom.

The convergence of the ecology and social justice movements places within our grasp the creation of a society that could only have been dreamed about in the past. This society will exist in harmony with nature and be at peace with itself.

BLACK ECOLOGY

Nathan Hare

THE EMERGENCE of the concept of ecology in American life is potentially of momentous relevance to the ultimate liberation of black people. Yet blacks and their environmental interests have been so blatantly omitted that blacks and the ecology movement currently stand in contradiction to each other.[1]

The legitimacy of the concept of black ecology accrues from the fact that: (1) the black and white environments not only differ in degree but in nature as well; (2) the causes and solutions to ecological problems are fundamentally different in the suburbs and ghetto (both of which human ecologists regard as "natural [or ecological] areas"; and (3) the solutions set forth for the "ecological crisis" are reformist and evasive of the social and political revolution which black environmental correction demands.

In the realm of white ecology, pollution "closes your beaches and prevents your youngsters from wading, swimming, boating, water-skiing, fishing, and other recreation close to home."[2] And, "we want clear water, for boating, and swimming, and fishing—and clean water just to look at."[3]

Similar involvement includes the planting of redwood trees, saving the American eagle, and redeeming terrestial beauty. Thus it is seen that ecologists aimed at the hearts and purse strings of industrialists and hit the eyeballs of the white bourgeoisie.

Ecology accordingly has come to refer for the most part to chemical and physical or esthetic conditions only, while professional ecologists themselves have been known to differ in their definition of ecology.

> ... the concept is borrowed from biology, where it means the study of relations between organisms and environment. In biological usage it includes relations between individual organisms and environment (autecology) and between groups and environment (synecology). In social science it is restricted to human synecology, that is, the study of relations between human groups (or populations) and their respective environments, especially their physical environments.[4]

A recent U.S. Department of Health, Education and Welfare report defines environment as "the aggregate of all the external conditions and influences affecting the life and development of an organism, human behavior, society, etc."[5] It is imperative therefore for us to understand how both the physical and social environments of blacks and whites have increasingly evolved as contrasts.

With the industrialization and urbanization of American society, there arose a relatively more rapid and drastic shift of blacks from Southern farms to Northern factories, particularly during periods when they were needed in war industries.[6] Moreover, urban blacks have been increasingly imprisoned in the physical and social decay in the hearts of major central cities, an imprisonment which most emphatically seems doomed to continue.[7] At the same time whites have fled to the suburbs and the exurbs, separating more and more the black and white worlds.[8] The "ecology crisis" arose when the white bourgeoisie, who have seemed to regard the presence of blacks as a kind of pollution, discovered that a sample of what they and their rulers had done to the ghetto would follow them to the suburb.

But there is a greater degree of all varieties of pollutants in the black ghetto, which also lies extremely exposed to the most final variety of environmental destruction imaginable—the "sneak atom bomb attack peril" this month reported by an authoritative study made by Great Britain's Institute for Strategic Studies.[9]

> Say Russia does drop a 10-megaton bomb on Washington, D.C. or Chicago, for example. Up to five miles from ground zero (the point of the explosion), nine out of ten of all inhabitants would be killed instantly and the rest seriously injured or victimized by radiation. All structures would be demolished. From 5 to 9.7 miles out, half of the inhabitants would be killed, a third of them injured, all others dazed, shocked, and sickened by radiation, and all buildings damaged beyond repair ... . In other words, this would just about take care of the Negro community.[10]

But the ecological ordeal of the black race does not have to wait for a nuclear attack; present conditions are deadly enough. The environmental crisis of whites (in both its physical and social aspects) already pales in comparison to that of blacks.

In addition to a harsher degree of industrial pollutants such as "smoke, soot, dust, fly ash, fumes, gases, stench, and carbon monoxide"[11]—which, as in the black ghetto, "if there is no wind or if breezes are blocked dispersal will not be adequate"[12]—the black ghetto contains a heavier preponderance or ratio, for instance, of rats and cockroaches. These creatures comprise an annoyance and "carry filth on their legs and bodies and may spread disease by polluting food. They destroy food and damage fabrics and bookbindings."[13] Blacks also are more exposed to accidents, the number four killer overall and number one in terms of working years lost by a community.[14]

> ... poverty amid affluence, urban squalor and decay, and alienation of young people pollute the environment as much as garbage and industrial smoke ... . A polluted political system which enables a handful of senile Southerners to dominate, through the seniority system, the law-making body of a supposedly free people is a political system which finds racism, poverty, and poisoned rivers equally congenial in its scheme of things.[15]

Moreover, "the ecological perspective directs attention to various kinds of phenomena. These include, among others: (1) the psychological behavior of persons (singly and in groups of various kinds) ... ."[16] Crime, insanity and other forms of social pathology pollute the central city environment.[17]

> It would be a tragic mistake to consider only the material costs of slums. The great expansion of slums in recent times has become a most serious social problem because the areas demoralize a large segment of the urban population.[18]

At the heart of this predicament, though not that alone, is the crowded conditions under which most black persons must live. Black spatial location and distribution not only expose blacks to more devastating and divergent environmental handicaps; they also affect black social and psychological adjustment in a number of subtle ways.

At certain levels of optimum density, flies in fruit jars have been known to die in droves, and rats in crowded places to attack and eat their young and otherwise behave in strange and aberrant ways. Frantz Fanon and others have patiently charted the way in which oppressed peoples so crowded turn upon themselves when, for whatever reason, they feel too weak to fight their oppressor.[19] Blacks accordingly are relatively more prone to be victims, contrary to popular belief, of all major crimes of violence as well as a number of other forms. Although it is true that blacks also exhibit higher rates of criminal activity, this merely stands in ecological succession to such groups as the Irish and Italians who in other eras inhabited the lower strata of the urban slums.[20] Only a minority of blacks are criminals; more are victims of crime. Due largely to existence in a criminally infested environment, blacks are about four times as likely to fall victim to forcible rape and robbery and about twice as likely to face burglary and aggravated assault.[21]

The social and psychological consequences of overcrowding are tangled and myriad in degree. To begin with, the more persons per unit of space the less important each individual there; also the noisier the place, other things equal, and the greater the probability of interpersonal conflict. Studies show that there is a greater hearing loss with age and that much of it is due to honking horns, loud engines and general traffic noise.[22] The importance of space to contentment also is suggested by the fact that in a survey of reasons for moving to the urban fringe, that of "less congested, more room" was twenty times more frequently given than the fact that the environment was "cleaner."[23]

The extent of black over-crowding may be seen in the fact that if population density were as great for the United States at large as it is for some blocks in Harlem, every person in the nation could live in one-half of New York city.[24] Using the yardstick of 1.5 persons per room, blacks are about four times as likely to be overcrowded as whites and they also are more often impelled to live "doubled up" with another family.[25] This necessity for doubling up imposes physical and psychological stress and affects self-perception and social behavior.[26] A study of working class blacks in Chicago revealed that most of them, owing to a lack of space for beds, slept less than five hours in a given night.[27]

But the residential pollution of blacks rests not alone in overcrowding and the greater prevalence of unsightly and unsanitary debris and commerical units such as factories. The very housing afforded blacks is polluted.[28] This fact is crucial when we consider that the word "ecology" was derived by a German biologist from the word "aikos" meaning "house." A house, like the clothes we wear, is an extension of one's self. It may affect "privacy, childrearing practices, and housekeeping or study habits."[29] Three of every ten dwellings inhabited by black families are dilapidated or without hot water, toilet, or bath.[30] Many more are clearly fire hazards.

The shortage of adequate housing and money for rent produces high rates of black mobility which have far-reaching effects on the black social environment.[31] It means that blacks will disproportionately live among strangers for longer periods of time and, in the case of children, attend school in strange classrooms.

The household and neighborhood environments of blacks are perhaps of greater detriment to black health. The ability to control temperature and humidity at will—climate control—in homes can affect the incidence of respiratory infections. Its impact on comfort and productivity in all seasons is without doubt.[32] Health as a community resource is invaluable.

> ... health is, aside from the personal comfort or pain accruing to a given individual, a natural resource for the black community or any other. Health not only affects demographic composition and change; it also affects the ability of individuals—and therefore the community—to play their social, political, and economic roles. Tied in with this assumption is the fact that the advance in health since the eighteenth century may be attributed mainly to improvements in the physical environment ... . Not only are the rates of mortality higher for blacks by each cause of death; there are some significant variations in degree. Blacks also, of course, are subject to higher rates of illness. Much of the differential causes of mortality revolve around communicable diseases, the narrowing of which has been the major factor associated with the decline in differential mortality rates by race.[33]

Throughout a person's life, both his probability of dying and the type of death he meets may be in large part a product of the kind of community in which he lives.[34] It is no coincidence in this context that the high rates of death for blacks are in the area of communicable diseases and nonmotor-vehicle accidents, mainly in industry. Blacks are more than twice as likely to die from pneumonia and influenza.[35] In the case of syphillis the death rate for blacks is about four times as high. The same is true of tuberculosis and of dysentery; and blacks die more from whooping cough and other communicable diseases.[36] The effect of all of this for even those who do not die is relatively more activity limitations on the job than whites.[37]

The life expectancy of blacks is almost ten years less than that of whites, and black infant and maternal mortality rates are at the level which whites exhibited twenty years ago. Black women are more than four times as likely to die of childbirth, and black children are about three times as likely to succumb to post-natal mortality. This is because (among other factors such as dietary deficiencies) black births are about twelve times as likely to occur in a setting in which there is an "attendant not in a hospital and not specified."[38]

Moreover, poor nutrition during pregnancy and in early childhood can retard the brain's development. Illnesses and inadequate medical care can combine with unsanitary conditions to effect physiological pollution. "The glazed eyes of children, legs that never grew straight, misshapen feet," and skin disorders are visible signs of this form of pollution.[39]

Yet there is alive today a neo-Malthusian fashion which blames "population explosion" for the ecology crisis.[40] Actually, the problem is not so much one of population explosion as population implosion, or "the increasing concentration of peoples on relatively small proportions both of the world's and America's land surface."[41] There is both an inadequate distribution of land and people and, more significantly, of people and resources. The United States accounts for only one-fifteenth of the world's population but controls at least three-fifths of its resources.[42] Within the United States three-fourths of the corporate wealth is controlled by about one percent of the people. Hence, one man's overpopulation is not so much a problem to him as is another man's overeating.

There is apparently something within the conditions of poverty that impels people to produce a larger number of children. Although the black birth rate is higher than that of whites, that is not true among women married to college trained men, where white women bear more children than do black women.[43]

No solution to the ecology crisis can come without a fundamental change in the economics of America particularly with reference to blacks. Although some of the ecological differentials between blacks and whites spring directly from racism and hence defy economic correlations,[44] many aspects of the black environmental condition are associated with basic economics. Blacks are employed in the most undesirable or polluted occupations,[45] lagging far behind their educational attainment.[46] About two-thirds work in unskilled and semi-skilled industries.[47] Aggravating, and associated with, the occupational effects on the black environment is the consistently low family income of blacks which must generally support larger families. Since the turn of the century, the family income of blacks has remained about half that of whites. Six in ten of all black children must grow up in poor families.[48] The figure is even higher for black families with a female head.[49] Unemployment is continually at least twice as high for blacks and has been shown to affect the rate of illegitimacy and marital separation, leaving many black familes fatherless.[50]

In addition to unemployment, the same technology which defaces the general society also displaces a disproportionate number of blacks occupationally, into the throes of underemployment. At the same time, the black mother is more likely to be taken out of the home environment to work.[51] Today the war in Viet nam continues to send many of the most vibrant black males disproportionately to die in a foreign land in battle with fellow peoples of oppression. This means that five or ten years from now, assuming that blacks do not reject monogamy, an already depleted black sex ratio will drop considerably and there will exist even a greater shortage of young black males for black women to marry. The result will be increased marital and family disorganization.

Thus the reformist solutions tendered by the current ecology movement emerge as somewhat ludicrous from the black perspective. For instance, automobiles are generally regarded to be the major source of air pollution.[52] This is compounded in the case of blacks by the relatively smaller space in which they must live and drive amid traffic congestion and junked cars. On top of this, white commuters from the suburbs and the outer limits of the central city drive into the central city for work or recreation and social contacts, polluting the black environment further. In every region of the country there has been a direct parallel between the increase in the number of cars and the growth of the suburban and fringe population.[53] Although automobile manufacturers are the chief profiteers, the contradiction of alien automobile polluters who daily invade and "foul the nest" of black urban residents remains.

Some of these commuters are absentee landlords who prevail as "ghetto litterbugs" by way of corrupt and negligent housing practices. Thus, blacks suffer the predicament wherein the colonizer milks dry the resources and labor of the colonized to develop and improve his own habitat while leaving that of the colonized starkly "underdeveloped."

> The problems of the ghetto are comparable to a colonized country. Middle city businesses and housing are owned and taxed by downtown and nothing is given in return except renewal programs that are determined by the needs of foreign interests and the transportation network that feeds downtown .... The job market is determined by the needs of foreign business geared to producing goods that middle city ghetto dwellers can't afford and often don't want.[54]

The real solution to the environmental crisis is the decolonization of the black race. Blacks in the United States number more than 25,000,000 people, comprising a kidnapped and captive nation surpassed in size by only twenty other nations in the entire world. It is necessary for blacks to achieve self-determination, acquiring a full black government and a multi-billion dollar budget so that blacks can better solve the more serious environmental crises of blacks. To do so blacks must challenge and confront the very foundations of American society. In so doing we shall correct that majority which appears to believe that the solution lies in decorating the earth's landscape and in shooting at the moon.

Notes

1. For example, in a 760-page volume, "being the record of a conference convened by the Conservation Foundation in April, 1965, at Warrenton, Virginia," only one-fifth of one page was devoted to the black race. (See F. Fraser Darling and John P. Milton, ed., *Future Environments of North America.* Garden City: Natural History Press, 1966). Likewise, in a 286-page volume, "a Dial Report on the deteriorating quality of the American environment," the index contains no entry for "Negro" or "Black." The book is entitled *Moment in the Sun,* but blacks received no "moment in the sun." Also, "suburbia" appears but there is no mention of the "slum" or the "ghetto." (See *Moment in the Sun.* New York: Dial Press. Robert Rienow and Leona Train Rienow, 1967.)

2. Izaak Walton League of America, *Clean Water.* Glenview: February, 1970, p. 1.

3. Federal Water Pollution Control Administration, *A New Era for America's Waters.* Washington, D.C.: U.S. Government Printing Office, 1967, p. 4.

4. Julius Gould and William L. Kolb, ed., *A Dictionary of the Social Sciences.* Glencoe: The Free Press, 1964, p. 215.

5. U.S. Department of Health, Education and Welfare, *Environmental Health Planning Guide,* Washingtin, D.C.: 1968, p. 1.

6. Nathan Hare, "Recent Trends in the Occupational Mobility of Negroes in the United States: An Intracohort Analysis," *Social Forces,* December, 1965. U.S. Bureau of the Census, *Statistical Abstract of the United States.* Washington, D.C.: U.S. Government Printing Office, 1969, p. 20, Table 20.

In 1910, when whites were about twice as urban as blacks, only 27 percent of blacks were urban. By 1960 blacks had grown more urban than whites; about four out of five of all blacks now live in urban places. See Bureau of Labor Statistics, *op. cit.,* p. 67 and Karl E. Taeuber and Alma F. Taeuber, *Negroes in Cities.* Chicago: Aldine Publishing Company, 1965.

7. Bureau of Labor Statistics, *The Negroes in the United States: Their Economic and Social Situation.* Washington, D.C.: U.S. Government Printing Office, June 1966, p. 69. *Statistical Abstract, op. cit.,* p. 18, Table 17. Scott Greer, *The Emerging City,* New York: Free Press, 1962, p. 82. Jeanne R. Lowe, *Cities in a Race with Time,* New York: Vintage Books, 1968, p. 283. George Schermer Associates, *More Than Shelter,* Washington, D.C.: U.S. Government Printing Office, 1968, pp. 20, 26, 51-53.

8. Bureau of the Census, *We the Black People of the United States.* Washington, D.C. U.S. Government Printing Office, 1970. Taeuber and Taeuber, *op. cit.*

9. *San Francisco Examiner,* April 11, 1970.

10. Nathan Hare, "Can Negroes Survive a Nuclear War?" *Negro Digest,* May, 1963, p. 29.

11. Kenneth G. Bueche and Morris J. Schur, *Air Pollution Control,* Boulder: Bureau of Governmental Research and Service, University of Colorado, 1963, p. 5.

12. *Air Pollution:* The Facts. Christmas Seal Associates National Tuberculosis Association, 1967, p. 4.

13. Entomology Research Division, *Cockroaches,* Washington, D.C.: U.S. Government Printing Office, Leaflet No. 430, October, 1969, p. 2.

14. Ralph Thomlinson, *Population Dynamics: Causes and Consequences of World Demographic Change.* New York: Random House, 1965, p. 109.

15. The Editors, "Action for Survival," *The Progressive,* April, 1970, pp. 4, 5.

16. Harold and Margaret Sprout, *The Ecological Perspective on Human Affairs.* Princeton: Princeton University Press, 1965, p. 8.

17. Egon Ernest Bergel, *Urban Sociology.* New York: McGraw-Hill Book Company, 1955, p. 420.

18. *Ibid.,* p. 421.

19. Martin Oppenheimer, *The Urban Guerilla.* Chicago: Quadrangle Books, 1969, p. 64.

20. Office of Policy Planning and Research, U.S. Department of Labor, *The Negro Family,* Washington, D.C. U.S. Government Printing Office, March, 1965, p. 25.

21. *Statistical Abstracts, op. cit.,* p. 140, Table 204.

22. Gary C. Smith, "Suggestions for the Schools," in Garrett De Bell, ed., *The Environmental Handbook.* New York: Ballantine Books, 1970.

23. Walter T. Martin, *The Rural-Urban Fringe*. Eugene: University of Oregon Press, 1953, p. 37. Svend Riemer, "Maladjustment to the Family Home," *American Sociological Review*, October, 1945, pp. 642-648.

24. Abram Kardiner and Lionel Ovesey, *The Mark of Oppression*. New York: Meridian Books, 1968, p. 52.

25. Division of Housing Research, *The* 1950 *Housing Situation in Charts*, Washington, D.C.: Housing and Home Finance Agency, June, 1952.

26. Alvin L. Scorr, *Slums and Social Insecurity*, Washington, D.C.: U.S. Government Printing Office, 1966, pp. 17, 18. George Schermer.

27. Allison Davis, "Motivation of the Underprivileged Worker," in William F. Whyte, ed., *Industry and Society*. New York: McGraw Hill, 1946, pp. 84-106.

28. Lowe, *op. cit.*, p. 237.

29. Scharr, *op cit.*, p. 8.

30. *We the Black People*, *op. cit.* See also Bureau of Labor Statistics, *op. cit.*, pp. 39, 40, 209.

31. Scharr, *op. cit.*, p. 86.

32. Thomlinson, *op. cit.*, p. 92.

33. Nathan Hare, "Does Separatism in Medical Care Offer Advantages for the Ghetto?" in John C. Norman, ed. *Medicine in the Ghetto*. Appleton-Century-Crofts, 1969, p. 44.

34. Wilbur Zilensky, *A Prologue to Population Geography*. Englewood Cliffs: Prentice-Hall, 1966, p. 42.

35. Marshall B. Clinard, *Sociology of Deviant Behavior*. New York: Holt, Rinehart and Winston, 1963, p. 42.

36. Bureau of Labor Statistics, *op cit.*, p. 223.

37. *Ibid.*, p. 227.

38. *Ibid.*, p. 222.

39. The President's Commission on Income Maintenance Programs, *Poverty Amid Plenty: The American Paradox*, Washington, D.C.: U.S. Government Printing Office, 1969, pp. 17, 18.

40. Paul R. Ehrlich, *The Population Bomb*. New York: Ballantine Books, 1970.

41. Patricia Leavey Hodge and Philip M. Hauser, *The Challenge of America's Metropolitan Population Outlook*—1960 to 1985. Washington, D.C.: U.S. Government Printing Office, 1968, p. 1.

Two-thirds of the people in the United States now live in 212 metropolitan areas comprising only one-tenth of the total land area; 53 percent are crowded into less than one percent of the land area.

42. Rienow and Rienow, *op. cit.*, p. 3.

43. Bureau of Labor Statistics, *op. cit.*

44. Taeuber and Taeuber, *op. cit.*, Robert H. Connery, ed. *Urban Riots: Violence and Change*. New York: Vintage Books, 1969, p. 112.

45. Scharr, *op. cit.*, pp. 68-69. Bureau of the Census. *Nonwhite Population by Race*, PC(2)-1C, Table 39.

46. Bureau of Labor Statistics, *op. cit.*, p. 31.

47. John Sirjamaki, *The Sociology of Cities*, New York: Random House, 1964, p. 247.

48. Schermer Associates, *op. cit.*, pp. 22, 34 President's Commission, *op. cit.*, pp. 14, 30, *Statistical Abstracts*, *op. cit.*, p. 324, p. 477.

49. Nathaniel Keith, Housing America's Low—and Moderate—Income Families. Washington, D.C.: U.S. Government Printing Office, 1968, pp. 1, 2.

50. Office of Policy Planning and Research, *op. cit.*, p. 22.

51. President's Commission, *op. cit.*, p. 36.

52. Ralph Nader, "The Profits in Pollution," *Progressive*, April, 1970, p. 19. Kenneth P. Cantor, "Warning: The Automobile is Dangerous ...", in Garrett De Bell, *op. cit.*, pp. 201, 202. Gary G. Smith, *op. cit.*, p. 303.

53. Cantor, *op. cit.*, p. 201. William E. Cole, *Urban Society*, Cambridge: Riverside Press, 1958, p. 167.

54. Berkeley People's Architecture, *op. cit.*, p. 240.

Further Reading

Bredemeier, H.C., and Getis, J., eds. 1973. *Environments, People, and Inequalities: Some Current Problems.* New York: Wiley.

Brodine, V. 1971. A Special Burden. *Environment* 13 (March): 22-33.

Brown, F.M. 1970. Black Survival. In *Earth Day,* ed. Environmental Action Foundation, pp. 95-105. New York: Arno.

Brown, L. 1990. The Illusion of Progress. In *State of the World 1990,* ed. L. Brown, C. Flavin, and S. Postel, pp. 3-16. New York: Norton.

Brown, L., and Wolf, E. 1986. Reversing Africa's Decline. In *State of the World 1986,* ed. L. Brown, C. Flavin, and S. Postel, pp. 177-194. New York: Norton.

Brown, L. and Young, J. 1990. Feeding the World in the Nineties. In *State of the World 1990,* ed. L. Brown, C. Flavin, and S. Postel, pp. 59-78. New York: Norton.

Counsell, S. 1987a. Brazil: Rough Road. *Not Man Apart* 17, No. 5 (September-October): 6.

Counsell, S. 1987b. U.S.: Rainforest Legislation: *Not Man Apart* 17, No. 6 (Nov./Dec.): 9.

Craig, P.P., and Berlin, E. 1971. The Air of Poverty. *Daffron, C. 1975. Using NEPA to Exclude the Poor. Environmental Affairs* 4 (Winter): 81-122.

Durning, A. 1990. Ending Poverty. In *State of the World 1990,* ed. L. Brown, C. Flavin, and S. Postel, pp. 135-153. New York: Norton.

Faramelli, N. 1971-72. "Perilous Links Between Economic Growth, Justice and Ecology. *Environmental Affairs* 1: 218-227.

______. 1972-73. Toying with the Environment and the Poor. *Environmental Affairs* 2: 469-505.

Freeman, A.M. 1973. Income Distribution and Environmental Quality. In *Pollution, Resources and the Environment,* ed. A.C. Enthoven and A.M. Freeman, pp. 100-107. New York: W.W. Norton.

Gould, S.J. 1990. The Golden Rule—a Proper Scale for Our Environmental Crisis. *Natural History* 9 (September): 34-90.

Grossman, R., and Daneker, G. 1977. *Jobs and Energy.* Washington, D.C.: Environmentalists for Full Employment.

Hallow, R.Z. 1969. The Blacks Cry Genocide. *The Nation* 208 (April 28): 535-537.

McCaull, J. 1976. Discriminatory Air Pollution. *Environment* 18 (March): 26-31.

Miller, G.T. 1975. Ecology and the Poor. In *Living in the Environment,* pp. E29-E33. Belmont, California: Wadsworth.

Myers, N. June 1987. *Not Far Afield: U.S. Interests and the Global Environment.* Washington, D.C.: World Resources Institute.

Neuhaus, R. 1971. *In Defense of People,* pp. 106-139; 152-181. New York: Macmillan.

Ostheimer, J.M., and Ritt, L.G. 1976. Environment and Black Americans. In *Environment, Energy, and Black Americans,* pp. 5-7; 10-12; 29-35. London: Sage.

Postel, S. 1984. Protecting Forests. In *State of the World 1984,* ed. L. Brown, pp. 74-94. New York: Norton.

Salwasser, H. 1986. Conserving a Regional Spotted Owl Population. In *Ecolological Knowledge and Environmental Problem Solving,* ed. G.H. Orians, et al. pp. 227-247. Washington, D.C.: National Academy Press.

Seelman, K. 1978. The Ethical Implications of Energy Production and Use. *Solar Age* 3: 16-21.

Shrader-Freschette, K.S. 1985a. *Risk Analysis and Scientific Method.* Boston, Mass.: Reidel.

Shrader-Frechette, K.S. 1985b. Environmental Ethics and Global Imperatives. In *The Global Possible: Resources, Development, and the New Century* ed. R. Repetto, pp. 97-130. New Haven, Connecticut: Yale University Press.

Shrader-Frechette, K.S. 1988. Agriculture, Ethics, and Restrictions on Property Rights. *Journal of Agricultural Ethics* 1, No. 1: 21 ff.

Smart, J.J.C., and Williams, B. 1973. Utilitarianism: For and Against. Cambridge: University Press.

Smith, J.N., ed. 1974. *Environmental Quality and Social Justice.* Wash., D.C.: Cons. Found.

Stein, J. 1977. Water for the Wealthy. *Environment* 19 (May): 6-14.

Train, R.E. Environment, Energy, Economic Equity, Can the World Balance These Vital Elements? *Environmental Affairs* VI (1977): 1-10.

Webb, G. 1989a. Global Effort Kills World Bank Loan to Brazil. *Not Man Apart* 19, 1:16.

Webb, G. 1989b. Peoples of the Forest Unite. *Not Man Apart* 19, No. 1 (February-May): 16.

Wiley, G. 1970. Ecology and the Poor. In *Earth Day,* ed. Environmental Action Foundation, pp. 233-236. New York: Arno.

World Wildlife Fund, WWF, 1990. *Madagascar Is a World Apart.* Washington, D.C.: World Wildlife Fund.

SECTION FOUR:

Some Contemporary Dilemmas

10

POPULATION, INDIVIDUAL FREEDOM AND THE COMMON GOOD

Introduction

IN 1968, in his classic work, *The Population Bomb,* Paul Ehrlich predicted that, with the population doubling times in poor nations as low as 19 years, it would be impossible for the countries to double food, energy, teachers, roads, and buildings in so short a time. Reviewing his earlier work in 1990, Ehrlich affirmed the truth of his depressing prediction. Western industrialized nations and China have made some reductions in fertility, but the poor nations are growing faster than ever, without a comparable increase in food production and manufacturing. Indeed, Africa's food production per person, for example, has fallen 11 percent since 1970. During the 1980s, global population increased by 842 million, and during the 1990s, it is projected to grow by 959 million. Yet, during the 1980s, incomes were falling in most of Africa and Latin America. To survive, many Third-World countries went deeply into debt. As a result, in places like Nigeria, Peru, and the Philippines, more than half their export earnings are going to service their loans.

Just as poor countries are caught in an impossible situation of debt, rising population, and falling income, individual families in developing nations face a similarly impossible situation. They need children to help with the work and to provide security to parents in their old age. By having more children, however, they must share less land, water, and food per person. The only escape from the trap is to have fewer children. Otherwise, trying to improve one's lot, without also using birth control, as a *Washington Post* columnist put it, is like trying to pour water uphill.

Population density is perhaps the one factor most clearly related to destruction of the environment. As one author put it, "the stork has passed the plow—the birth rate has exceeded the ability of the land to support life. Surprisingly, the population problem is much more serious in the USA than in underdeveloped countries. According to biologist Wayne Davis, it would take 500 Indians to have the same adverse effect on the environment as one affluent American with his driveways, leaded gas, strip-mined coal, junk cars, and discarded cans and bottles.

The ethical questions raised by the population explosion go to the heart of political philosophy. How does one reconcile individual with collective interests? How ought one to balance personal freedom to choose family size with societal responsibility toward feeding and housing all its citizens? How does one preserve the environment for present and future inhabitants while maximizing basic human rights, including those of reproduction? Can morality be decided solely in terms of inalienable rights or must future consequences have a bearing on what is called 'good' and 'bad'? How does one balance traditional ethical norms with social and environmental well-being? Does the end (an ecosystem capable of supporting a high quality of life) ever justify the means (involuntary control of procreation)?

One set of answers to the questions posed by the problem of overpopulation has been given by the United Nations in its "Declaration on Social Progress." In this document UN members explain that *countries* are responsible for improvement of the environment, equitable distribution of income, and socioeconomic progress, but that *individuals* have the "exclusive right" to determine the number and spacing of their children. Consistent with the UN Declaration on Human Rights, the General Assembly has agreed that planetary well-being can be achieved in an ethical manner only if basic human rights, such as procreation, are not violated.

Garrett Hardin, Professor Emeritus of Human Ecology at the University of California at Santa Barbara attacks the conclusions of the UN Declaration. In the first reading he argues that breeding is an exclusive parental right only if each family is dependent on its own resources and only if children of improvident parents starve to death. Hardin sees a fundamental contradiction in the UN theory that *society* has a responsibility to provide for the well-being of citizens, but that *individuals* have the right to control the number of offspring. He argues that one cannot talk about "exclusive rights" of humans to reproduce, because "the morality of an act is a function of the state of the system at the time it is performed." Hardin's chief ethical thesis is that complete, individual, reproductive liberty destroys societal freedom to enjoy an environment to support it.

In the second selection Daniel Callahan, a philosopher and director of the Institute for Society, Ethics, and the Life Sciences, in New York, also discusses freedom. Callahan, however, comes to a conclusion opposite to that of Hardin. Arguing that rights can never be subservient to social or environmental goals, he maintains that reproductive liberty is a higher priority than economic, social, or demographic utility. He concludes that freedom of choice is the most important ethical consideration relevant to the population problem.

One of the reasons why the population problem appears so insoluble is that it is a "commons" problem. In commons problems, individuals serve their own interests—by having more children or by overfishing or overgrazing, for example—at the expense of everyone else who then faces depleted total resources. A key aspect of the commons problems is that the "gainers," those who use their power to diminish common resources and to gain from their depletion, often are not exactly the same as the losers, those who are powerless to avoid the hurt caused by the resource depletion. (Of course, everyone is a loser when the environmental commons is diminished, but the "gainers" have a smaller overall loss.) This inequity, between the gainers and the losers, raises the ethical issue of the degree to which the environmental losers—the powerless persons hurt by a diminished commons—have the right to react, to force the gainers to change their antisocial behavior.

In addition to the population problem, environmental ethics faces a number of other commons difficulties that raise the same moral issues regarding inequity, coercion, and the disjunction of "gainers" and "losers." Some of these commons problems include acid rain, river water pollution, global warming, ozone depletion, and others.

Acid rain occurs because sulfur and nitrogen oxides emitted during the burning of fossil fuels typically "come down" in one form or another—often causing acid deposition responsible for pervasive and long-term environmental damage, especially destruction of forests. Because of wind and weather problems, all but a few countries export at least half of the sulfur dioxide they emit. At one extreme, Scandinavian countries have few emissions, but 75 to 90 percent of their sulfur deposits are imported. At the other extreme, the United Kingdom is the largest sulfur emitter in Western Europe, but it exports two-thirds of its emissions. Less polluted air would benefit everyone, but most countries do not want to reduce their emissions unless others do the same, because they believe that they will be "losers" rather than net "gainers."

A similar situation occurs with global warming. Combusion of fossil fuels releases greenhouse gases, like CO_2, that are warming the globe and that will cause rising sea levels, floods, droughts, and widespread famines. A major ethical issue is whether and how it might be justifiable to coerce the "gainers" into reducing greenhouse emissions that are harming the net "losers."

In the case of ozone depletion, numerous companies release chlorofluorocarbons, CFCs, that reduce the ozone layer of the atmosphere, thereby allowing more harmful ultraviolet rays to penetrate. This penetration causes several environmentally damaging effects, among them skin cancer and reduction in phytoplankton activity. For example, the US Environmental Protection Agency (EPA) estimates that, as a result of CFCs, an additional 3 to 15 million people born before 2075 will contract skin cancer because of ozone depletion. The same ethical issues of inequity, coercion, and the disjunction of gainers and losers occur in the ozone case as in other commons problems.

Readings

1. Hardin, G. The Tragedy of the Commons. *Science* 162 (December 13): 1243-1248. Copyright 1968 by the American Association for the Advancement of Science. Reprinted by permission of the Association and Garrett Hardin.

2. Callahan, D. 1972. Ethics and Population Limitation. *Science* 175 (February 4): 487-94. Copyright 1972 by the American Association for the Advancement of Science. Reprinted by permission of the Association and Daniel Callahan.

THE TRAGEDY OF THE COMMONS

Garrett Hardin

AT THE END of a thoughtful article on the future of nuclear war, Wiesner and York[1] concluded that: "Both sides in the arms race are ... confronted by the dilemma of steadily increasing military power and steadily decreasing national security. *It is our considered professional judgment that this dilemma has no technical solution.* If the great powers continue to look for solutions in the area of science and technology only, the result will be to worsen the situation."

I would like to focus your attention not on the subject of the article (national security in a nuclear world) but on the kind of conclusion they reached, namely that there is no technical solution to the problem. An implicit and almost universal assumption of discussions published in professional and semipopular scientific journals is that the problem under discussion has a technical solution. A technical solution may be defined as one that requires a change only in the techniques of the natural sciences, demanding little or nothing in the way of change in human values or ideas of morality.

In our day (though not in earlier times) technical solutions are always welcome. Because of previous failures in prophecy, it takes courage to assert that a desired technical solution is not possible. Wiesner and York exhibited this courage; publishing in a science journal, they insisted that the solution to the problem was not to be found in the natural sciences. They cautiously qualified their statement with the phrase, "It is our considered professional judgment. ..." Whether they were right or not is not the concern of the present article. Rather, the concern here is with the important concept of a class of human problems which can be called "no technical solution problems," and, more specifically, with the identification and discussion of one of these.

It is easy to show that the class is not a null class. Recall the game of tick-tack-toe. Consider the problem, "How can I win the game of tick-tack-toe?" It is well known that I cannot, if I assume (in keeping with the conventions of game theory) that my opponent understands the game perfectly. Put another way, there is no "technical solution" to the problem. I can win only by giving a radical meaning to the word "win." I can hit my opponent over the head; or I can drug him; or I can falsify the records. Every way in which I "win" involves, in some sense, an abandonment of the game, as we intuitively understand it. (I can also, of course, openly abandon the game—refuse to play it. This is what most adults do.)

The class of "No technical solution problems" has members. My thesis is that the "population problem," as conventionally conceived, is a member of this class. How it is conventionally conceived needs some comment. It is fair to say that most people who anguish over the population problem are trying to find a way to avoid the evils of overpopulation without relinquishing any of the privileges they now enjoy. They think that farming the seas or developing new strains of wheat will solve the problem—technologically. I try to show here that the solution they seek cannot be found. The population problem cannot be solved in a technical way, any more than can the problem of winning the game of tick-tack-toe.

Population, as Malthus said, naturally tends to grow "geometrically," or, as we would now say, exponentially. In a finite world this means that the per capita share of the world's goods must steadily decrease. Is ours a finite world?

A fair defense can be put forward for the view that the world is infinite; or that we do not know that it is not. But, in terms of the practical problems that we must face in the next few generations with the foreseeable technology, it is clear that we will greatly increase human misery if we do not, during the immediate future, assume that the world available to the terrestrial human population is finite. "Space" is no escape.[2]

A finite world can support only a finite population; therefore, population growth must eventually equal zero. (The case of perpetual wide fluctuations above and below zero is a trivial variant that need not be discussed.) When this condition is met, what will be the situation of mankind? Specifically, can Bentham's goal of "the greatest good for the greatest number" be realized?

No—for two reasons, each sufficient by itself. The first is a theoretical one. It is not mathematically possible to maximize for two (or more) variables at the same time. This was clearly stated by von Neumann and Morgenstern,[3] but the principle is implicit in the theory of partial differential equations, dating back at least to D'Alembert (1717-1783).

The second reason springs directly from biological facts. To live, any organism must have a source of energy (for example, food). This energy is utilized for two purposes: mere maintenance and work. For man, maintenance of life requires about 1,600 kilo-calories a day ("maintenance calories"). Anything that he does over and above merely staying alive will be defined as work, and is supported by "work calories" which he takes in. Work calories are used not only for what we call work in common speech; they are also required for all forms of enjoyment, from swimming and automobile racing to playing music and writing poetry. If our goal is to maximize population it is obvious what we must do: We must make the work calories per person approach as close to zero as possible. No gourmet meals, no vacations, no sports, no music, no literature, no art .... I think that everyone will grant, without argument or proof, that maximizing population does not maximize goods. Bentham's goal is impossible.

In reaching this conclusion I have made the usual assumption that it is acquisition of energy that is the problem. The appearance of atomic energy has led some to question this assumption. However, given an infinite source of energy, population growth still produces an inescapable problem. The problem of the acquisition of energy is replaced by the problem of its dissipation, as J.H. Fremlin has so wittily shown.[4] The arithmetic signs in the analysis are, as it were, reversed; but Bentham's goal is still unobtainable.

The optimum population is, then, less than the maximum. The difficulty of defining the optimum is enormous; so far as I know, no one has seriously tackled this problem. Reaching an acceptable and stable solution will surely require more than one generation of hard analytical work—and much persuasion.

We want the maximum good per person; but what is good? To one person it is wilderness, to another it is ski lodges for thousands. To one it is estuaries to nourish ducks for hunters to shoot; to another it is factory land. Comparing one good with another is, we usually say, impossible because goods are incommensurable. Incommensurables cannot be compared.

Theoretically this may be true; but in real life incommensurables *are* commensurable. Only a criterion of judgment and a system of weighting are needed. In nature the criterion is survival. Is it better for a species to be small and hideable, or large and powerful? Natural selection commensurates the incommensurables. The compromise achieved depends on a natural weighting of the values of the variables.

Man must imitate this process. There is no doubt that in fact he already does, but unconsciously. It is when the hidden decisions are made explicit that the arguments begin. The problem for the years ahead is to work out an acceptable theory of weighting. Synergistic effects, nonlinear variation, and difficulties in discounting the future make the intellectual problem difficult, but not (in principle) insoluble.

Has any cultural group solved this practical problem at the present time, even on an intuitive level? One simple fact proves that none has: there is no prosperous population in the world today that has, and has had for some time, a growth rate of zero. Any people that has intuitively identified its optimum point will soon reach it, after which its growth rate becomes and remains zero.

Of course, a positive growth rate might be taken as evidence that a population is below its optimum. However, by any reasonable standards, the most rapidly growing populations on earth today are (in general) the most miserable. This association (which need not be invariable) casts doubt on the optimistic assumption that the positive growth rate of a population is evidence that it has yet to reach its optimum.

We can make little progress in working toward optimum population size until we explicitly exorcize the spirit of Adam Smith in the field of practical demography. In economic affairs, *The Wealth of Nations* (1776) popularized the "invisible hand," the idea that an individual who "intends only his own gain," is, as it were, "led by an invisible hand to promote ... the public interest."[5] Adam Smith did not assert that this was invariably true, and perhaps neither did any of his followers. But he contributed to a dominant tendency of thought that has ever since interfered with positive action based on rational analysis, namely, the tendency to assume that decisions reached individually will, in fact, be the best decisions for an entire society. If this assumption is correct it justifies the continuance of our present policy of laissez-faire in reproduction. If it is correct we can assume that men will control their individual fecundity so as to produce the optimum population. If the assumption is not correct, we need to reexamine our individual freedoms to see which ones are defensible.

Tragedy of Freedom in a Commons

The rebuttal to the invisible hand in population control is to be found in a scenario first sketched in a little-known pamphlet[6] in 1833 by a mathematical amateur named William Forster Lloyd (1794-1852). We may well call it "the tragedy of the commons," using the word "tragedy" as the philosopher Whitehead used it[7]: "The essence of dramatic tragedy is not unhappiness. It resides in the solemnity of the remorseless working of things." He then goes on to say, "This inevitableness of destiny can only be illustrated in terms of human life by incidents which in fact involve unhappiness. For it is only by them that the futility of escape can be made evident in the drama."

The tragedy of the commons develops in this way. Picture a pasture open to all. It is to be expected that each herdsman will try to keep as many cattle as possible on the commons. Such an arrangement may work reasonably satisfactorily for centuries because tribal wars, poaching, and disease keep the numbers of both man and beast well below the carrying capacity of the land. Finally, however, comes the day of reckoning, that is, the day when the long-desired goal of social stability becomes a reality. At this point, the inherent logic of the commons remorsely generates tragedy.

As a rational being, each herdsman seeks to maximize his gain. Explicitly or implicitly, more or less consciously, he asks, "What is the utility *to me* of adding one more animal to my herd?" This utility has one negative and one positive component.

1. The positive component is a function of the increment of one animal. Since the herdsman receives all the proceeds from the sale of the additional animal, the positive utility is nearly + 1.

2. The negative component is a function of the additional overgrazing created by one more animal. Since, however, the effects of overgrazing are shared by all the herdsmen, the negative utility for any particular decision-making herdsman is only a fraction of -1.

Adding together the component partial utilities, the rational herdsman concludes that the only sensible course for him to pursue is to add another animal to his herd. And another; and another ... . But this is the conclusion reached by each and every rational herdsman sharing a commons. Therein is the tragedy. Each man is locked into a system that compels him to increase his herd without limit—in a world that is limited. Ruin is the destination toward which all men rush, each pursuing his own best interest in a society that believes in the freedom of the commons. Freedom in a commons brings ruin to all.

Some would say that this is a platitude. Would that it were! In a sense, it was learned thousands of years ago, but natural selection favors the forces of psychological denial.[8] The individual benefits as an individual from his ability to deny the truth even though society as a whole, of which he is a part, suffers. Education can counteract the natural tendency to do the wrong thing, but the inexorable succession of generations requires that the basis for this knowledge be constantly refreshed.

A simple incident that occurred a few years ago in Leominster, Massachusetts, shows how perishable the knowledge is. During the Christmas shopping season the parking meters downtown were covered with plastic bags that bore tags reading: "Do not open until after Christmas. Free parking courtesy of the mayor and city council." In other words, facing the prospect of an increased demand for already scarce space, the city fathers reinstituted the system of the commons. (Cynically, we suspect that they gained more votes than they lost by this retrogressive act.)

In an approximate way, the logic of the commons has been understood for a long time, perhaps since the discovery of agriculture or the invention of private property in real estate. But it is understood mostly only in special cases which are not sufficiently generalized. Even at this late date, cattlemen leasing national land on the western ranges demonstrate no more than an ambivalent understanding, in constantly pressuring federal authorities to increase the head count to the point where overgrazing produces erosion and weed-dominance. Likewise, the oceans of the world continue to suffer from the survival of the philosophy of the commons. Maritime nations still respond automatically to the shibboleth of the "freedom of the seas." Professing to believe in the "inexhaustible resources of the oceans," they bring species after species of fish and whales closer to extinction.[9]

The National Parks present another instance of the working out of the tragedy of the commons. At present, they are open to all, without limit. The parks themselves are limited in extent—there is only one Yosemite Valley—whereas population seems to grow without limit. The values that visitors seek in the parks are steadily eroded. Plainly, we must soon cease to treat the parks as commons or they will be of no value to anyone.

What shall we do? We have several options. We might sell them off as private property. We might keep them as public property, but allocate the right to enter them. The allocation might be on the basis of wealth, by the use of an auction system. It might be on the basis of merit, as defined by some agreed-upon standards. It might be by lottery. Or it might be on a first-come, first-served basis, administered to long queues. These, I think, are all the reasonable possibilities. They are all objectionable. But we must choose—or acquiesce in the destruction of the commons that we call our National Parks.

Pollution

In a reverse way, the tragedy of the commons reappears in problems of pollution. Here it is not a question of taking something out of the commons, but of putting something in—sewage, or chemical, radioactive, and heat wastes into water; noxious and dangerous fumes into the air; and distracting and unpleasant advertising signs into the line of sight. The calculations of utility are much the same as before. The rational man finds that his share of the cost of the wastes he discharges into the commons is less than the cost of purifying his wastes before releasing them. Since this is true for everyone, we are locked into a system of "fouling our own nest," so long as we behave only as independent, rational, free-enterprisers.

The tragedy of the commons as a food basket is averted by private property, or something formally like it. But the air and waters surrounding us cannot readily be fenced, and so the tragedy of the commons as a cesspool must be prevented by different means, by coercive laws or taxing devices that make it cheaper for the polluter to treat his pollutants than to discharge them untreated. We have not progressed as far with the solution of this problem as we have with the first. Indeed, our particular concept of private property, which deters us from exhausting the positive resources of the earth, favors pollution. The owner of a factory on the bank of a stream—whose property extends to the middle of the stream—often has difficulty seeing why it is not his natural right to muddy the waters flowing past his door. The law, always behind the times, requires elaborate stitching and fitting to adapt it to this newly perceived aspect of the commons.

The pollution problem is a consequence of population. It did not much matter how a lonely American frontiersman disposed of his waste. "Flowing water purifies itself every 10 miles," my grandfather used to say, and the myth was near enough to the truth when he was a boy, for there were not too many people. But as population became denser, the natural chemical and biological recycling processes became overloaded, calling for a redefinition of property rights.

How to Legislate Temperance?

Analysis of the pollution problem as a function of population density uncovers a not generally recognized principle of morality, namely: *the morality of an act is a function of the state of the system at the time it is performed.*[10] Using the commons as a cesspool does not harm the general public under frontier conditions, because there is no public; the same behavior in a metropolis is unbearable. A hundred and fifty years ago a plainsman could kill an American bison, cut out only the tongue for his dinner, and discard the rest of the animal. He was not in any important sense being wasteful.

Today, with only a few thousand bison left, we would be appalled at such behavior.

In passing, it is worth noting that the morality of an act cannot be determined from a photograph. One does not know whether a man killing an elephant or setting fire to the grassland is harming others until one knows the total system in which his act appears. "One picture is worth a thousand words," said an ancient Chinese; but it may take 10,000 words to validate it. It is as tempting to ecologists as it is to reformers in general to try to persuade others by way of the photographic shortcut. But the essence of an argument cannot be photographed; it must be presented rationally—in words.

That morality is system-sensitive escaped the attention of most codifiers of ethics in the past. "Thou shalt not ..." is the form of traditional ethical directives which make no allowance for particular circumstances. The laws of our society follow the pattern of ancient ethics, and therefore are poorly suited to governing a complex, crowded, changeable world. Our epicyclic solution is to augment statutory law with administrative law. Since it is practically impossible to spell out all the conditions under which it is safe to burn trash in the back yard or to run an automobile without smog-control, by law we delegate the details to bureaus. The result is administrative law, which is rightly feared for an ancient reason—*Quis custodiet ipsos custodes?*—"Who shall watch the watchers themselves?" John Adams said that we must have "a government of laws and not men." Bureau administrators, trying to evaluate the morality of acts in the total system, are singularly liable to corruption, producing a government by men, not laws.

Prohibition is easy to legislate (though not necessarily to enforce); but how do we legislate temperance? Experience indicates that it can be accomplished best through the mediation of administrative law. We limit possibilities unnecessarily if we suppose that the sentiment of *Quis custodiet* denies us the use of administrative law. We should rather retain the phrase as a perpetual reminder of fearful dangers we cannot avoid. The great challenge facing us now is to invent the corrective feedbacks that are needed to keep custodians honest. We must find ways to legitimate the needed authority of both the custodians and the corrective feedbacks.

Freedom To Breed Is Intolerable

The tragedy of the commons is involved in population problems in another way. In a world governed soley by the principle of "dog eat dog"—if indeed there ever was such a world—how many children a family had would not be a matter of public concern. Parents who bred too exuberantly would leave fewer descendants, not more, because they would be unable to care adequately for their children. David Lack and others have found that such a negative feedback demonstrably controls the fecundity of birds.[11] But men are not birds, and have not acted like them for millenniums, at least.

If each human family were dependent only on its own resources; *if* the children of improvident parents starved to death; *if*, thus, overbreeding brought its own "punishment" to the germ line—*then* there would be no public interest in controlling the breeding of families. But our society is deeply committed to the welfare state,[12] and hence is confronted with another aspect of the tragedy of the commons.

In a welfare state, how shall we deal with the family, the religion, the race, or the class (or indeed any distinguishable and cohesive group) that adopts overbreeding as a policy to secure its own aggrandizement?[13] To couple the concept of freedom to breed with the belief that everyone born has an equal right to the commons is to lock the world into a tragic course of action.

Unfortunately this is just the course of action that is being pursued by the United Nations. In late 1967, some 30 nations agreed to the following[14]:

> The Universal Declaration of Human Rights describes the family as the natural and fundamental unit of society. It follows that any choice and decision with regard to the size of the family must irrevocably rest with the family itself, and cannot be made by anyone else.

It is painful to have to deny categorically the validity of this right; denying it, one feels as uncomfortable as a resident of Salem, Massachusetts, who denied the reality of witches in the seventeenth century. At the present time, in liberal quarters, something like a taboo acts to inhibit criticism of the United Nations. There is a feeling that the United Nations is "our last and best hope," that we shouldn't find fault with it; we shouldn't play into the hands of the archconservatives. However, let us not forget what Robert Louis Stevenson said: "The truth that is suppressed by friends is the readiest weapon of the enemy." If we love the truth we must openly deny the validity of the Universal Declaration of Human Rights, even though it is promoted by the United Nations. We should also join with Kingsley Davis[15] in attempting to get Planned Parenthood-World Population to see the error of its ways in embracing the same tragic ideal.

Conscience Is Self-Eliminating

It is a mistake to think that we can control the breeding of mankind in the long run by an appeal to conscience. Charles Galton Darwin made this point when he spoke on the centennial of the publication of his grandfather's great book. The argument is straightforward and Darwinian.

People vary. Confronted with appeals to limit breeding, some people will undoubtedly respond to the plea more than others. Those who have more children will produce a larger fraction of the next generation than those with more susceptible consciences. The difference will be accentuated, generation by generation.

In C.G. Darwin's words: "It may well be that it would take hundreds of generations for the progenitive instinct to develop in his way, but if it should do so, nature would have taken her revenge, and the variety *Homo contraceptens* would become extinct and would be replaced by the variety *Homo progeni tivus*"[16]

The argument assumes that conscience or the desire for children (no matter which) is hereditary—but hereditary only in the most general formal sense. The result will be the same whether the attitude is transmitted through germ cells, or exosomatically, to use A.J. Lotka's term. (If one denies the latter possibility as well as the former, then what's the point of education?) The argument has here been stated in the context of the population problem, but it applies equally well to any instance in which society appeals to an individual exploiting a commons to restrain himself for the general good—by means of his conscience. To make such an appeal is to set up a selective system that works toward the elimination of conscience from the race.

Pathogenic Effects of Conscience

The long-term disadvantage of an appeal to conscience should be enough to condemn it; but has serious short-term disadvantages as well. If we ask a man who is exploiting a commons to desist "in the name of conscience," what are we saying to him? What does he hear?—not only at the moment but also in the wee small hours of the night when, half asleep, he remembers not merely the words we used but also the nonverbal communication cues we gave him unawares? Sooner or later, consciously or subconsciously, he senses that he has received two communications, and that they are contradictory: (i) (intended communication) "If you don't do as we ask, we will openly condemn you for not acting like a responsible citizen"; (ii) (the unintended communication) "If you *do* behave as we ask, we will secretly condemn you for a simpleton who can be shamed into standing aside while the rest of us exploit the commons."

Everyman then is caught in what Bateson has called a "double bind." Bateson and his co-workers have made a plausible case for viewing the double bind as an important causative factor in the genesis of schizophrenia.[17] The double bind may not always be so damaging, but it always endangers the mental health of anyone to whom it is applied. "A bad conscience," said Nietzsche, "is a kind of illness."

To conjure up a conscience in others is tempting to anyone who wishes to extend his control beyond the legal limits. Leaders at the highest level succumb to this temptation. Has any President during the past generation failed to call on labor unions to moderate voluntarily their demands for higher wages, or to steel companies to honor voluntary guidelines on prices? I can recall none. The rhetoric used on such occasions is designed to produce feelings of guilt in noncooperators.

For centuries it was assumed without proof that guilt was a valuable, perhaps even an indispensable, ingredient of the civilized life. Now, in this post-Freudian world, we doubt it.

Paul Goodman speaks from the modern point of view when he says: "No good has ever come from feeling guilty, neither intelligence, policy, nor compassion. The guilty do not pay attention to the object but only to themselves, and not even to their own interests, which might make sense, but to their anxieties."[18]

One does not have to be a professional psychiatrist to see the consequences of anxiety. We in the Western would are just emerging from a dreadful two-centuries-long Dark Ages of Eros that was sustained partly by prohibition laws, but perhaps more effectively by the anxiety-generating mechanisms of education. Alex Comfort has told the story well in *The Anxiety Makers*[19]; it is not a pretty one.

Since proof is difficult, we may even concede that the results of anxiety may sometimes, from certain points of view, be desirable. The larger question we should ask is whether, as a matter of policy, we should ever encourage the use of a technique the tendency (if not the intention) of which is psychologically pathogenic. We hear much talk these days of responsible parenthood; the coupled words are incorporated into the titles of some organizations devoted to birth control. Some people have proposed massive propaganda campaigns to instill responsibility into the nation's (or the world's) breeders. But what is the meaning of the word responsibility in this context? Is it not merely a synonym for the word conscience? When we use the word responsibility in the absence of substantial sanctions we are not trying to browbeat a free man in a commons into acting against his own interest? Responsibility is a verbal counterfeit for a substantial *quid pro quo*. It is an attempt to get something for nothing.

If the word responsibility is to be used at all, I suggest that it be in the sense Charles Frankel uses it.[20] "Responsibility," says this philosopher, "is the product of definite social arrangements."

Notice that Frankel calls for social arrangements—not propaganda.

Mutual Coercion Mutually Agreed-Upon

The social arrangements that produce responsibility are arrangements that create coercion, of some sort. Consider bank-robbing. The man who takes money from a bank acts as if the bank were a commons. How do we prevent such action? Certainly not by trying to control his behavior solely by a verbal appeal to his sense of responsibility. Rather than rely on propaganda we follow Frankel's lead and insist that a bank is not a commons; we seek the definite social arrangements that will keep it from becoming a commons. That we thereby infringe on the freedom of would-be robbers we neither deny nor regret.

The morality of bank-robbing is particularly easy to understand because we accept complete prohibition of this activity. We are willing to say "Thou shalt not rob banks," without providing for exceptions. But temperance also can be created by coercion. Taxing is a good coercive device. To keep downtown shoppers temperate in their use of parking space we introduce parking meters for short periods, and traffic fines for longer ones. We need not actually forbid a citizen to park as long as he wants to; we need merely make it increasingly expensive for him to do so. Not prohibition, but carefully biased options are what we offer him. A Madison Avenue man might call this persuasion; I prefer the greater candor of the world coercion.

Coercion is a dirty word to most liberals now, but it need not forever be so. As with the four-letter words, its dirtiness can be cleansed away by exposure to the light, by saying it over and over without apology or embarassment. To many, the word coercion implies arbitrary decisions of distant and irresponsible bureaucrats; but this is not a necessary part of its meaning. The only kind of coercion I recommend is mutual coercion, mutually agreed upon by the majority of the people affected.

To say that we mutually agree to coercion is not to say that we are required to enjoy it, or even to pretend we enjoy it. Who enjoys taxes? We all grumble about them. But we accept compulsory taxes because we recognize that voluntary taxes would favor the conscienceless. We institute and (grumblingly) support taxes and other coercive devices to escape the horror of the commons.

An alternative to the commons need not be perfectly just to be preferable. With real estate and other material goods, the alternative we have chosen is the institution of private property coupled with legal inheritance. Is this system perfectly just? As a genetically trained biologist I deny that it is. It seems to me that, if there are to be differences in individual inheritance, legal possession should be perfectly correlated with biological inheritance—that those who are biologically more fit to be the custodians of property and power should legally inherit more. But genetic recombination continually makes a mockery of the doctrine of "like father like son" implicit in our laws of legal inheritance. An idiot can inherit millions, and a trust fund can keep his estate intact. We must admit that our legal system of private property plus inheritance is unjust—but we put up with it because we are not convinced, at the moment, that anyone has invented a better system. The alternative of the commons is too horrifying to contemplate. Injustice is preferable to total ruin.

It is one of the peculiarities of the warfare between reform and the status quo that it is thoughtlessly governed by a double standard. Whenever a reform measure is proposed it is often defeated when its opponents triumphantly discover a flaw in it. As Kingsley Davis has pointed out,[21] worshippers of the status quo sometimes imply that no reform is possible without unanimous agreement, an implication contrary to historical fact. As nearly as I can make out, automatic rejection of proposed reforms is based on one of two unconscious assumptions: (i) that the status quo is perfect; or (ii) that the choice we face is between reform and no action; if the proposed reform is imperfect, we presumably should take no action at all, while we wait for a perfect proposal.

But we can never do nothing. That which we have done for thousands of years is also action. It also produces evils. Once we are aware that the status quo is action, we can then compare its discoverable advantages and disadvantages with the predicted advantages and disadvantages of the proposed reform, discounting as best we can for our lack of experience. On the basis of such a comparison, we can make a rational decision which will not involve the unworkable assumption that only perfect systems are tolerable.

Recognition of Necessity

Perhaps the simplest summary of this analysis of man's population problems is this: the commons, if justifiable at all, is justifiable only under conditions of low-population density. As the human population has increased, the commons has had to be abandoned in one aspect after another.

First we abandoned the commons in food gathering, enclosing farm land and restricting pastures and hunting and fishing areas. These restrictions are still not complete throughout the world.

Somewhat later we saw that the commons as a place for waste disposal would also have to be abandoned. Restrictions on the disposal of domestic sewage are widely accepted in the Western world; we are still struggling to close the commons to pollution by automobiles, factories, insecticide sprayers, fertilizing operations, and atomic energy installations.

In a still more embryonic state is our recognition of the evils of the commons in matters of pleasure. There is almost no restriction on the propagation of sound waves in the public medium. The shopping public is assaulted with mindless music, without its consent. Our government is paying out billions of dollars to create supersonic transport which will disturb 50,000 people for every one person who is whisked from coast to coast 3 hours faster. Advertisers muddy the airwaves of radio and television and pollute the view of travelers. We are a long way from outlawing the commons in matters of pleasure. Is this because our Puritan inheritance makes us view pleasure as something of a sin, and pain (that is, the pollution of advertising) as the sign of virtue?

Every new enclosure of the commons involves the infringement of somebody's personal liberty. Infringements made in the distant past are accepted because no contemporary complains of a loss. It is the newly proposed infringements that we vigorously oppose; cries of "rights" and "freedom" fill the air. But what does "freedom" mean? When men mutually agreed to pass laws against robbing, mankind became more free, not less so. Individuals locked into the logic of the commons are free only to bring on universal ruin; once they see the necessity of mutual coercion, they become free to pursue other goals. I believe it was Hegel who said, "Freedom is the recognition of necessity."

The most important aspect of necessity that we must now recognize, is the necessity of abandoning the commons in breeding. No technical solution can rescue us from the misery of overpopulation. Freedom to breed will bring ruin to all. At the moment, to avoid hard decisions many of us are tempted to propagandize for conscience and responsible parenthood. The temptation must be resisted, because an appeal to independently acting consciences selects for the disappearance of all conscience in the long run, and an increase in anxiety in the short.

The only way we can preserve and nurture other and more precious freedoms is by relinquishing the freedom to breed, and that very soon. "Freedom is the recognition of necessity"—and it is the role of education to reveal to all the necessity of abandoning the freedom to breed. Only so, can we put an end to this aspect of the tragedy of the commons.

References

1. J.B. Wiesner and H.G. York, *Scientific American* 211 (No. 4), 27 (1964).

2. G. Hardin, *J. Hered.* 50, 68 (1959); S. von Hoernor, *Science* 137, 18 (1962).

3. J. von Neumann and O. Morgenstern, *Theory of Games and Economic Behavior* (Princeton University Press, Princeton, New Jersey, 1947), p. 11.

4. J.H. Fremlin, *New Scientist.* No. 415 (1964), p. 285.

5. A. Smith, *The Wealth of Nations* (Modern Library, New York, 1937), p. 423.

6. W.F. Floyd, *Two Lectures on the Checks to Population* (Oxford University Press, Oxford, England, 1833), reprinted (in part) in *Population, Evolution, and Birth Control,* G. Hardin, Ed. (Freeman, San Francisco, 1964), p. 37.

7. A.N. Whitehead, *Science and the Modern World* (Mentor, New York, 1948), p. 17.

8. G. Hardin, Ed. *Population, Evolution, and Birth Control* (Freeman, San Francisco, 1964), p. 56.

9. S. McVay, *Scientific American,* 216 (No. 8), 13 (1966).

10. J. Fletcher, *Situation Ethics* (Westminster, Philadelphia, 1966).

11. D. Lack, *The National Regulation of Animal Numbers* (Clarendon Press, Oxford, 1954).

12. H. Girvetz, *From Wealth to Welfare* (Stanford University Press, Stanford, California, 1950).

13. G. Hardin, *Perspective Biological Medicine,* 6, 366 (1963).

14. U. Thant, *Int. Planned Parenthood News,* No. 168 (February 1968), p. 3.

15. K. Davis, *Science,* 158, 730 (1967).

16. S. Tax, Ed., *Evolution after Darwin* (University of Chicago Press, Chicago, 1960), vol. 2, p. 469.

17. G. Bateson, D.D. Jackson, J. Haley, J. Weakland, *Behavioral Science,* 1, 251 (1956).

18. P. Goodman, *New York Review of Books* 10 (8), 22 (23 May 1968).

19. A. Comfort, *The Anxiety Makers* (Nelson, London, 1967).

20. C. Frankel, *The Case for Modern Man* (Harper, New York, 1955), p. 203.

21. J.D. Roslansky, *Genetics and the Future of Man* (Appleton-Century-Crofts, New York, 1966), p. 177.

ETHICS AND POPULATION LIMITATION

Daniel Callahan

THROUGHOUT ITS HISTORY, the human species has been preoccupied with the conquest of nature and the control of death. Human beings have struggled to survive, as individuals, families, tribes, communities, and nations. Procreation has been an essential part of survival. Food could not have been grown, families sustained, individuals supported, or industry developed without an unceasing supply of new human beings. The result was the assigning of a high value to fertility. It was thought good to have children: good for the children themselves, for the parents, for the society, and for the species. While it may always have been granted that extenuating circumstances could create temporary contraindications to childbearing, the premise on which the value was based endured intact. There remained a presumptive right of individual procreation, a right thought to sustain the high value ascribed to the outcome: more human beings.

That the premise may now have to be changed, the value shifted, can only seem confounding. As Erik Erikson has emphasized, it is a risky venture to play with the "fire of creation," especially when the playing has implications for almost every aspect of individual and collective life.[1] The reasons for doing so would have to be grave. Yet excessive population growth presents such reasons—it poses critical dangers to the future of the species, the ecosystem, individual liberty and welfare, and the structure of social life. These hazards are serious enough to warrant a reexamination and, ultimately, a revision of the traditional value of unrestricted procreation and increase in population.

The main question is the way in which the revision is to proceed. If the old premise—the unlimited right of and need for procreation—is to be rejected or amended, what alternative premises are available? By what morally legitimate social and political processes, and in light of what values, are the possible alternatives to be evaluated and action taken? These are ethical questions, bearing on what is taken to constitute the good life, the range and source of human rights and obligations, the requirements of human justice and welfare. If the ethical problems of population limitation could be reduced to one overriding issue, matters would be simplified. They cannot. Procreation is so fundamental a human activity, so wide-ranging in its personal and social impact, that controlling it poses a wide range of ethical issues. My aim here is primarily to see what some of the different ethical issues are, to determine how an approach to them might be structured, and to propose some solutions.

With a subject so ill-defined as "ethics and population limitation," very little by way of common agreement can be taken for granted. One needs to start at the "beginning," with some basic assertions.

Facts and Values

There would be no concern about population limitation if there did not exist evidence that excessive population growth jeopardizes present and future welfare. Yet the way the evidence is evaluated will be the result of the values and interests brought to bear on the data. Every definition of the "population problem" or of "excessive population growth" will be value-laden, expressive of the ethical orientations of those who do the defining. While everyone might agree that widespread starvation and malnutrition are bad, not everyone will agree that crowding, widespread urbanization, and a loss of primitive forest areas are equally bad. Human beings differ in their assessments of relative good and evil. To say that excessive population growth is bad is to imply that some other state of population growth would be good or better—for example, an "optimum level of population." But as the demographic discussion of an optimum has made clear, so many variables come into play that it may be possible to do no more than specify a direction: "the desirability of a lower *rate* [italics added] of growth."[2]

If the ways in which the population problem is defined will reflect value orientations, these same definitions will have direct implications for the ways in which the ethical issues are posed. An apocalyptic reading of the demographic data and projections can, not surprisingly, lead to coercive proposals. Desperate problems are seen to require desperate and otherwise distasteful solutions.[3] Moreover, how the problem is defined, and how the different values perceived to be at stake are weighted, will have direct implications for the priority given to population problems in relation to other social problems. People might well agree that population growth is a serious issue, but they might (and often do) say that other issues are comparatively more serious.[4] If low priority is given to population problems, this is likely to affect the perception of the ethical issues at stake.

Why Ethical Questions Arise

Excessive population growth raises ethical questions because it threatens existing or desired human values and ideas of what is good. In addition, all or some of the possible solutions to the problem have the potential for creating difficult ethical dilemmas. The decision to act or not to act in the face of the threats is an ethical decision. It is a way of affirming where the human good lies and the kinds of obligations individuals and societies have toward themselves and others. A choice in favor of action will, however, mean the weighing of different options, and most of the available options present ethical dilemmas.

In making ethical choices, decisions will need to be made on (i) the human good and values that need to be served or promoted—the ends; (ii) the range of methods and actions consistent and coherent with those ends—the means; and (iii) the procedure and rationale to be used in trying to decide both upon ends and means and upon their relation to each other in specific situations—the ethical criteria for decision-making. A failure to determine the ends, both ultimate and proximate, can make it difficult or impossible to choose the appropriate means. A failure to determine the range of possible means can make it difficult to serve the ends. A failure to specify or articulate the ethical criteria for decision-making can lead to capricious or self-serving choices, as well as to the placing of obstacles in the way of rational resolution of ethical conflicts.

In the case of ethics and the population problem, both the possibilities and the limitations of ethics become apparent. In the face of a variety of proposals to solve the population problem, some of them highly coercive, a sensitivity to the ethical issues and some greater rigor in dealing with them is imperative. The most fundamental matters of human life and welfare are at stake. Yet because of the complexity of the problem, including its variability from one nation or geographical region to the next, few hard and fast rules can be laid down about what to do in a given place at a given time.

Still, since some choices must be made (and not to choose is to make a choice as well), the practical ethical task will be that of deciding upon the available options. While I will focus on some of the proposed options for reducing birthrates, they are not the only ones possible. Ralph Potter has discussed some others.[5]

> It has generally been assumed that policy must be primarily, if not exclusively, concerned with bringing about a decline in the rate of population increase through a reduction in the birthrate. But there are other choices. It is generally considered desirable but impossible to increase resources at a sufficient pace and through an adequate duration to preserve the present level of living for all within an expanding population. It is generally considered possible but undesirable to omit the requirement that all persons have access to that which is necessary for a good life. There is still the option of redefining what is to be considered necessary for a good life or of foregoing some things necessary for a good life in order to obtain an equitable distribution in a society that preserves the autonomy of parents to determine the size of their families.

A useful way of posing the issue practically is to envision the ethical options ranked on a preferential scale, from the most desirable to the least desirable. For working purposes, I will adopt as my own the formulation of Kenneth E. Boulding: "A moral, or ethical, proposition is a statement about a rank order of preferences among alternatives, which is intended to apply to more than one person."[6] Ethics enters at that point when the preferences are postulated to have a value that transcends individual tastes or inclinations. Implicitly or explicitly, a decision among alternatives becomes an ethical one when it is claimed that one or another alternative *ought* to be chosen—not just by me, but by others as well. This is where ethics differs from tastes or personal likings, which, by definition, imply nonobligatory preferences that are applicable to no more than one person (even if the tastes are shared).

General Ethical Issues

I will assume at the outset that there is a problem of excessive population growth, a problem serious for the world as a whole (with a 2 percent annual growth rate), grave for many developing nations (where the growth rate approaches 3 percent per annum), and possibly harmful for the developed nations as well (with an average 1 percent growth rate). The threats posed by excessive population growth are numerous: economic, environmental, agricultural, political, and sociopsychological. There is considerable agreement that something must be done to meet these threats. For the purpose of ethical analysis, the first question to be asked is, "in trying to meet these threats, what human ends are we seeking to serve?" Two kinds of human ends can be distinguished—proximate and ultimate.

Among the important proximate ends being sought in attempts to reduce birthrates in the developing countries are a raising of literacy rates, a reduction in dependency ratios, the elimination of starvation and malnutrition, more rapid economic development, and an improvement in health and welfare services; among these ends in the developed countries are a maintenance or improvement of the quality of life, the protection of nonrenewable resources, and the control of environmental pollution. For most purposes, it will be sufficient to cite goals of this sort. But for ethical purposes, it is critical to consider not just proximate, but ultimate ends as well. For it is legitimate to ask of the specified proximate ends what ultimate human ends they are meant to serve. Why is it important to raise literacy rates? Why is it necessary to protect nonrenewable resources? Why ought the elimination of starvation and malnutrition to be sought? For the most part, these are questions that need not be asked or that require no elaborate answers. The ethical importance of such questions is that they force us to confront the goals of human life. Unless these goals are confronted at some point, ethics cannot start or finish.

Philosophically, solving the population problem can be viewed as determining at the outset what final values should be pursued. The reason, presumably, that a reduction in illiteracy rates is sought is that it is thought valuable for human beings to possess the means of achieving knowledge. The elimination of starvation and malnutrition is sought because of the self-evident fact that human beings must eat to survive. The preservation of nonrenewable resources is necessary in order that human life may continue through future generations. There is little argument about the validity of these propositions, because they all presuppose some important human values: knowledge, life, and survival of the species, for instance. Historically, philosophers have attempted to specify what, in the sense of "the good," human beings essentially seek. What do they, in the end, finally value? The historical list of values is long: life, pleasure, happiness, knowledge, freedom, justice, and self-expression, among others.

This is not the place to enter into a discussion of all of these values and the philosophical history of attempts to specify and rank them. Suffice it to say that three values have had a predominant role, at least in the West: freedom, justice, and security-survival. Many of the major ethical dilemmas posed by the need for population limitation can be reduced to ranking and interpreting these three values. Freedom is prized because it is a condition for self-determination and the achievement of knowledge. Justice, particularly distributive justice, is prized because it entails equality of treatment and opportunity and an equitable access to those resources and opportunities necessary for human development. Security-survival is prized because it constitutes a fundamental ground for all human activities.

Excessive population growth poses ethical dilemmas because it forces us to weight and rank these values in trying to find solutions. How much procreative freedom, if any, should be given up in order to insure the security-survival of a nation or a community? How much security-survival can be risked in order to promote distributive justice? How much procreative freedom can be tolerated if it jeopardizes distributive justice?

Ethical dilemmas might be minimized if there were a fixed agreement on the way the three values ought to be ranked. One could say that freedom is so supreme a value that both justice and security-survival should be sacrificed to maintain it. But there are inherent difficulties in taking such a position. It is easily possible to imagine situations in which a failure to give due weight to the other values could result in an undermining of the possibility of freedom itself. If people cannot survive at the physical level, it becomes impossible for them to exercise freedom of choice, procreative or otherwise. If the freedom of some is unjustly achieved at the expense of the freedom of others, then the overall benefits of freedom are not maximized. If security-survival were given the place of supremacy, situations could arise in which this value was used to justify the suppression of freedom or the perpetuation of social injustice. In that case, those suppressed might well ask, "Why live if one cannot have freedom and justice?"

For all of these reasons it is difficult and perhaps unwise to specify a fixed and abstract rank order of preference among the three values. In some circumstances, each can enter a valid claim against the others. In the end, at the level of abstractions, one is forced to say that all three values are critical; none can permanently be set aside.

The Primacy of Freedom

In the area of family planning and population limitation, a number of national and international declarations have given primacy to individual freedom. The Declaration of the 1968 United Nations International Conference on Human Rights is representative[7-8]: "... couples have a basic human right to decide freely and responsibly on the number and spacing of their children and a right to adequate education and information in this respect." While this primacy of individual freedom has been challenged,[9] it retains its position, serving as the ethical and political foundation of both domestic and foreign family planning and population policies. Accordingly, it will be argued here that (i) the burden of proof for proposals to limit freedom of choice (whether on the grounds of justice or security-survival) rests with those who make the proposals, but that (ii) this burden can, under specified conditions, be discharged if it can be shown that a limitation of freedom of choice in the name of justice or security-survival would tend to maximize human welfare and human values. This is only to say that, while the present international rank order of preference gives individual freedom primacy, it is possible to imagine circumstances that would require a revision of the ranking.

One way of approaching the normative issues of ranking preferences in population limitation programs and proposals is by locating the key ethical actors, those who can be said to have obligations. Three groups of actors can be identified: individuals (persons, couples, families), the officers and agents of voluntary (private-external) organizations, and the government officials responsible for population and family planning programs. I will limit my discussion here to individuals and governments. What are the ethical obligations of each of the actors? What is the right or correct course of conduct for them? I will approach these questions by first trying to define some general rights and obligations for each set of actors and then by offering some suggested resolutions of a number of specific issues.

I begin with individuals (persons, couples, families) because, in the ranking of values, individual freedom of choice has been accorded primacy by some international forums—and it is individuals who procreate. What are the rights and obligations of individuals with regard to procreation?

Individuals have the right voluntarily to control their own fertility in accordance with their personal preferences and convictions (7, p. 15). This right logically extends to a choice of methods to achieve the desired control and the right to the fullest possible knowledge of available methods and their consequences (medical, social, economic, and demographic, among others).

Individuals are obligated to care for the need and respect the rights of their existing children (intellectual, emotional, and physical); in their decision to have a child (or another child), they must determine if they will be able to care for the needs and respect the rights of the child-to-be. Since individuals are obliged to respect the rights of others, they are obliged to act in such a way that these rights are not jeopardized. In determining family size, this means that they must exercise their own freedom of choice in such a way that they do not curtail the freedom of others. They are obliged, in short, to respect the requirements of the common good in their exercise of free choice.[10] The source of these obligations is the rights of others.

The role of governments in promoting the welfare of their citizens has long been recognized. It is only fairly recently, however, that governments have taken a leading role in an antinatalist control of fertility.[11] This has come about by the establishment, in a number of countries, of national family planning programs and national population policies. While many countries still do not have such policies, few international objections have been raised against the right of nations to develop them. So far, most government population policies have rested upon and been justified in terms of an extension of freedom of choice. Increasingly, though, it is being recognized that, since demographic trends can significantly affect national welfare, it is within the right of nations to adopt policies designed to reduce birthrates and slow population growth.

A preliminary question must, therefore, be asked. Is there any special reason to presume or suspect that governmental intervention in the area of individual procreation and national fertility patterns raises problems which, in *kind*, are significantly different from other kinds of interventions? To put the question another way, can the ethicopolitical problems that arise in this area be handled by historical and traditional principles of political ethics, or must an entirely new ethic be devised?

I can see no special reason to think that the formation of interventionist, antinatalist, national population policies poses any unique *theoretical* difficulties. To be sure, the perceived need to reduce population growth is historically new; there exists no developed political or ethicopolitical tradition dealing with this specific problem. Yet the principle of governmental intervention in procreation-related behavior has a long historical precedent: in earlier, pronatalist population policies, in the legal regulation of marriage, and in laws designed to regulate sexual behavior. It seems a safe generalization to say that governments have felt (and generally have been given) as much right to intervene in this area as in any other where individual and collective welfare appears to be at stake. That new forms of intervention may seem to be called for or may be proposed (that is, in an anti- rather than pronatalist direction) does not mean that a new ethical principle is at issue. At least, no such principle is immediately evident.

Yet, if it is possible to agree that no new principles are involved, it is still possible to argue that a further extension of an old principle—the right of government intervention into procreation-related behavior—would be wrong. Indeed, it is a historical irony that, after a long international struggle to establish individuals' freedom of choice in controlling their own fertility, that freedom should immediately be challenged in the name of the population crisis. Irony or not, there is no cause to be surprised by such a course of events. The history of human liberty is studded with instances in which, for a variety of reasons, it has been possible to say that liberty is a vital human good and yet that, for the sake of other goods, restriction of liberty seems required. A classical argument for the need of a government is that a formal and public apparatus is necessary to regulate the exercise of individual liberty for the sake of the common good.

In any case, the premise of my discussion will be that governments have as much right to intervene in procreation-related behavior as in other areas of behavior affecting the general welfare. This right extends to the control of fertility in general and to the control of individual fertility in particular. The critical issue is the way in which this right is to be exercised—its conditions and limits—and that issue can only be approached by first noting some general issues bearing on the restriction of individual freedom of choice by governments.

Governments have the right to take those steps necessary to insure the preservation and promotion of the common good—the protection and advancement of the right to life, liberty, and property. The maintenance of an orderly and just political and legal system, the maintenance of internal and external security, and an equitable distribution of goods and resources are also encompassed within its rights. Its obligations are to act in the interests of the people, to observe human rights, to respect national values and traditions, and to guarantee justice and equality. Since excessive population growth can touch upon all of these elements of national life, responses to population problems will encompass both the rights and the obligations of governments. However, governmental acts should represent collective national decisions and be subject to a number of stipulations.

I now recapitulate the points made so far and summarize some propositions, which I then use to suggest solutions to some specific ethical issues.

1. General moral rules (i) individuals have the right to freedom of procreative choice, and they have the obligation to respect the freedom of others and the requirements of the common good; (ii) governments have the right to take those steps necessary to secure a maximization of freedom, justice, and security-survival, and they have the obligation to act in such a way that freedom and justice are protected and security-survival enhanced.

2. Criteria for ethical decision-making: (i) one (individual, government, organization) is obliged to act in such a way that the fundamental values of freedom, justice, and security-survival are respected; (ii) in cases of conflict, one is obliged to act in such a way that any limitation of one or more of the three fundamental values—a making of exceptions to the rules concerning these values—continues to respect the values and can be justified by the promise of increasing the balance of good over evil.

3. Rank order of preference: (i) those choices of action that ought to be preferred are those that accord primacy to freedom of choice; (ii) if conditions appear to require a limitation of freedom, this should be done in such a way that the direct and indirect harmful consequences are minimized and the chosen means of limitation are just—the less the harm, the higher the ranking.

Some Specific Ethical Issues

Since it has already been contended that individual freedom of choice has primacy, the ethical issues to be specified here will concentrate on those posed for governments. This focus will, in any event, serve to test the limits of individual freedom.

Faced with an excessive population growth, a variety of courses are open to governments. They can do nothing at all. They can institute, develop, or expand voluntary family planning programs. They can attempt to implement proposals that go "beyond family planning."[12]

Would it be right for governments to go beyond family planning if excessive population growth could be shown to be a grave problem? This question conceals a great range of isues. Who would decide if governments have this right? Of all the possible ways of going beyond family planning, which could be most easily justified and which would be the hardest to justify? To what extent would the problem have to be shown to be grave? As a general proposition, it is possible ethically to say that governments would have the right to go beyond family planning. The obligation of governments to protect fundamental values could require that they set aside the primacy of individual freedom in order to protect justice and security-survival. But everything would depend on the way they proposed to do so.

Would it be right for governments to establish involuntary fertility controls? These might include (if technically feasible) the use of a mass "fertility control agent," the licensing of the right to have children, compulsory temporary or permanent sterilization, or compulsory abortion.[12] Proposals of this kind have been put forth primarily as "last resort" methods, often in the context that human survival may be at stake. "Compulsory control of family size is an unpalatable idea to many," the Ehrlichs have written, "but the alternatives may be much more horrifying ... human survival seems certain to require population control programs ... ." (3, p. 256). Their own suggestion is manifestly coercive: "If ... relatively uncoercive laws should fail to bring the birthrate under control, laws could be written that would make the bearing of a third child illegal and that would require an abortion to terminate all such pregnancies" (3, p. 274).

That last suggestion requires examination. Let us assume for the moment that the scientific case has been made that survival itself is at stake and that the administrative and enforcement problems admit of a solution. Even so, some basic ethical issues would remain. "No one," the United Nations has declared, "shall be subjected to torture or to cruel, inhuman, or degrading treatment or punishment" (13, Article 5). It is hard to see how compulsory abortion, requiring governmental invasion of a woman's body, could fail to qualify as inhuman or degrading punishment. Moreover, it is difficult to see how this kind of suggestion can be said to respect in any way the values of freedom and justice. It removes free choice altogether, and in its provision for an abortion of the third child makes no room for distributive justice at all; its burden would probably fall upon the poorest and least educated. It makes security-survival the prime value, but to such an extent and in such a way that the other values are ignored altogether. But could not one say, when survival itself is at stake, that this method would increase the balance of good over evil? The case would not be easy to make (i) because survival is not the only human value at stake; (ii) because the social consequences of such a law could be highly destructive (for example, the inevitably massive fear and anxiety about third pregnancies that would result from such a law); and (iii) because it would be almost impossible to show that this is the *only* method that would or could work to achieve the desired reduction in birthrates.

Would it be right for governments to develop "positive" incentive programs, designed to provide people with money or goods in return for a regulation of their fertility? These programs might include financial rewards for sterilization, for the use of contraceptives, for periods of nonpregnancy or nonbirth, and for family planning bonds or "responsibility prizes" (12, p. 2). In principle, incentive schemes are noncoercive; that is, people are not forced to take advantage of the incentive. Instead, the point of an incentive is to give them a choice they did not previously have.

Yet there a number of ethical questions about incentive plans. To whom would they appeal most? Presumably, their greatest appeal would be to the poor, those who want or need the money or goods offered by an incentive program; they would hold little appeal for the affluent, who already have these things. Yet if the poor desperately need the money or goods offered by the incentive plan, it is questionable whether, in any real sense, they have a free choice. Their material needs may make the incentive seem coercive to them. Thus, if it is only or mainly the poor who would find the inducements of an incentive plan attractive, a question of distributive justice is raised. Because of their needs, the poor have less choice than the rich about accepting or rejecting the incentive; this could be seen as a form of exploitation of poverty. In sum, one can ask whether incentive schemes are or could be covertly coercive, and whether they are or could be unjust.[14] If so, then while they may serve the need for security-survival, they may do so at the expense of freedom and justice.

At least three responses seem possible. First, if the need for security-survival is desperate, incentive schemes might well appear to be the lesser evil, compared with more overtly coercive alternatives. Second, the possible objections to incentive schemes could be reduced if, in addition to reducing births, they provided other benefits as well. For instance, a "family planning bond" program would provide the additional benefit of old-age security[15]. Any one of the programs might be defended on the grounds that those who take advantage of it actually want to control births in any case (if this can be shown). Third, much could depend upon the size of the incentive benefits. At present, most incentive programs offer comparatively small rewards; one may doubt that they offer great dilemmas for individuals or put them in psychological straits. The objection to such programs on the grounds of coercion would become most pertinent if it can be shown that the recipients of an incentive benefit believe they have no real choice in the matter (because of their desperate poverty or the size of the benefit); so far, this does not appear to have been the case.[16]

While ethical objections have been leveled at incentive programs because of some experienced corrupt practices in their implementation, this seems to raise less serious theoretical issues. Every program run by governments is subject to corruption; but there are usually ways of minimizing it (by laws and review procedures, for instance). Corruption, I would suggest, becomes a serious theoretical issue only when and if it can be shown that a government program is *inherently* likely to create a serious, inescapable, and socially damaging system of corruption. This does not appear to be the case with those incentive programs so far employed or proposed.

Would it be right for governments to institute "negative" incentive programs? These could take the form of a withdrawal of child or family allowances after a given number of children, a withdrawal of maternity benefits after a given number, or a reversal of tax benefits, to favor those with small families (12, p. 2). A number of objections to such programs have been raised. They are directly coercive in that they deprive people of free choice about how many children they will have by imposing a penalty on excess procreation; thus they do not attach primary importance to freedom of choice. They can also violate the demands of justice, especially in those cases where the burden of the penalties would fall upon those children who would lose benefits available to their siblings. And the penalties would probably be more onerous to the poor than to the rich, further increasing the injustice. Finally, from quite a different perspective, the social consequences of such programs could be most undesirable. They could, for instance, worsen the health and welfare of those mothers, families, and children who would lose needed social and welfare benefits. Moreover, such programs could be patently unjust in those places where effective contraceptives do not exist (most places at preent). In such cases, people would be penalized for having children whom they could not prevent with the available birth control methods.

It is possible to imagine ways of reducing the force of these objections. If the penalties were quite mild, more symbolic than actual [as Garrett Hardin has proposed[17]], the objection from the viewpoint of free choice would be less; the same would apply to the objection from the viewpoint of justice. Moreover, if the penalty system were devised in such a way that the welfare of children and families would not be harmed, the dangerous social consequences would be mitigated. Much would depend, in short, upon the actual provisions of the penalty plan and the extent to which it could minimize injustice and harmful social consequences. Nonetheless, penalty schemes raise serious ethical problems. It seems that they would be justifiable only if it could be shown that security-survival was at stake and that, in their application, they would give due respect to freedom and justice. Finally, it would have to be shown that, despite their disadvantages, they promised to increase the balance of good over evil—which would include a calculation of the harm done to freedom and justice and a weighing of other, possibly harmful, social consequences.

An additional problem should be noted. Any penalty or benefit scheme would require some method of governmental surveillance and enforcement. Penalty plans, in particular, would invite evasion—for example, hiding the birth of children to avoid the sanctions of the scheme. This likelihood would be enhanced among those who objected to the plan on moral or other grounds, or who believed that the extra children were necessary for their own welfare. One does not have to be an ideological opponent of "big government" to imagine the difficulties of trying to ferret out violators or the lengths to which some couples might go to conceal pregnancies and births. Major invasions of privacy, implemented by a system of undercover agents, informants, and the like, would probably be required to make the scheme work. To be sure, there are precedents for activities of this kind (as in the enforcement of income tax laws), but the introduction of further governmental interventions of this kind would raise serious ethical problems, creating additional strains on the relationship between the government and the people. The ethical cost of an effective penalty system would have to be a key consideration in the development of any penalty program.

Would it be right for governments to introduce antinatalist shifts in social and economic institutions? Among such shifts might be a raising of marriage ages, manipulation of the family structure away from nuclear families, and bonuses for delayed marriage (12, pp. 2-3). The premise of these proposals is that fertility patterns are influenced by the context in which choices are made and that some contexts (for example, higher female employment) are anti- rather than pronatalist. Thus instead of intervening directly into the choices women make, these proposals would alter the environment of choice; freedom of individual choice would remain. The attractiveness of these proposals lies in their noninterference with choice; they do not seem to involve coercion. But they are not without their ethical problems, at least in some circumstances. A too-heavy weighting of the environment of choice in an antinatalist direction would be tantamount to an interference with freedom of choice—even if, technically, a woman could make a free choice. In some situations, a manipulation of the institution of marriage (for example, raising the marriage age) could be unjust, especially if there exist no other social options for women.

The most serious problems, however, lie in the potential social consequences of changes in basic social institutions. What would be the long-term consequences of a radical manipulation of family structure for male-female relationships, for the welfare of children, for the family? One might say that the consequences would be good or bad, but the important point is that they would have to be weighed. Should some of them appear bad, they would then have to be justified as entailing a lesser evil than the continuation of high birthrates. If some of the changes promised to be all but irreversible once introduced, the justification would have to be all the greater. However, if the introduction of shifts in social institutions had some advantages in addition to antinatalism—for instance, greater freedom for women, a value in its own right—these could be taken as offsetting some other, possibly harmful, consequences.

Would it be right for the government of a developed nation to make the establishment of a population control program in a developing nation a condition for extending food aid?[18-19] This would be extremely difficult to justify on ethical grounds. At the very least, it would constitute an interference in a nation's right to self-determination.[20] Even more serious, it would be a direct exploitation of one nation's poverty in the interests of another nation's concept of what is good for it; and that would be unjust. Finally, I would argue that, on the basis of Article 3 of the "Universal Declaration of Human Rights,"[21] a failure to provide needed food aid would be a fundamental violation of the right to life (when that aid could, without great cost to the benefactor nation, be given). The argument that such aid, without an attendant population control program, would only make the problem worse in the long run, is defective. Those already alive, and in need of food, have a right to security-survival. To willfully allow them to die, or to deprive them of the necessities of life, in the name of saving even more lives at a later date cannot be justified in the name of a greater preponderance of good over evil. There could be no guarantee that those future lives would be saved, and there would be such a violation of the rights of the living (including the right to life) that fundamental human values would be sacrificed.

Would it be right for a government to institute programs that go beyond family planning—particularly in a coercive direction—for the sake of future generations? This is a particularly difficult question, in great part because the rights of unborn generations have never been philosophically, legally, or ethically analyzed in any great depth.[22] On the one hand, it is evident that the actions of one generation can have profound effects on the options available to future generations. And just as those living owe much of their own welfare to those who preceded them (beginning with their parents), so, too, the living would seem to have obligations to the unborn. On the other hand though, the living themselves do have rights—not just potential, but actual. To set aside these rights, necessary for the dignity of the living, in favor of those not yet living would, I think, be to act arbitrarily.

A general solution might, however, be suggested. While the rights of the living should take precedence over the rights of unborn generations, the living have an obligation to refrain from actions that would endanger future generations' enjoyment of the same rights that the living now enjoy. This means, for instance, that the present generation should not exhaust nonrenewable resources, irrevocably pollute the environment, or procreate to such an extent that future generations will be left with an unmanageably large number of people. All of these obligations imply a restriction of freedom. However, since the present generation does have the right to make use of natural resources and to procreate, it must be demonstrated (not just asserted) that the conduct of the present generation poses a direct threat to the rights of future generations. In a word, the present generation cannot be deprived of rights on the basis of vague speculations about the future or uncertain projections into the future.

Do governments have the right unilaterally to introduce programs that go beyond family planning? It is doubtful that they do. Article 21 of the "Universal Declaration of Human Rights"[13] asserts that "Everyone has the right to take part in the government of his country, directly or through freely chosen representatives. ... The will of the people shall be the basis of the authority of government." There is no evident reason that matters pertaining to fertility control should be exempt from the requirements of this right. By implication, not only measures that go beyond family planning, but family planning programs as well require the sanctions of the will of the people and the participation of the people in important decisions.

A Ranking of Preferences

The preceding list of specific issues by no means exhausts the range of possible ethical issues pertaining to governmental action; it is meant only to be illustrative of some of the major issues. Moreover, the suggested solutions are only illustrative. The complexities of specific situations could well lead to modifications of them. That is why ethical analysis can rarely ever say exactly what ought to be done in *x* place at *y* time by *z* people. It can suggest general guidelines only.

I want now to propose some general ethical guidelines for governmental action, ranking from the most preferable to the least preferable.

1. Given the primacy accorded freedom of choice, governments have an obligation to do everything within their power to protect, enhance, and implement freedom of choice in family planning. This means the establishment, as the first order of business, of effective voluntary family planning programs.

2. If it turns out that voluntary programs are not effective in reducing excessive population growth, then governments have the right, as the next step, to introduce programs that go beyond family planning. However, in order to justify the introduction of such programs, it must be shown that voluntary methods have been adequately and fairly tried, and have nonetheless failed and promise to continue to fail. It is highly doubtful that, at present, such programs have "failed"; they have not been tried in any massive and systematic way.[23]

3. In choosing among possible programs that go beyond family planning, governments must first try those which, comparatively, most respect freedom of choice (that is, are least coercive). For instance, they should try "positive" incentive programs and manipulation of social structures before resorting to "negative" incentive programs and involuntary fertility controls.

4. Further, if circumstances force a government to choose programs that are quasi- or wholly coercive, they can justify such programs if, and only if, a number of prior conditions have been met (i), if, in the light of the primacy of free choice, a government has discharged the burden of proof necessary to justify a limitation of free choice—and the burden of proof is on the government (this burden may be discharged by a demonstration that continued unrestricted liberty poses a direct threat to distributive justice or security-survival; and (ii) if, in light of the right of citizens to take part in the government of their country, the proposed limitations on freedom promise, in the long run, to increase the options of free choice, decisions to limit freedom are collective decisions, the limitations on freedom are legally regulated and the burden falls upon all equally, and the chosen means of limitations respect human dignity, which will here be defined as respecting those rights specified in the United Nations' "Universal Declaration of Human

Rights."[13] The end—even security-survival—does not justify the means when the means violate human dignity and logically contradict the end.

As a general rule, the more coercive the proposed plan, the more stringent should be the conditions necessary to justify and regulate the coercion. In addition, one must take account of the possible social consequences of different programs, consequences over and above their impact on freedom, justice, and security-survival. Thus, if it appears that some degree of coercion is required, that policy or program should be chosen which (i) entails the least amount of coercion, (ii) limits the coercion to the fewest possible cases, (iii) is most problem-specific, (iv) allows the most room for dissent of conscience, (v) limits the coercion to the narrowest possible range of human rights, (vi) threatens human dignity least, (vii) establishes the fewest precedents for other forms of coercion, and (viii) is most quickly reversible if conditions change.

While it is true to say that social, cultural, and political life requires, and has always required, some degree of limitation of individual liberty—and thus some coercion—that precedent does not, in itself, automatically justify the introduction of new limitations.[24] Every proposal for a new limitation must be justified in its own terms—the specific form of the proposed limitation must be specifically justified. It must be proved that it represents the least possible coercion, that it minimizes injustice to the greatest extent possible, that it gives the greatest promise of enhancing security-survival, and that it has the fewest possible harmful consequences (both short- and long-term).

Freedom and Risk-Taking

The approach I have taken to the ethics of population limitation has been cautionary. I have accepted the primacy of freedom of choice as a given not only because of its primacy in United Nations and other declarations, but also because it is a primary human value. I have suggested that the burden of proof must lie with those proposals, policies, or programs that would place the primacy elsewhere. At the same time, I have laid down numerous conditions necessary to discharge the burden of proof. Indeed, these conditions are so numerous, and the process of ethical justification so difficult, that the possibility of undertaking decisive action may seem to have been excluded. This is a reasonable concern, particularly if time is short. Is it reasonable to give the ethical advantage to freedom of choice?[25] Does this not mean that a great chance is being taken? Is it not unethical to take risks of that sort, and all the more so since others, rather than ourselves, will have to bear the burden if the risk-taking turns out disastrously? In particular, would it not be irresponsible for governments to take risks of this magnitude?

Three kinds of responses to these questions are possible. First, as mentioned, it can and has been argued that freedom of choice has not been adequately tested. The absence of a safe, effective, and inexpensive contraceptive has been one hindrance, particularly in developing countries; it is reasonable to expect that such a contraceptive will eventually be developed. The weakness of existing family planning programs (and population policies dependent upon them) has, in great part, been the result of inadequate financing, poor administration, and scanty research and survey data. These are correctable deficiencies, assuming that nations give population limitation a high priority. If they do not give population limitation a high priority, it is unlikely that more drastic population policies could be successfully introduced or implemented. Very little effort has been expended anywhere in the world to educate people and persuade them to change their procreation habits. Until a full-scale effort has been made, there are few good grounds for asserting that voluntary limitation will be ineffective.

Second, while the question of scientific-medical-technological readiness, political viability, administrative feasibility, economic capability, and assumed effectiveness of proposals that would go beyond family planning is not directly ethical in nature, it has important ethical implications. If all of these categories seem to militate against the practical possibility of instituting very strong, immediate, or effective coercive measures, then it could become irresponsible to press for or support such measures. This would especially be the case if attention were diverted away from what could be done, for example, an intensification of family planning programs.

Third, primacy has been given to freedom of choice for ethical reasons. Whether this freedom will work as a means of population limitation is a separate question. A strong indication that freedom of choice will be ineffective does not establish grounds for rejecting it. Only if it can be shown that the failure of this freedom to reduce population growth threatens other important human values, thus establishing a genuine conflict of values, would the way be open to remove it from the place of primacy. This is only another way of asserting that freedom of choice is a right, grounded in a commitment to human dignity. The concept of a "right" becomes meaningless if rights are wholly subject to tests of economic, social, or demographic utility, to be given or withheld depending upon their effectiveness in serving social goals.

In this sense, to predicate human rights at all is to take a risk. It is to assert that the respect to be accorded human beings ought not to be dependent upon majority opinion, cost-benefit analysis, social utility, governmental magnanimity, or popular opinion. While it is obviously necessary to adjudicate conflicts among rights, and often to limit one right in order to do justice to another, the pertinent calculus is that of rights, not of utility. A claim can be entered against the primacy of one right only in the name of one or more other important rights. The proper route to a limitation of rights is not directly from social facts (demographic, economic, and so on) to rights, as if these facts were enough in themselves to prove the case against a right. The proper route is from showing that the social facts threaten rights, and in that way, to showing that a limitation of one right may be necessary to safeguard or enhance other rights. To give primacy to the right of free choice is to take a risk. The justification for the risk is the high value assigned to the right, a value that transcends simply utilitarian considerations.

References and Notes

1. E.H. Erikson, *Insight and Responsibility* (Norton, New York, 1964), p. 132.

2. B. Berelson, in *Is There an Optimum Level of Population?*, S.F. Singer, Ed. (McGraw-Hill, New York, 1971), p. 305.

3. See, for instance, P.R. Ehrlich and A.H. Ehrlich, *Population, Resources, Environment Issues in Human Ecology* (Freeman, San Francisco, 1970), pp. 321-324.

4. A 1967 Gallup Poll, for example, revealed that, while 54 percent of those surveyed felt that the rate of American population growth posed a serious problem, crime, racial discrimination and poverty were thought to be comparatively more serious social problems. J.F. Kanther, *Stud. Fam. Plann.* No. 30 (May 1968), p. 6.

5. R.B. Potter, Jr., in *Freedom, Coercion and the Life Sciences*, L. Kass and D. Callahan, eds., in press.

6. K.E. Boulding, *Amer. Econ. Rev.* 59, 1 (March 1969).

7. *Final Act of the International Conference on Human Rights* (United Nations, New York, 1968), p. 15.

8. "Declaration on Population: The World Leaders Statement," *Stud. Fam. Plann.* No. 26 (January 1968), p. 1.

9. For instance, not only has Garrett Hardin, in response to the "The World Leaders' Statement"[8], denied the right of the family to choose family size, he has also said that "If we love the truth we must openly deny the validity of the Universal Declaration of Human Rights, even though it is promoted by the United Nations" [*Science 162,* 1246 (1968)]. How literally is one to take this statement? The declaration, after all, affirms such rights as life, liberty, dignity, equality, education, privacy, and freedom of thought. Are none of these rights valid? Or, if those rights are to remain valid, why is only the freedom to control family size to be removed from the list?

10. See A.S. Parkes, in *Biology and Ethics,* F.J. Ebling, ed. (Academic Press, New York, 1969), pp. 109-116.

11. In general, "antinatalist" means "attitudes or policies directed toward a reduction of births," and "pronatalist" means "attitudes or policies directed toward an increase in births."

12. See B. Berelson, *Stud. Fam. Plann.* No. 38 (February 1969), p. 1.

13. "Universal Declaration of Human Rights," in *Human Rights: A Compilaton of International Instruments of the United Nations* (United Nations, New York, 1967).

14. See E. Pohlman and K.G. Rao, *Licentiate 17,* 236 (1967).

15. See, for instance, R.G. Ridker, *Stud. Fam. Plann.* No. 43 (June 1969), p. 11.

16. The payments made in six different family planning programs are listed in *Incentive Payments in Family Planning Programmes* (International Planned Parenthood Federation, London, 1969), pp. 8-9.

17. G. Hardin, *Fam. Plann. Perspect. 2,* 26 (June 1970).

18. See, for example, W.H. Davis, *New Repub.* (20 June 1970), p. 19.

19. P.R. Ehrlich, *The Population Bomb* (Ballantine Books, New York, 1968), pp. 158-173.

20. See the "International Covenant on Economic, Social and Cultural Rights," Article 1, section 1, paragraph 1, in *Human Rights: A Compilation of International Instruments of the United Nations* (United Nations, New York, 1967), p. 4: "All people have the right to self-determination. By virtue of that right they freely pursue their economic, social and cultural development."

21. "Everyone has the right to life, liberty and the security of person."[13]

22. One of the few recent discussions on the obligation to future generations is in M.P. Golding [*UCLA Law Review 15,* 457 (February 1968)].

23. See D. Nortman, in *Reports on Population/Family Planning* (Population Council, New York, December 1969), pp. 1-48. Judith Blake is pessimistic about the possibilities of family planning programs [*J. Chronic Dis. 18,* 1181 (1965)]. See also J. Blake [*Science 164,* 552 (1969)] and the reply of O. Harkavy, F.S. Jaffe, S.M. Wishik [*Ibid. 165,* 367 (1969)].

24. See E. Pohlman, *Eugen. Quart. 13,* 122 (June 1966): "the spectre of 'experts' monkeying around with such private matters as family size desires frightens many people as being too 'Big Brotherish.' But those involved in eugenics, or psychotherapy, or child psychology, or almost any aspect of family planning are constantly open to the charge of interfering in private lives, so that the charge would not be new ... . Of course, many injustices have been done with the rationale of being 'for their own good.' But the population avalanche may be used to justify—perhaps rationalize—contemplation of large-scale attempts to manipulate family size desires, even rather stealthily." This mode of reasoning may explain how some people will think and act, but it does not constitute anything approaching an ethical justifi cation.

25. P.R. Ehrlich (*19,* pp. 197-198) argues that the taking of strong steps now to curb population growth is the wiser and safer gamble than doing nothing or too little. This seems to me a reasonable enough position, up to a point. That point would come when the proposed steps would seriously endanger human dignity; an ethic of survival, at the cost of other basic human values, is not worth the cost.

26. This article is an abridgment of an "Occasional Paper" [*Ethics and Population Limitation* (Population Council, New York, 1971)] and was written while the author was a staff associate at the Population Council in 1969-70. I would particularly like to thank Bernard Berelson for his suggestions and criticisms.

Further Reading

Adelstein, M.E., and Pival, J.G. 1972. *Ecocide and Population.* New York: St. Martin's Press.

Allison, A., ed. 1970. *Population Control.* Harmondsworth, England: Penguin.

Berelson, B. 1979. Beyond Family Planning. *Science* 163 (February 7): 533-543.

Blake, J. 1969. Population Policy for Americans: Is the Government Being Misled? *Science* 164 (May 2): 522-529.

Borrie, W.D. 1973. *Population, Environment, and Society.* Aukland: Aukland University Press.

Brown, C.R., Newland, K., and Stokes, B. 1976. Twenty-Two Aspects of the Population Problem. *The Futurist* 10:5 (October) pp. 238-245.

Brown, L. 1985. Stopping Population Growth. In *State of the World 1984,* ed. L. Brown et al., pp. 20-34. New York: Norton.

Callahan, D., ed. 1971. *The American Population Debate,* Garden City, New York: Doubleday Anchor.

Cépéde, M. 1964. *Population and Food.* New York: Sheed and Ward.

Chandler, W. 1986. Investing in Children. In *State of the World 1986,* ed. L. Brown and E. Wolf, etc., pp. 159-178. New York: Norton.

Chrisman, R. 1970. Ecology Is a Racist Shuck. *Scanlan's Monthly* 1 (August): 46-49.

Christiansen, D. 1977. Ethics and Compulsory Population Control. *The Hastings Center Report* 7 (February): 30-33.

Colinvaux, P.A. 1970. *The Environment of Crowded Men.* New York: MSS Educational Publishing Company.

Davis, W.H. 1971. Overpopulated America. In *Environ/Mental,* ed. P. Shepard and D. McKinley. New York: Houghton Mifflin.

Durning, A. 1990. Ending Poverty. In *State of the World 1990,* ed. L. Brown, et al., pp. 131-153. New York: Norton.

Dyck, A.J. 1977. Alternative Views of Moral Priorities in Population Policy. *BioScience* 27 (April): 272-276.

Ehrlich, P.R. 1968. *The Population Bomb.* New York: Ballantine.

———. 1970. *Population, Resources, Environment.* San Francisco: W.H. Freeman.

———.and Ehrlich, A. 1990. Growing, Growing, Gone. *Sierra* 75, No. 2 (March/April): 36-40.

———., and Holdren, J. 1974. Population and Panaceas: A Technological Perspective. In *Science, Technology, and Freedom,* ed. W.H. Truitt and T.W. Solomons. Atlanta: Houghton Mifflin.

Forrester, J.W. 1976. Population Vs. Standard of Living. *The Futurist* 10:5 (October): pp. 246-250.

Hardin, G. 1972. *Exploring New Ethics for Survival.* New York: Viking.

———., ed. 1969. *Population, Evolution, and Birth Control.* San Francisco: W.H. Freeman.

Hare, N. 1970. Black Ecology. *The Black Scholar* 2 (April): 2-8.

Hart, H.H., ed. 1973. *Population Control: For and Against.* New York: Hart.

Howard, W.E. 1969. The Population Crisis Is Here *Now. BioScience* 19 (September): 779-784.

Hutchinson, E.P. 1967. *The Population Debate.* Boston: Houghton Mifflin.

Leisner, R.S. and Kormondy, E.J., eds. 1971. *Population and Food.* Dubuque, Iowa: W.C. Brown.

Lopp, R.E. 1973. *The Logarithmic Century.* Englewood Cliffs, New Jersey: Prentice-Hall.

McHarg, I.L. 1974. Man: Planetary Disease. In *Environment and Society,* ed. R.T. Roelofs, J.N. Crowley, and D.L. Hardesty. Englewood Cliffs, New Jersey: Prentice-Hall.

Mappes, T.A., and Zembaty, J.S. 1977. *Social Ethics: Morality and Social Policy,* pp. 332-375. New York: McGraw-Hill.

Marden, P. 1975. *Population, Environment, and the Quality of Life.* New York: Wiley.

Micklin, M. 1973. *Population, Environment, and Social Organization.* Hinsdale, Illinois: Dryden Press.

Miller, F., and Sartorius, R. 1979. Population Policy and Public Goods. *Philosophy and Public Affairs* 8 (Winter): 148-174.

Miller, G.T. 1975. *Living in the Environment,* pp. 7-21; 105-121; E16-E33. Belmont, California: Wadsworth.

Morrison, P. 1979. Beyond the Baby Boom. *The Futurist* 13 (April): 131-135.

Mudd, S., ed. 1964. *The Population Crisis and the Use of World Resources.* Bloomington, Indiana: Indiana University Press.

Myrdal, G. 1940. *Population.* Cambridge, Massachusetts: Harvard University Press.

Nelson, J.L. 1972. *Population and Survival.* Englewood Cliffs, New Jersey: Prentice-Hall.

Neuhaus, R. 1971. *In Defense of People.* New York: Macmillan

Oppenheimer, V.C. 1971. *Population.* New York: Foreign Policy Association.
Parry, H.B., ed. 1974. *Population and Its Problems.* Oxford: Clarendon Press.
Petersen, W. 1969. *Population.* New York: Macmillan.
Pohlman, E., ed. 1973. *Population: A Clash of Prophets.* New York: New American Library.
Pope, C. 1971. Family Planning Is Inherently Genocidal. True? False? *Environmental Action* 3 (May 15): 13-16.
Postel, S. 1984. *Air Pollution, Acid Rain, and the Future of Forests.* Washington, D.C.: Worldwatch Institute.
Postel, S. 1986. *Altering the Earth's Chemistry.* Washington, D.C.: Worldwatch Institute.
Ravenholt, R.T. 1976. Winning the Battle Against Overpopulation. *The Futurist* 10:2 (April) pp. 64-68.
Reid, S.T., Lyon, D.L., eds. 1972. *Population Crisis.* Glenview, Illinois: Scott, Foresman.
Reining, P., and Tinker, I., eds. 1975. *Population: Dynamics, Ethics, and Policy.* Washington, D.C.: American Association for the Advancement of Science.
Ridker, R.G., ed. 1976. *Population and Development.* Baltimore: Johns Hopkins Press.
Rosenfield, A. 1977. The Ethics of Supervising Family Planning in Developing Nations. *The Hastings Center Report* 7 (February): 25-29.
Ross, D.N., ed. 1976. *The Challenge of Overpopulation and Food Shortages.* New York: The Conference Board.
Salas, R.M. 1978. World Population Growth. *The Futurist* 12:5 (October): 276-282.
Shimm, M.G., ed. 1961. *Population Control.* New York: Oceana.
Spengler, J.J. 1975. *Population and America's Future.* San Francisco: W.H. Freeman.
Steinberg, I.S. 1974. *Population and Frustration.* Columbus, Ohio: Ohio State University Press.
Todd, W.M. 1971-72. Population Policy. *Environmental Affairs* 1: 716-734.
United Nations. 1973. Declaration on Social Progress and Development. In *Human Rights: A Compilation of International Instruments of the United Nations,* pp. 97-102. New York: United Nations.
US Department of the Interior. 1966. *The Population Challenge.* Washington, D.C.: Government Printing Office.
Waldron, I. 1973. *Environment and Population.* New York: Holt, Rinehart and Winston.
Warwick, D.P. 1974. Ethics and Population Control. *The Hastings Center Report* 4 (June): 1-4.
Wogaman, J.P. 1973. *The Population Crisis and Moral Responsibility.* Washington, D.C.: Public Affairs Press.
Wrong, D.H. 1967. *Population and Society.* New York: Random House.

11

THE PROBLEM OF PESTICIDES

Introduction

WHEN the Union Carbide plant in Bhopal, India accidentally vented a lethal chemical used in making pesticides—methyl isocyanate—in 1984, the death toll reached 3,500. Surprisingly, a study released in 1990 by the US Environmental Protection Agency (EPA) revealed that, since 1980, there have been 15 chemical spills in the USA that exceeded Bhopal in both volume and toxicity. The EPA report suggests that not enough is being done to protect USA victims of pesticides and other toxic chemicals.

One example of our not doing enough to protect ourselves involves the case of dioxin. When scientists discovered in the 1960s that the pesticide DDT was causing a decline in the bird and fish populations and building up in human tissue, the government banned it. Approximately 10 years later, when we discovered that PCBs were accumulating in humans and other animals and were likely carcinogenic, Congress banned PCBs, despite the protests of Monsanto. Every since the 1980s, however, scientists have been discovering similar harmful effects about a class of chemicals called "dioxins" that are used in pesticides, refining, and other industries. Even in the 1990s, however, government has allowed dioxin pollution to continue. The reason, say many environmentalists, is that dioxin has many powerful industrial friends who are dependent upon it. They claim that environmental and human health is being given second priority to economic interests.

Employment of pesticides has been a political, ethical, and public policy issue at least since 1962 when Rachel Carson published *Silent Spring,* her classic attack on chemical control of pests. Like most other environmental problems, the question of whether to use chemical pesticides is a difficult one, because it requires a philosophical—not merely technological—answer. By what ethical principles does one decide the equitable balance between the interests of those who profit from use of pesticides and those who bear the health costs? How does one reconcile possibly irreparable damage to the ecosystem and to certain species of animals with the benefits that accrue from chemical control of weeds, insects, and disease? Is it ethically possible to use the concept of personal liberty to defend industrial-agricultural activities which create economic prosperity but which may also cause unintended biological and environmental destruction? Although the answers to questions such as these require ethical justification, they are complicated by a number of actual or technological issues about which there is much dispute. Some of these largely scientific questions must be answered before ethical solutions to the pesticide problem can be discovered. Can one use increased numbers of natural predators to achieve biological rather than chemical control of pests? Must chemicals be used in ever increasing amounts if they destroy the nontarget predators that might otherwise keep pest population in check? Do organisms ever become immune to these toxins? What are the long-term biological effects of pesticidal residues?

The technological and ethical questions raised by the pesticide issue are basically problems of tradeoff. How much pollution ought one to bear in order to obtain high crop yields or protection from insect-borne diseases? Obviously we wish to feed ourselves and to be protected from medical risks. The difficulty is knowing when these benefits have been bought at too great an environmental cost. In other words, which is the greater good, long-term ecological and environmental stability or short-term protection from disease? If we say that the former is of more worth, then we might be condemning innocent persons to malaria or to starvation. Alternatively, if we claim that the latter is of greater value, then we might be acting contrary to the best interests of wildlife and future generations of human beings.

In the USA alone, more than two billion dollars are spent annually for chemical pesticides. Advocates of their use point out that they have saved the lives of millions of people by increasing the amount of food available and by controlling insect-borne disease. Human illnesses partially controlled in this manner include malaria, bubonic plague, typhus, and sleeping sickness. Moreover, without some form of pest control, the US Department of Agriculture estimates that we would have a 30-percent drop in food production. This in turn would mean either that food prices would have to increase by 50 percent or that several hundred million acres of new land would need to be brought under cultivation.

Proponents of pesticides claim that their use is even more essential in developing nations than in the developed world.

Because the population in third-world countries is doubling much faster than their ability to raise food, advocates say that the chemicals can make the difference betwen life and death. Ever since 1980, for example, per-capita food production has been declining in Africa. Over the short-term, broad-spectrum chemical pesticides often represent an inexpensive, quick "technological fix" for the problem of impending famine.

Opponents of pesticides point out, however, that even in developing nations use of the chemicals raises serious ethical problems. Among them is the fact that persons in third-world nations often are unable to give genuine free, informed consent to the risk of using pesticides; for example, they cannot read the warning labels on the packages. As a consequence, the World Health Organization (WHO) claims that 40,000 persons—mostly in developing nations—are killed annually by pesticide poisoning. Moreover, many of the deaths are caused by chemicals that are banned (as unsafe) for use in the USA and other western nations. Nearly one-third (29 percent) of all USA pesticide exports are banned for use at home. Their shipment abroad raises the question of the ethical limits on our duties to insure equal protection of all persons, even those in other countries. Opponents of pesticides also point out that the chemicals have global mobility and increase in concentration as they move up the food chain. This means that the costs and benefits of their use will not be borne by the same people, and that chemical effects on nontarget wildlife and marine organisms can contribute to extinction of certain species and to the instability of ecosystems.

The readings in this chapter present divergent views on the pesticide problem. The first selection is by H.F. Kraybill, a researcher at the National Cancer Institute in Bethesda, Maryland. In his article, Kraybill discusses the carcinogenic properties of pesticides; he subscribes to the thesis that there are "acceptable" doses of these chemicals. He also argues that the government is not warrranted in calling certain pesticides carcinogens when their cancer-causing activity has been confirmed only in lower animals and not in humans. He concludes that, whatever direct evidence of hazards to persons is incomplete, one should employ the experimental approach and allow the pesticides in question to be used because there is "societal and economic need" for them.

Kraybill's reasoning is criticized in the second selection, one I wrote especially for this book. I believe that his acceptance of a threshold for pesticidal injury is incorrect on both methodological and ethical grounds. After arguing that Kraybill's conclusions sanction numerous violations of equity and due process, I present two further points, viz., that there are workable alternatives to massive use of pesticides, and that widespread employment of these chemicals is not needed for medical, agricultural, or economic reasons.

Part of the solution to the pesticide controversy may be to realize that, as one congressman put it, the problem is not one-sided, not another "warm puppy issue." One cannot simply opt for the solution which safeguards the environment or which best protects the interests of the pesticide industry. Regardless of the position adopted, it is likely that *someone* will be hurt, either farm owners, the hungry, the victims of disease or, on the other hand, farm workers, wildlife, and future generations. The task of the ethician is to arrive at the most equitable distribution of costs in a situation where there are no zero-cost alternatives.

Readings

1. Kraybill, H.F. 1975. Pesticide Toxicity and the Potential for Cancer: A Proper Perspective. *Pest Control* 43 (December): 10-16. Copyright ©1971 *Pest Control Magazine*, Harvest Publishing Company, 9800 Detroit Avenue, Cleveland, Ohio. Reprinted by permission of *Pest Control* Magazine.

2. Shrader-Frechette, K.S. 1979. Pesticide Toxicity: An Ethical Perspective. Written especially for inclusion in this volume.

PESTICIDE TOXICITY AND THE POTENTIAL FOR CANCER

H.F. Kraybill, Ph.D.

MAN RECEIVES A multiplicity of environmental insults, which may be biological, chemical and physical. Some are of natural origin over which he may have no control. To name but a few, they are UV radiation, mold toxins, plant and bacterial toxins, natural radioactivity, and metallic or metalloid elements of geological origin. Our technological era has introduced new and added toxicants from synthetic industrial and agricultural chemicals, which may be found in the biosphere (i.e., water, air, and food). The universe reportedly has four million chemicals, with an annual increase of several thousand new ones.[1] Pesticides are a class of agricultural chemicals on which many toxicity studies have been conducted over the past decades.

Significance of Toxicity

Toxicity, like hazard, is a relative factor in consideration of adverse effects or responses upon animal or plant life. Absolute safety is unattainable. Facetiously, pathologists emphasize that living is indeed a hazardous process. We make decisions and adjustments daily to mitigate, or hopefully eliminate, the hazards in our living or environment. Certain nomenclature is used in defining the type of hazard or toxicity. First of all, one considers acute toxicity, which is the immediate or short-term adverse effect from exposure to a toxic agent such as a pesticide. Pesticides are chemicals which are toxic and are intended as such to eliminate pests. Beyond the short-term or acute effect evidenced in accidental poisoning to pesticides is the subchronic toxic effect which, in experimental animals, would mean three to six months; and for man a matter of months or longer. Finally, there is the more subtle and pernicious effect of a toxic chemical after a long period of exposure, usually referred to as chronic toxicity. Clinically, certain diseases are referred to as chronic since the illness is protracted and the symptoms observable over a span of months and years. The adverse biomedical response from a toxic agent on a chronic basis can be reflected in certain disease processes, one of which is cancer. Our attention is directed to this unwanted, debilitating effect which has been described as a "hundred diseases" because of its complexity, its site of action (target organs), and its variable mechanism under imposed conditions.

Basically, pesticidal chemicals are intended to kill pests and not people. Unfortunately, in our application of pesticides we proliferate them into the atmosphere, the waterways, our domestic animals and wildlife, and our food and diet as part of the food chain. Lastly, man in his exposure to the environment and in the work-place, such as pest control operations, formulating, aerial spraying, and manufacture, receives inadvertently the stress from pesticides which may, as in the case with DDT, result in a body burden (deposition in the body fat and other organs). Considerable attention has been focused on the residues of pesticides permitted after the crop is harvested and as the food moves from the processor to the consumer. Regulations have been developed to protect the ultimate consumer from ingesting unacceptable levels of these toxic chemicals. As long as we expect the maximum in food and fiber production, and pesticides are the only feasible control procedures for pests, one can expect a small finite level of these chemicals in foods and hence the principle of tolerance. When that pesticide chemical is declared a carcinogen (cancer producing agent) a safety or public health decision must be made.

In these considerations, especially for pesticide exposure and its relation to possible cancer induction, the focus, as indicated, is upon the general population exposure; but another equally significant consideration is the larger potential exposure received by workers in the production and application of pesticides. This component of the population is at the highest risk. Experimental studies are designed to accumulate toxicity data to assess the risk from pesticide exposure for the general population and for the worker. One must hasten to add that although cancer may appear as one of the awesome biomedical effects, other noncarcinogenic episodes such as permanent nerve damage or heritable genetic effects called mutations must not be overlooked.

Evaluation of the carcinogenic risk of various chemicals is achieved mostly through *in vivo* procedures (i.e., testing their chronic toxicity in experiments with animals). The route of exposure is usually through feeding a known amount in an experimental diet, through skin absorption tests (skin painting), or by inhalation studies; the latter to simulate exposures that pest control operators, formulators, and others may experience. More recently, rapid bioassay procedures such as *in vitro* techniques have been used where organ cell cultures and bacterial systems are subjected to the pesticide chemical, and cell transformations are observed as indicators of the neoplastic or cancer process. These rapid tests are not yet developed to the point where they can replace the typical two-year feeding experiment with rats, mice, or hamsters. Other species, of course, have been employed such as dogs, fish, rabbits, guinea pigs, and monkeys. Repeated observations, or epidemiological studies, may be made on man when he has been exposed to pesticides. It is not acceptable today from a medico-legal standpoint to purposefully administer a chemical to a human test subject to observe adverse biomedical effects. If man is inadvertently exposed to a pesticide then he can be observed. Such is the case with numerous industrial chemicals where groups of workers are followed retrospectively (case histories on exposures and symptoms recorded back over previous years), or prospectively (observations projected to years beyond the present exposure incident). Some of the best clues and indications of carcinogenicity of chemicals come from such studies. They represent the ideal for unequivocal evidence. They are observable effects on man, the subject of interest, and do not require extrapolation from an experimental animal which may or may not be comparable to man in risk assessment relative to cancer. Animal experiments are a compromise, a reference point to predict probable hazard.

The National Cancer Institute, through its carcinogenicity screening program, evaluates approximately 450 chemicals in a two-year period. Of the various chemicals categorized by use classes, pesticides represent about 17 percent of the total bioassayed (Table I). This would indicate that pesticides are an important class for safety evaluation.[2] Of the pesticides tested for carcinogenic activity, insecticides represent the largest class by percentage distribution[2] as shown in Table II. Of the insecticides tested, the organochlorine compounds (e.g., chlordane, heptachlor, dieldrin, DDT, lindane, etc.) are well represented in the bioassay program and probably present the most problems in toxicological testing. They usually show the most activity relevant to carcinogenic response and are invariably predominant, as a group, in the bioassay with the mouse. Rats are responders to some of these pesticides, but not as frequent. The bioassay, as conducted, is a screening procedure for activity and should not be considered as a comprehensive toxicological study unless it includes a plan or design to acquire data on body burden and metabolic fate, and a dose-response curve reflecting the incidence of tumors or cancers in the mouse or rat at various dose levels. Findings from the test groups are compared to incidence data on a control group of animals (untreated) so that a statistical evaluation can be made to arrive at an estimate of safety or risk. Such evaluative studies are expensive and time-comsuming, but are useful in a regulatory or decision-making process.

Over the years, it has become obvious to most in the pesticide industry that findings from chronic toxicity studies, including carcinogenicity assessment, have led to regulatory actions by the appropriate federal and state agencies. Table III shows some chemical pesticides reported as carcinogenic in various test species. Some of these have been banned while actions are pending on others.

Principles Employed in Assessment of Risk

Prior to rendering a decision on the carcinogenic risk of a chemical upon man, one must muster forth all the relevant facts to put the problem into proper perspective. The following guidelines should be helpful in this process.

A. Body Retention—Body Burden

It is usually assumed that a chemical which has the proclivity for retention in organs and tissues, as is the case with some organochlorine pesticides, automatically imposes a greater hazard than those readily excreted. This is not a basis for inferring either benefit or hazard. Similarily, the fact that an excretion rate is high does not necessarily imply the existence of a "no effect" level. The latter could prevail if the excretion of the chemical occurs independent of the concentration.

There is a background level of naturally occurring environmental toxicants (mycotoxins, polycyclic aromatic hydrocarbons, ets.), which may react with other chemicals or be retained in the tissues. This background level may be below the threshold described later. Hence, there is always in existence a baseline upon which chemicals introduced into the environment are really an added risk.

B. Threshold Levels—Safe Doses

Webster defines threshold as "that point at which a stimulus is strong enough to be perceived or produce a response or the entrance or beginning point of something." Some investigators believe that there is no threshold for a carcinogen or a mutagen or at least there is no scientific method available to establish such a threshold. This thesis is advanced, drawing a parallelism from radiation carcinogenesis to chemical carcinogenesis. A chemical differs from radiation hits in that it comes in contact with receptors in the intact animal and, having gone through certain biochemical reactions stepwise, is represented by a variety of metabolic transformations. Also, claims have been made as to the irreversibility of the toxic effect and that other additives in the milieu could synergize the action at the cell level. This concept, in part, is being challenged by newer ones. Falk[4], supported by others, has indicated that one single molecule may not cause a permanent change, which either alone or with other agents acting additively or cumulatively over a lifetime could produce a cancer. Roe[5] confirmed this concept when he noted that animals exposed to very low doses of a carcinogen did not respond as expected when challenge was interrupted for a few months before a promoter was administered. It was concluded that the cells originally insulted had either disappeared or would no longer respond. A basic unit of the cell DNA (deoxyribonucleic acid) is considered as the target for interaction of chemicals on the cell, thus leading to cell transformation. Mutagenesis and carcinogenesis are considered as the results of some alteration of the DNA. Studies with bacterial and human cells have shown that there is a repair of genetic damage. Therefore, the reaction of a carcinogen or a mutagen with DNA is not necessarily irreversible.[6-7] Hence, "no effect" thresholds must exist since a one molecule stress on a cell does not endow irreversibility.

To illustrate the point further, selenium, for example, has been shown to be carcinogenic at a high dose level; but, at a lower dose level it is required, like some other metals and metalloids, as an essential element. Again a "zero tolerance" is inapplicable for such events.

Using liver cells as a model, Dinman[8] has developed a theoretical approach on the probability of involvement of different chemical elements in a biological reaction with varying concentration. He very clearly shows that there is a finite level for an essential element. There is no activity below this level. Considerably below the threshold, there is indeed an adverse effect on the physiology, or a disease state, due to a deficiency. Thus, foreign substances or chemicals must exert an effect as to reactivity similar to the essential elements. Using this analogy, some investigators have calculated a threshold concentration for a carcinogen. For a 70 kg man, assuming 6×10^{13} cells in the body, a calculation was made that a threshold could be 10^4 molecules per cell or 8.6×10^{15} molecules/kg or 1×10^{-8} molar concentration.[9] Restated, the threshold for a carcinogen falls within the range of minimal concentrations of endogenous essential chemicals such as the hormones and vitamins. Herschler[3] has applied these principles to calculate the threshold for a carcinogen. Assuming a chemical carcinogen at molecular weight of 500 and administering this chemical to an adult with normal food intake, he calculated that one could keep below the threshold with a concentration of the chemical at 5 ppb in the total food ingested (Table IV). This provides some idea of the magnitude of the probable threshold and the requirements for a compliance action to technically and economically achieve a low-level residue or exposure.

From a physiological viewpoint, the body, in reacting to an insult, attempts to a maintain a homeostasis within finite limits with exogenous and endogenous components at the cell level. Not exceeding this finite level, whether the insult is a stimulus or inhibition, there is usually a null point for an observed toxic effect.[10] A carcinogen as a chemical reacts biochemically and pharmacologically, as any chemical would, in its transformation or metabolism within the body. At low levels—below the threshold—one can then anticipate that equilibrium is maintained; the cells or organs are not swamped; the probability for multiple hits on target DNA is not achieved; and repair mechanisms may be operable. Eckhardt[11] has maintained that sensitivity of analytical methods automatically establishes a tolerance for all substances, including a carcinogen. He feels that this is a correct procedure and admits that zero is not a precise value but is an undefined tolerance.

Some believe that carcinogenicity is a "black or white" phenomenon, so to speak; that is, a chemical is or is not a carcinogen. This concept is based on the fact that a tumor which appears in an experimental animal after administration of a certain chemical reflects cellular changes at the molecular levels. However, in some instances a cellular reaction is started but fails to progress to tumor formation within the lifespan of the animal, which could be dependent on the dose administered (i.e., dose too high or too low). While it may be difficult, as previously stated, to ascertain a "no effect" dose, one should think in terms of an acceptable risk level.[12] Druckery and others[13] have certainly shown that dose-response relationships exist, both with respect to tumor formation and the induction period. We have previously cited the case with selenium.

C. Safety Margin for Susceptible Members of the Population

Experimentally, in the evaluation of the carcinogenicity of chemicals, the investigator frequently selects an inbred susceptible strain of rodent which will respond to the challenge of the chemical within a range of doses. This approach is designed to simulate the probable role of susceptible individuals in the human population with potential variant thresholds. As laudable as this objective may be, the heterogeneity of the human population with their disease profiles and variant exposures during a lifetime may present a different picture which may not allow for extrapolation.[4] Of course, much has been written on the difficulties of extrapolating toxicity data from one species, such as a rodent, to man where metabolic patterns and biochemical behavior on response are different. The same difficulty prevails in extrapolating a "no adverse effect" in an animal to man.

D. Time to Tumor Formation

Some chemicals of high carcinogenic activity, when administered continuously to animals, will induce tumors at a short interval, the time interval being dose-dependent. If that same chemical is only given once or infrequently the time to tumor formation may be advanced, sometimes approaching the life span of the animals. If the chemical has a demonstrated low carcinogenic activity (potency), studies with small numbers of animals and low doses of the chemical may not reveal tumor formation within the lifetime of the animal. By increasing the size of the animal population, which is costly, one might perceive a tumor incidence.

Despite the above, there may be chemicals of low toxicity, which even when administered at inordinately high doses, may still have a time to tumor formation that exceeds the life time of the animal. In comparison to man, this could mean that the induction period would fall far beyond a life span of 100 years, or as one writer has emphatically stated "beyond the age of Methuselah."

E. Metabolic Overloading

Experimentally, it is common practice to challenge the animal with a dose schedule that far exceeds the exposure levels received by the general population. In many instances, the dose to the animal even exceeds the calculated exposure level for man in the work-place. The reasons for this procedure are obvious—it is a compromise between challenging a small group of animals with a high dose to induce a positive response, or having thousands of animals, which is not always feasible, dosed at levels more nearly approximating the exposure levels of a human population and their associated tumor incidence.

In some instances this procedure biases the tests in favor of highly toxic compounds (i.e., low toxicity compounds are tested at doses of 1,000, 10,000 or 20,000 times the dose of a high toxicity compound, while environmental and occupational exposure for man could be at the same levels for both classes of compounds). In fact, the test dose recommended for some studies has exceeded the industrial capacity for production of a compound needed in the tests. Inequality in dosing, or this bias, is in favor of biodegradable compounds. Frequently, the overdosing also swamps the capacity of enzymes in the liver required for detoxification of the chemical, a situation probably not encountered in man's real life exposure to the chemical. The only comparable situation in man is acute poisoning, sometimes resulting in death. The major question then becomes whether one can really extrapolate such abnormal stresses in the animal to any comparable exposure situation in man. The differences in cellular environment and response to such non-comparative stimuli under these conditions create problems for scientific assessment.

F. Epidemiological Verification of Experimental Studies

It is axiomatic that the spectrum of chemicals observed to induce cancer in man were also subsequently shown to be carcinogens in experimental animals. The one exception to the proven human carcinogen—arsenic shown to be a carcinogen in some occupational studies—has not yet been corroborated in animal studies.[14] Not many chemicals, however, showing carcinogenicity in animals have been proven out in man. There are many reasons why correlations are not always evident, but this should not be surprising since there are species variations, strain variations and sex variations relevant to response and susceptibility to toxic agents. Certainly, genetic factors, immunological competence, nutrition, enzyme alterations, and traumatic injury to organs and tissues all may play a role to enhance susceptibility to toxicants.

Man, of course, is exposed to multiple insults and the numerous cases of additive effects of synergism should be kept in mind in assessment of risk. A few examples are: asbestos exposure plus cigarette smoke, diet variants plus carcinogens, etc.; all of these are considered in the context of environmental cancer.

Roe[15] in his review entitled "Carcinogenesis and Sanity" has emphasized the need for distinguishing between primary effects of a chemical as to whether it is a carcinogen or is merely acting synergistically as a cocarcinogen. This situation could well apply to DDT, but more will be discussed on this matter later.

The matter of extrapolation of carcinogenicity findings in rodents to man repeatedly comes up. The best posture to assume is one of concern, but statements are always equivocal. For example, the relevance of findings on aldrin and dieldrin, demonstrated as a tumorigen in the mouse with considerable variation among strains, has been troublesome to some. As Deichmann[16] has stated "the effects in mice in general are not, without reservation, predictive of other mammalian species, including other rodents, as well as man." Quite frequently a regulatory agency, in the decision-making process on safety, may declare a chemical as an "imminent hazard." This may raise some controversy in scientific circles about an agent that could conceivably have, at best, a long-term effect. If the end point in man is cancer, although the induction time could range from 11 to 30 years as was the case with vinyl chloride, then the concern is obvious and the decision to rule on the side of safety is apparent and justified.

Not all cases, however, are as clearly warranted as was the case with vinyl chloride. For example, DDT has only been established as a carcinogen in the mouse and not in other species. It is now believed that not DDT but DDE could be the etiologic agent.[18] Be that as it may, there is no evidence thus far that DDT is a carcinogen to man although man has been exposed to DDT since 1940. Studies on man, of course, are quite limited. Since DDT is ubiquitous, it is practically impossible to find a cohort study group which is free of DDT in their bodies and thus free of exposure.

DDT, like many of the organochlorine compounds, usually targets on the liver and biliary passages, with ancillary sites such as gall bladder and the extrahepatic gall ducts. Using the data from the American Cancer Society, the Bureau of Census, Vital Statistics of the Department of Health, Education, and Welfare,[19] Deichmann[16] attempted to show that deaths from liver disease, including malignant neoplasms of the liver and biliary passages, were declining from 1940 to 1972 (Table V). In 1968, the ICDA[20] redesignated their coding of liver neoplasms and the ancillary cancers of contiguous organs so that calculated rates reflected in Table V, given by the National Cancer Institute Field Studies and Statistical Program, reflect the lower calculated rates for the years specified.[21]

The liver death rates listed by Deichmann (1975) collated from various sources is an all-inclusive, nonspecific group which does not delineate liver neoplasms. The change in ICDA codes makes interpretation of data prior to 1968 rather difficult since it does not separate out the specific rates for target organs such as liver, extrahepatic and intrahepatic liver neoplasms, gall bladder and bile duct and primary and secondary site neoplasms.

One could generalize that for aggregate cancer deaths for these organ sites (probable target organs for DDT and other organochlorine pesticides), the cancer incidence, in this case on a mathematical basis, would not appear to have increased as significantly as might be presumed. However, for a population group at high risk in some geographical regions where there is high exposure, more liver cancers could conceivably be noted and might, in those regions, actually show an increase after 35 years. The thesis advanced of a declining incidence for liver cancer, or at least a static rate in the United States, is a very tempting argument for negative causal relationships for organochlorine pesticides and human cancer. However, one must await a more definitive analysis of all data by the National Cancer Institute to ascertain whether there are regions where the cancer specific rates show a different trend than those given for liver cancer in the preceding discussion.

Discussion

The decision-making process on the potential cancer hazard to man from exposure to pesticides referenced to previously reported carcinogenesis studies in experimental animals is a complex matter. For this reason, the preceding principles cited, which one must consider in risk assessment, are intended to facilitate a scientific appraisal. One must look at the data and findings with a critical overview. Perhaps more questions are being raised than there are answers.

There is at this time no universal agreement on approaches to appraisement of the carcinogenic potential of an environmental chemical. However, one detects that some of the firmly held views of the past are now being challenged. Toxicologists and experimental oncologists are thinking in terms of safe doses and thresholds, the aspect of added risk, and the quantification of this risk as it is balanced against societal benefits. Obviously, a visceral reaction toward a zero exposure and elimination of all chemicals with some toxic or carcinogenic activity is no longer tenable and indeed unwarranted if mankind is to live in balance with its environment. The principles cited are intended to provide the essential inquisitiveness for those who have the responsibility for permitting the safe and continued use of agents that will maintain our food and fiber production in a world of advanced technology necessary for survival.

Admittedly, the mechanism of carcinogenesis is not fully understood and, therefore, a carcinogen, as a toxic agent, may evoke a unique biomedical response. However, one must not lose sight of the fact that these chemicals exert effects on metabolic pathways whereby they may be degraded and excreted. As described earlier, below a certain level these chemicals in the milieu may exert an effect at the cellular level similar to that of hormones or vitamins. Some chemicals shown to have some carcinogenic activity at high levels, such as selenium, at low levels are physiologically essential in maintenance of the hematopoietic (blood forming) system.

As in any toxicity investigation, some experimental approaches for screening a chemical for carcinogenicity are in the extreme and violate certain principles of biochemistry and pharmacology. Fortunately, these studies are being challenged in design, such as the tendency to overload an animal metabolic system with a chemical or drug which would not be countenanced if one were to bioassay a seemingly inocuous chemical as a vitamin or an essential mineral. Homeostasis must be maintained, organs must not become refractory to the overload, and different metabolic pathways must not be taken if one expects to induce neoplasia that is relevant to environmental stresses which man actually receives. In one study on DDT, the experimental doses used to prove carcinogenic activity in the mouse were 853 times the dietary exposure for man and about three times the highest exposure man would receive in the workplace.[22]

We are indeed being confronted with an expanding number of chemicals which, in various test systems, are proven as either carcinogenic or mutagenic. *Appropriateness of the test procedure,* a Proviso delineated in the Delaney Amendment to the Food, Drug and Cosmetic Act, should be duly considered by inquiring scientists. Certainly, one does not wish to generate blind phobias or unwarranted concerns in the general public which, far too often, precipitate some regulatory action mediated by political intervention without the benefit of sound scientific deliberation and judgment.

Roe,[13] for example, has indicated that one should clarify the distinction between a carcinogen and cocarcinogen before the list of compounds to be excluded from the environment becomes so long as to render the task utterly impossible. He concludes that "if cancer may be induced by an agent in a variety of species and following a variety of routes of administration then the likelihood is that the agent is a true carcinogen." (Vinyl chloride, benzidene, and especially benzo(a)pyrene would be unequivocal.) He goes on to state that "if cancer can only be induced in one species or under special circumstances, or if the effect observed is only an increase in the incidence of a type of tumor that arises commonly in the test species, or there is no clear relationship between dose and effect then the likelihood is that where tumors appear, the agent is acting only as a cocarcinogen or promoter." One could probably name quite a few candidates that fit this class which have been misclassified as carcinogens.

Care then must be taken in the planning of experiments to insure that they are useful in extrapolation of data to man. This is necessary in that experimentally derived reports serve as a warning and should be valid because of their national importance and socioeconomic impact.

Therefore, a general view emerges that exclusion of environmental chemicals with indications of carcinogenicity occurs in the absence of technological and societal pressures for balancing out risk and benefit. Rephrased, if there is a substitute for the indicted pesticide, even if inadequately tested for carcinogenicity, the restriction or banning is easy. Thus the debate is of no great consequence and the scientific appraisal or reappraisal is essentially non-existent. Some chemical pesticides may fall into this class. However, where the societal and economic need is great, the case has merit and the benefit side of the equation is readily apparent. With time, more societal decisions will be evident, more risk-benefit analyses will be made, and some pesticides could be retained by virtue of their benefits. Reasonableness dictates that there is no absolute safety. Indeed, if good controls are instituted for mitigation or elimination of exposures then the safety of pesticide use increases. To enjoy the benefits of chemical pesticides which may be needed for some time to come, one must quantify the total cost to society of the risk versus benefit. These latter decisions may not be made by scientists but scientists should play a role in providing good evidence on the biomedical aspects of the chemical pesticide. Ultimately, one may envision a colloquium of governmental, industrial and academic scientists collaborating in the scientific decision-making process on carcinogenicity. It is too awesome a task to be left to those with an emotional response or to one agency having narrow programmatic interests. We are entering a new era in scientific interchange and an era of new conceptual approaches in the appraisement of environmental carcinogens, including pesticides.

Summary

The basis for considering the potential carcinogenic hazard of any pesticide derives from studies in laboratory animals. *None of the pesticide chemicals thus far have been shown to be carcinogenic to man. Experimental studies are of paramount importance and so important that they demand critical appraisal as to the validity of the findings.* Experimental work is a means to an end and, lacking evidence on safety or hazard to man from direct observations on man, the experimental approach is used to furnish guidance in assessment of health risks. The experimental procedure must, however, have credibility in design; otherwise this means, or approach, assumes greater importance than the end.

As discussed earlier, one can establish an exposure level in an experimental population where no tumors appear. This, as has been shown, is conditioned statistically by the numbers of animals used in the test, the species or strain used, or other factors. Despite these contingencies, the dose may be insufficient to initiate a change in cellular constituents which could progress to tumor formation in the life span of the animal. This leads one to believe, along with other rationalizations, that at very low levels it is conceivable that carcinogens will not evoke tumors.

Perhaps there is no safe dose to be determined for a pesticide, a drug, an industrial chemical, a natural toxin or other classes of environmental chemicals, but this does not mean a safe dose level does not exist insofar as carcinogenicity.

The assessment of carcinogenesis is, according to some, a statistical phenomenon; i.e., measuring tumor incidence of a test group against the incidence for a control group. We should not inevitably rely on statistical tests of validity which may curb our imagination. We should inquire as to the biological aspects of the events observed. One should not become preoccupied with the response to a pesticide at a high dose challenge but should perceive the possibility of lack of response at a low dose level. The adaptive response, for example, at a low dose should not necessarily be misconstrued as injury but as a beneficial defense mechanism.

One should not exhaust one's energy or resourcefulness in the inexorable task of looking for a "zero" exposure, the "no effect level," but should view the problem in the context of an "acceptable level of risk." This concept should hold for a carcinogenic or noncarcinogenic event.

We recognize that some pesticides have been indicted as a carcinogenic risk to the mouse; others as a risk to the rat, dog, or trout, and a few as a risk to two species. Since we have no direct evidence in man in which to seek refuge, our decision-making process then is based on a possible interpretation and extrapolation from the animal experience.

We must, therefore, resort to the best scientific judgment in the appraisal of the risk in this area of uncertainties. One must appraise the validity of the observations derived experimentally by considering carefully such factors as body burden of the pesticide, the possibility of a threshold, the degree of added risk, the margin of safety for susceptible members of the population, the time frame for tumor formation, the degree of metabolic overloading (overstating of the risk) and the possibility of any epidemiological leads.

In the final analysis, one is faced with a societal decision. Fortified with all the scientific evidence as to safety versus hazard, one ultimately must, through designated bodies, establish what degree of risk, if any, the society can accept if benefits are to be accrued from the use of a pesticide. Specifically, will the national population accept, in the case of a chemical carcinogen, an increase of one cancer per million of population, or 100 million of population? The mechanism for evaluation is based on the concept that there is not in this universe a zero order of risk. Evaluation of risk is a continuous process in our living.

In this complex area of carcinogenesis, wherein we do not yet fully understand the mechanism of action, we must continually keep an open mind so that we can perceive and follow up the unexpected. Having such perception, we should recognize as Campbell[23] has stated "that the main reason people put truth on such a high pedestal is to avoid the danger of accidentally running into it."

Quite a few concepts and opinions are expressed in carcinogenesis evaluation, but as the old saying goes, "there's nothing like a little experience to upset a theory."

References

1. Kraybill, H.F. and Richardson, H.L. (1973). Orientation on a coordination program on environmental carcinogenesis. Bull. of the Soc. of Pharmacological and Environmental Pathologists, No. 4, pp. 18-22.

2. Kraybill, H.F. (1975). Research priorities —Interagency Coordination—Chemical Carcinogens. Proceedings of Environmental Protection Agency Conference on Health Effects of Pesticides, July 14-15, 1975, Arlington, Virginia. (In press).

3. Herschler, D. (1974). New approaches to a definition of threshold values for "irreversible" toxic effects. Arch. Toxicol. 32: 63-67.

4. Falk, H.L. (1975). Considerations of risks versus benefits. Environmental Health Perspectives, Vol. 11, pp. 1-5.

5. Roe, F.S.C. (1972). On the persistence of tumor initiation and the acceleration of tumor progression in mouse skin tumorigenesis. Int. J. Cancer, *9:* 264.

6. Morgan, K., Hastings, P.J. and von Borstel, R.C. (1973). A potential hazard: Explosive production of mutations by induction of mutators. Environ. Health Perspec. *6:* 207.

7. Cleaver, J.E. (1968). Defective repair replication of DNA in xeroderma pigmentosum. Nature *218:* 652.

8. Dinman, D.B. (1972). "Non-concept" of "no threshold": Chemicals in the environment. Science *175:* 495.

9. Friedman, L. (1972). Problems of evaluating the health significance of the chemicals present in food. Fifth Inter. Congress Pharmacol., San Francisco, Calif. July 27, 1972.

10. Stokinger, H.E. (1971). Sanity in research and evaluation of environmental health. Science *174:* 662.

11. Eckhardt, R.E. (1973). The Delaney Amendment. Prevent. Med. *2:* 156-158.

12. Munn, A. (1972). Extrapolation of experimental carcinogenicity to the reality of industrial exposure. Ann. Occup. Hyg. *15:* 75-79, Pergamon Press, 1972. Great Britain.

13. Druckery, H. (1967). Union International Contra Cancer Monograph. Series 7.

14. Kraybill, H.F. (1975). Carcinogenicity of arsenic and arsenic compounds. Proc. of Natl. Conf. on Health Effects of Occupational Lead and Arsenic Exposure. February 24-25, 1975, Chicago, Illinois (In press).

15. Roe, F.S.C. (1968). Carcinogenesis and sanity. Food Cosmet. Tox. *6:* 485.

16. Diechmann, W.B. (1975). The Market Basket: Food for thought. Cummings Memorial Lecture, Am. Ind. Hyg. Assoc. Journ., June, 1975, pp. 411-429.

17. Walker, A.I.T., Thorpe, E. and Stevenson, D.E. (1972). The toxicology of Dieldrin (HEOD). I. Long term oral toxicity studies mice. Fd. Cosmet. Toxicol. *11:* 415.

18. International Agency for Research on Cancer (1974). IARC Monographs on the evaluation of carcinogenic risk of chemicals to man—some organochlorine pesticides. Vol. 5, page 109. Int. Agency for Research on Cancer, Lyon, France.

19. Bureau of the Census (1943). Vital statistics of the United States, 1900-1940. U.S. Department of Commerce.

20. Department of Health, Education and Welfare, Public Health Service, National Center for Vital Statistics (1968). International Classification of Diseases (ICDA) adapted for use in the United States, 8th Revision, Government Printing Office, Washington, D.C.

21. Department of Health, Education and Welfare, Public Health Service, National Cancer Institute. (1975). Calculations on liver cancer rates in the USA by age adjustment. Furnished by the author by Field Studies and Statistics Program, October 2, 1975.

22. Kraybill, H.F. (1975). Conceptual approaches in non-occupational environmental carcinogenesis in Modern Advances in Toxicology: Environmental Cancer, eds. H.F. Kraybill and M.A. Mehlman. Hemisphere Publishing Company, Washington, D.C. (In press).

23. Campbell, J. (1966). Editorial in Analog. *77*:178.

TABLE I

Relative Distribution of Chemicals by Classes Evaluated for Carcinogenicity

	Percent of Total Bioassayed
Pharmaceuticals	20.8
Pesticides	17.4
Industrial Chemicals (Organic)	15.2
Metallo Compounds	6.7
Natural Food Products	5.7
Food Chemicals	1.6
Tobacco Ingredients	0.8
Environmental Agents (General)	0.2
Miscellaneous (Structural Analogues, Multiple Uses)	31.6
	100.0%

Reference: DHEW, PHS, NIH, National Cancer Institute, Division of Cancer Cause and Prevention, Carcinogenesis Annual Report FY 1975, Summary Report of Bioassay Operations Segment.

TABLE II

Distribution of Pesticides by Classes Evaluated for Carcinogenicity

	Percent of Total Bioassayed
Insecticides	52.1
Fungicides	25.5
Herbicides	18.4
Bactericides	2.0
Molluscicides	1.0
Acaricides	1.0
	100.0%

Reference: DHEW, PHS, NIH, National Cancer Institute, Division of Cancer Cause and Prevention, Carcinogenesis Annual Report FY 1975, Summary Report of Bioassay Operations Segment.

TABLE III

Some Chemical Pesticides Reported as Carcinogenic in Various Species

Chemical	Application	Species Tested
1. Arsenic Oxide	Insecticide	Man[1]
2. Carbarsone	Amebicide	Trout
3. Aramite	Miticide	Dog-Rat[2]
4. Isopropyl-n-Phenyl Carbamate	Herbicide	Mouse-Rat
5. Beta-Propiolactone	Fumigant	Rat-Mouse
6. Benzene Hexachloride (BHC)	Insecticide	Mouse
7. Dieldrin - Aldrin	Insecticide	Mouse
8. DDT	Insecticide	Mouse-Trout[3]
9. Ethylene Thiourea	Fungicide	Rat-Mouse
10. Aminotriazole	Herbicide	Rat-Trout[4]
11. Heptachlor	Insecticide	Mouse[5]
12. Chlordane	Insecticide	Mouse[5]

1. Occupational studies. 2. Banned. 3. Banned except export for malaria control. 4. Banned except for on right of way. 5. Regulatory actions in 1975.

TABLE IV

Estimation of a Threshold for Carcinogen

Part I	
Body Weight of Reference Man	70 KG
Number of Cells in Body	6×10^{13} Cells
Avogadros Number	6.02×10^{23} Molecules/Mole
Threshold Value/Cell	10^{4} Molecules
Threshold Value Molecules/kg	8.6×10^{15}
Molar Concentration/kg	1×10^{-8}
Part II	
Assuming M. Wt. of Chemical	500
Calculated Threshold	5 ppb

Friedman, L. (1972); Herschler, D. (1974); Dinman, D.B. (1972)

TABLE V

Liver Death Rate (1939-72) and Age Adjusted Cancer Rates for Neoplasms of Liver, Gall Bladder and Bile Duct (Rate/100,000 in Total United States)

Year	Liver Deaths	Liver and Intrahepatic[1]	Liver Unspecified[2]	Gall Bladder Bile Duct[3]	All Sites Intra and Extrahepatic[4]
1939	9.6				
1940	7.4				
1949	6.7				
1950	6.5				
1960	5.4				
1968		1.06	1.4	2.2	5.3
1969		1.06	1.4	2.0	5.3
1970	4.7	1.08	1.3	1.9	5.2
1971	4.7	1.09	1.3	1.9	5.2
1972	4.7				

ICDA Codes:[1] = 155;[2] = 197.8;[3] = 156;[4] = 155, 156, 197.7, and 197.8. Data from Deichmann (1975); Bureau of Census (1943); and National Cancer Institute (1975)

PESTICIDE TOXICITY: AN ETHICAL PERSPECTIVE

K.S. Shrader-Frechette

Introduction

THERE IS A SAYING, among members of one of the Bantu tribes of southwest Africa, that a person should not shoot a bird resting on his own head.[1] We humans, however, have shot many birds on our heads. We have ignored our interdependence with other members of our ecosystem, and we have forgotten that the health and well-being of all inhabitants of the planet are intimately interconnected.

Several years ago, for example, the World Health Organization used DDT to kill off malaria-carrying mosquitos in Borneo. The chemicals killed the mosquitos but not the roaches, who accumulated the pesticide in their bodies. When the long-tailed lizards, called geckoes, ate the roaches, the DDT in their prey caused disorders in the nervous systems of the lizards. This caused the geckoes to become sluggish and to fall victim to the village cats who ate them. The cats died from the DDT in the lizards. Rats, carrying the threat of an epidemic of plague, moved in from the Borneo forests, and cats had to be flown into the villages to catch the rats. The cats controlled the rat population, but the roofs of the peoples' huts began caving in. The lizards, now vanished, had formerly eaten the caterpillars that ate the roof thatching.[2]

The Pesticide Problem

Because pesticides have an effect on all living things, not just pests, their use continues to raise both ecological and ethical questions. Most of these issues focus on the morality of accepting a given risk (e.g., chemically induced cancer or poisoning), in order to obtain a certain benefit (e.g., protection from disease or crop destruction). The problem, of course, is knowing what levels and what types of risks are acceptable. Is a risk morally desirable, for example, if it causes no harm to present persons but, because of its cumulative genetic effects, threatens the well-being of future generations? In this same vein, how does one balance the rights of current individuals to food (allegedly enhanced by pesticidal protection) with the rights of future descendants to an intact genetic endowment (jeopardized by exposure to pesticides)?[3] Or, how does one balance the rights of humans to protection from disease and crop pests with the alleged rights of nontarget animals inadvertently killed by these chemicals?[4] Likewise how does one determine whether the risks from pesticides ought to be borne equitably? Is it ethically permissible for one subset of people (e.g., agricultural workers) to bear a health hazard from the toxins so that all persons may enjoy fruits and vegetables of a particular quality?

Ethical questions such as these are becoming more and more important, since chemical production is increasing every year; world pesticide use for 1984 is expected to increase 50 percent from its 1974 level. In North America alone, this increase is expected to be 25 percent, despite the fact that North America already accounts for approximately 40 percent of crop pesticide use in the world.[5] Since increasing employment of chemical pest controls leads to greater degrees of human and environmental hazards, the stakes in this ecological problem are becoming larger and larger. On the one hand, the UN Food and Agriculture Organization wants to reduce food losses to pests by 20 to 30 percent and to achieve a fourfold increase in world food supplies.[6] On the other hand, thousands of people each year are poisoned or injured by pesticides.[7] Likewise, industry spokespersons maintain that government regulation is effective in insuring consumer safety,[8] while environmentalists charge that the chemicals are one of the most immediately hazardous pollutants known, even when used as directed.[9]

With Rachel Carson, I believe that the immediate solution to the pesticide problem is not to eliminate these chemicals completely but rather to use them in a more discriminating manner.[10] To support this point of view, I suggest that ethics demands greater use of available means of biological, as opposed to chemical, control and that justice requires one to admit that existing policies of pest management violate principles of equity and due process. Since these conclusions cannot be substantiated without coming into direct conflict with persons who favor continuing (and even greater) use of pesticides, I will criticize a number of theses widely accepted by proponents of chemical pest control.[11]

Some of the most important of these points, used to support large-scale employment of pesticides, are the following.

1. Although chemical pest control is hazardous, it is realistic to accept the minimal degree of risk it presents, since absolute safety is unattainable in any sector of life.

2. Since pesticides have not been proved to be carcinogenic in humans, it is not clearly unethical to employ them.

3. Because there may be a threshold for injurious effects of chemicals, their use is not necessarily contrary to human well-being.

4. Many pesticides (e.g., DDT) have been wrongly banned as a result of believing that they were carcinogens when in actuality they are cocarcinogens.

5. Data on pesticide toxicity incorrectly suggest that the chemicals are more dangerous than they really are, since test animals receive exorbitant doses of the chemicals being evaluated.

6. Because there are no workable alternatives to pesticides, their employment is necessary.

7. Use of chemicals is needed to insure food, fiber, and human survival.

8. The benefits of using pesticides outweigh the costs.

In the next few pages, I will critique these eight major tenets used to justify current levels of pesticide use. In doing so, I hope to illuminate some of the ethical questions involved in debate over pest control.

"The pesticide problem" is highly complex; it is difficult to conceive of any simple solution to it without somehow jeopardizing the interests of one of the parties to the debate. Despite the difficulty of knowing precisely which particular policy ought to be followed, I argue, first, that the current notion of "acceptable risk" followed by many scientists and lawmakers errs as a standard for allowable hazards. On utilitarian principles, it fails because it is insensitive to considerations of equal protection and due process, and hence ignores the requirements upon public policy set by the Bill of Rights.

I argue, secondly, that it is both illogical and unethical to claim that certain levels of pesticide exposure are acceptable so long as they are shown only to be cocarcinogenic or only to cause animal, rather than human, cancer. Such reasoning commits the fallacy of the appeal to ignorance and wrongly presupposes the morality of experimenting on human beings. Instead, I explain why animal studies are necessary for assessing the risks involved in human exposure to pesticides. As such, animal experiments should be done on the basis of large dose-response studies, so as to provide scientific data for guaranteeing equal protection for persons who might be especially sensitive, even to low levels of pesticides. Otherwise, I argue, chemical-exposure policies might wrongly be guided by utilitarian, rather than egalitarian, ethics.

Any critique of current levels of pesticide use is unrealistic, unless it addresses the problem of how to cope with pest management when chemical controls are reduced. For this reason, I will be discussing alternatives to the employment of pesticides. There is clear evidence that chemicals are not the only workable and economically feasible means of pest control. I propose that, if both chemical and biological forms of management are possible and workable, then one is morally bound to choose the form which presents the least amount of risk to human well-being. I argue, further, that—although use of pesticides leads to conservation of land and energy resources—it often destroys the balance of nature by killing nontarget organisms, inducing pest resistance, and therefore intensifying the original problem. Finally, I explain why one of the most serious pesticidal injuries to the balance of nature is to human genetic integrity. I argue that extensive use of these chemicals violates the rights of future generations by jeopardizing human and animal genetic endowment.

In the closing pages of this section I investigate pesticide use in terms of its consequent costs and benefits. I argue that widespread and massive use of the chemicals is not necessarily justified on the grounds that they help to provide food, fiber, and survival for the human race. Rather, I maintain that quite different "pesticide-ethics" ought to be employed for developed and underdeveloped nations. Following the calculations of the US Department of Agriculture, I also indicate how a developed country (the US) could cut pesticide use by 50 percent without a substantial loss either in profits or crop production. Finally, I explain why pesticide use is often wrongly said to provide more good than evil to society. This erroneous statement, I argue, is based on the fact that US cost-benefit analyses of pesticide employment typically ignore social hazards (e.g., carcinogenesis) and parameters which are difficult to quantify (e.g., environmental and aesthetic losses). In closing, I suggest that, once cost-benefit studies are done in a truly comprehensive way, it will be difficult to argue that present policies of pesticide use are morally acceptable.

A number of scientists firmly believes that present levels of pesticide use are well within the limits of moral acceptibility. They also maintain that current chemical methods of pest control are based on policies which grossly overstate the risks of pesticides and wrongly understate their benefits. In fact, they suggest that the US government was wrong in banning DDT in 1972.[12]

As a result of its "stability in the environment, toxicity to wildlife and nontarget organisms, and adverse physiological effects on humans," the director of the US Environmental Protection Agency banned DDT in 1972 from crop and non-health-related uses.[13] Since that time, one of the "highest priorities" of "the pesticide interests" has been "to get DDT back into the mainstraim," says one well known entomologist.[14] Acceptance of DDT, say the environmentalists, is acceptance of "the pest-control status quo" as opposed to newer and safer methods of biological and integrated control.[15] Hence, the current pesticide problem is not only the focal point for discussion of what levels of risks may be imposed ethically on society, but also the framework for political debate over whether one of the cheapest and most widely used pesticides, DDT, ought to be allowed to be used more freely.

Pesticide Toxicity and Acceptable Risks

Let us examine, in detail, each of the reasons employed to justify widespread use of pesticides. One often cited argument is that, although such chemicals are hazardous, it is realistic to accept the minimal degree of risk they present, since absolute safety is unattainable in any sector of life.[16]

As far as it goes, this argument is very persuasive. Of course "living is indeed a hazardous process,"[17] and of course it is impossible for anyone anywhere to attain a position of absolute safety. What is correct about this approach is that it recognizes the pragmatic point that risks and benefits must be balanced, and that zero risk is an unattainable goal. What is deficient is that it discusses the moral acceptability of a hazard only in terms of the "degree of risk."[18] Surely the magnitude of a technological danger is not alone sufficient grounds for judging its permissibility. Let us see why this is so.

Two other important parameters affecting the acceptability of a hazard are how equitably the risk is distributed, and whether it is taken on voluntarily or imposed involuntarily. Clearly merely "scientific evidence as to safety versus hazard" is not sufficient, as has been suggested,[19] to determine what is an acceptable risk. Even if it could be proved that the carcinogenic (cancer-causing) and mutagenic (inducing genetic damage) hazard of pesticides were very slight, one would still have to show that this risk were equitably distributed in order to substantiate the claim that it were acceptable.

Suppose, for example, that a number of families live downwind of a large cotton crop regularly sprayed with a pesticide. Government scientists indicate that 63 percent of all pesticides are applied by aircraft,[20] that aerial sprays typically reach their targets in amounts less than 50 percent of that distributed,[21] that chemical application efficiency is undercut primarily by wind drifts, and that no more than 2 percent of applied amounts of pesticide are effective.[22] For all these reasons, it is plausible to assume that families living downwind of crops regularly subject to aerial pesticide application are likely to receive a much higher dose of these chemicals than are most Americans. In fact, statistical studies indicate that, in some farm counties in California, half of the children annually show one or more signs of pesticide poisoning.[23] If this is true, then the costs and benefits of the chemicals are not borne equitably. Whereas all or nearly all Americans may be said to benefit from efficient growing of cotton, for example, a small subset of this group (e.g., those living downwind of aerially treated crops) may be said to bear a greater cost in the form of increased health risk. If all Americans have rights to equal protection by virtue of the Fifth and Fourteenth Amendments, then the acceptability of the pesticide risk is, at least in part, clearly a function of the equity with which people are protected from, or compensated for, it.

While the previous considerations do not indicate, *a priori,* that use of these chemicals is unethical, they do show why arguments based on the magnitude of a hazard do not constitute sufficient grounds for determining the moral acceptability of a particular risk. A plausible case could likewise be made to show that reasoning of this type ignores other important parameters, in addition to equity. Although I will not go into these analogous considerations, it is clear that the judgment of the moral acceptability of a risk is also a function of whether the hazard is assumed involuntarily or voluntarily, whether it has immediate or delayed effects, whether it is a common or a dread hazard, whether it affects average or especially sensitive people, and whether the danger will be likely to be heightened by human misuse and error. None of these parameters has been considered by many scientists (e.g., Kraybill) who have argued in favor of widespread use of pesticides, although all of them conceivably could affect the validity of the judgment that employment of these chemicals represents a morally acceptable risk.[24] Because of the incompleteness of this often used argument, the conclusion that continuing and widespread employment of pesticides presents an acceptable risk cannot be said to be justified ethically on the grounds usually provided for it.

Pesticide Toxicity and Human Carcinogenesis

Two further reasons, frequently employed to justify extensive use of pesticides, focus on their alleged carcinogenic effects. It has been argued that, even though the chemicals cause cancer in animals, there is no clear proof that they do so in humans.[25] Moreover, a threshold for pesticidal injury may exist.[26]

One obvious problem with both these arguments is that they are implicit commissions of the fallacy of the appeal to ignorance. This logical error is committed whenever one assumes that because X has not been proved to be the case, therefore X is not the case, or because X has not been disproved, therefore X is the case. Both types of conclusions are invalid, because both rest on premises specifying no facts, but on premises stating an ignorance of relevant facts. From ignorance about situation Y, nothing follows about the nature of situation Y. Likewise, from the fact that there is no proof of pesticide-induced human cancer, it does not follow that there is no pesticide-induced human cancer. And from the fact that a threshold for pesticidal injury has not been disproved, it does not follow that a threshold for pesticidal injury exists. As numerous researchers admit, pesticide-induced carcinogenesis in humans and injury thresholds for pesticides are both open questions.[27] Hence, at best, it is logically misleading to use either of these considerations as grounds for claiming that certain modes of employing pesticides are morally acceptable. Let us look more closely at each of these arguments.

Apart from the appeal to ignorance, there are several other difficulties inherent in arguing that even though it is obvious that pesticides cause cancer in animals, it is not known whether they have the same effects on humans, and therefore not certain that pesticides ought to be used in a more discriminating fashion. The most obvious ethical flaw in this reasoning is that it assumes that it is morally acceptable to experiment on human beings. If one is unwilling to accept data on animal carcinogenesis as grounds for conclusions about human carcinogenesis, then the only option left is to accept human experimentation at the laboratory level or to allow it full-scale at the public level. Either case is highly questionable on moral grounds. One university professor, speaking before a Congressional subcommittee, called such experimentation "the human guinea-pig approach."[28]

Another problem, with the failure to accept data on animal carcinogenesis as sufficient grounds for protecting humans from particular pesticides, is that such a position is inconsistent with established scientific and legislative policy regarding carcinogenesis. Specifically the view is inconsistent with the Delaney Principle. Part of an amendment to the Food and Drug Act introduced in 1958 by Congressman James J. Delaney, this principle stipulates that "no [food] additive shall be deemed to be safe if it is found ... to induce cancer in man or animal."[29]

The scientific reasons for accepting data on animal carcinogenesis as grounds for inferring human carcinogenesis, and therefore for accepting the Delaney Principle, are not difficult to spell out. For one thing, provided that (a) the animal tests are used on species with similarities relevant to man, (b) that an adequate number and variety of experiments are performed, (c) that the exposure and dose are appropriate, and (d) that the extrapolation from animal to man is mathematically and metabolically valid, then the fact that certain substances "induce cancers in animals warrants the expectation that they will induce cancers in men."[30] In fact, experiments on animals may be preferable to those on humans for several reasons. First, "experiments can be performed that would be unacceptably hazardous to human subjects." Secondly, large control groups can be maintained more easily in animals than in humans; thirdly, very large numbers of subjects can be used.[31]

In the most famous and widely cited studies on effects of DDT on humans, only 51 prisoners were administered DDT for a period of several months. In this same test, only 5 men received the pesticide for as long as 18 months.[32] Obviously experiments done on animals could have been accomplished both with larger numbers of subjects and for a longer period of time. As it is, however, the Department of Health, Education, and Welfare has revealed that all DDT studies on humans have been done with inordinately small sample spaces and for periods of less than four years. For this reason, HEW has declared that the thesis that DDT does not cause cancer in humans is "open to question."[33]

Both HEW's conclusion and the verification of the length of DDT experiments are in implicit conflict with claims made by proponents of the chemical. One scientist, for example, maintains that "man has been exposed to DDT since 1940" and that "there is no evidence thus far" that it is a carcinogen to persons, even though it has been proved carcinogenic to animals.[34]

The obvious flaw in the reasoning here is the assumption that human exposure to DDT since 1940 is the same thing as controlled human experimentation on DDT since 1940. Clearly, if no studies were done on long-term exposure, then the long-term exposure, of itself, constitutes no grounds for concluding that the substance in question is not carcinogenic. Chronic health effects, such as cancers, are notoriously difficult to attribute to a certain cause. Cancers do not wear clear experimental "tags" indicating that they were caused by background radiation, weapons-testing fallout, DDT, or our parents' genetic endowment. Hence, without experimental controls, exposure to substances causing chronic health effects proves nothing. As one Environmental Protection Agency scientist put it: "It is often very difficult to attribute a chronic health effect in humans to the correct cause. This is due to the lack of experimental controls."[35]

Another problem with placing too much weight on short-term cancer experiments is that toxic results of pesticides are well known to be delayed for several reasons. For one thing, the delay period (between exposure to a hazard and the onset of its effect) of some known human carcinogens is very long. The latency period for lung cancers caused by exposure to radioactive ores, for example, is as high as 20 years; for skin cancers caused by ultraviolet exposure, the latency period is as great as 40 years. Obviously, therefore, the acute aspects of pesticide poisoning are not the most damaging, since chemical carcinogens have widespread and long latent effects.[36] Rachel Carson pointed out this fact almost two decades ago when she observed that many cancer patients have had a history of exposure to toxic chemicals and that nearly all these effects did not reveal themselves until many years later.[37]

A second reason why poisonous results of pesticides are usually delayed is that these chemicals are stored in human body fat. Fat-soluble pesticide residues thus "accumulate to high levels" and "cause toxic effects when the fatty tissues are mobilized"; hence demonstrable harmful consequences of pesticides are usually delayed for many years.[38]

Another reason why pesticidal effects are not immediately evident is that only the last member of a food chain receives the full burden of the toxicity. In a celebrated California incident, for example, the chemicals entered a lake, were concentrated in the bodies of fish, and only later proved lethal to the land grebes who fed on the fish. More will be said later about delayed food chain effects. What is relevant here, however, is that only long-term studies often reveal cancer hazards to man. This is a result of his being the terminal member of many food chains in which water, plants, and animals concentrate carcinogens.[39]

Pesticide Toxicity and the Threshold for Injury

Just as there are a number of methodological considerations (e.g., length of carcinogenesis experiments) which undercut the conclusion that pesticides are not known to cause cancer in man, so also there are numerous reasons for questioning the thesis that a threshold for chemical injury may exist. For one thing, scientists who proposed in the past that there was a threshold for acute or chronic injury from radiation have now been proved wrong. We have known for more than a decade that any amount of ionizing radiation, however small, puts the person exposed in a higher risk category for carcinogenic and mutagenic effects. Hundreds of thousands of people were exposed indiscriminately to fallout from atmospheric testing and to unnecessary X-rays, and many of them have contracted cancer, because it was once believed that there were safe doses of radiation. Owing to our past mistakes with alleged radiation thresholds, and in light of the fact that many pesticides, like radiation, have effects which are carcinogenic, mutagenic, persistent, and cumulative, one ought to proceed very carefully before suggesting that there is a threshold for chemical injury. If this belief is wrong, countless persons could pay for the error with their lives, health, and genetic integrity.

In spite of the potential for harm suggested by the threshold hypothesis, many scientists have claimed that such a threshold may exist, and then used this belief as grounds for arguing in favor of increased employment of pesticides. Let us examine the ethical assumptions implicit in the argument and, on this basis, determine whether the threshold hypothesis provides morally acceptable grounds for use of the toxins.

Even though the existence of thresholds for effects of chemical carcinogens "has never been firmly established,"[40] let us assume with a number of scientists that such a threshold does exist for certain pesticides. If we grant this point (for purposes of argument), are we thereby committed to accepting the ethical conclusion allegedly following from it? Are we committed to the belief that it is morally acceptable to subject persons to pesticides in amounts below the threshold dose? If we examine the assumptions implicit in the argument, I think it will be clear that this conclusion does not follow.

The basic reason why the establishment of a threshold for pesticidal damage does not guarantee the ethical acceptability of allowing population doses below threshold is that it is virtually impossible to know when a given individual has received such a dose. As Rachel Carson explained in her famous work, *Silent Spring,* although experimental situations are controlled, in real life humans never know how much of given pesticides they are receiving.[41] One person may eat five or ten times the "average" amount of a particular food and thereby ingest five or ten times the "normal" amount of a particular chemical. Or, as was mentioned earlier, a person would receive additional doses as a result of the facts that 63 percent of all pesticides are applied by aircraft, that aerial sprays typically reach their targets in amounts less than 50 percent of what is distributed, and that pesticide application efficiency is undercut primarily by wind drifts.[42] Numerous authors have documented deaths and injuries among humans as a result of what Carson calls "indiscriminate" aerial spraying.[43] Hence it is reasonable to assume that the body burden of pesticides might be increased significantly for those persons affected by aerial spraying.

Other factors determining an individual's dose of a chemical are synergistic effects. Although a particular dose may be below an alleged threshold, two or more pesticidal contaminants, or a single pesticidal toxin plus some naturally occurring event, may cause a far more serious consequence than the sum of their individual effects.[44] This means, practically speaking, that the synergistic results of a subthreshold dose of a contaminant often cause injuries far in excess of those theoretically calculated to occur. As Rachel Carson puts it, any "supposedly 'safe dose'" could be enough to tip the scales that are "already loaded" with allegedly safe doses of other chemicals.[45] Likewise, temperature combined with a subthreshold dose of a pesticide, might cause excessive or unpredicted injuries. The metabolism of an aquatic organism, for instance, doubles for every 10° C increase in temperature, and this metabolic increase renders any amount of pesticide ingested much more toxic. As one Environmental Protection Agency scientist explained, synergistic effects of pesticides have not been completely investigated by anyone; therefore, he says, current "static bioassays" give results about chemical dosage which are "of little use in predicting the effect in natural systems."[46] What this means is that any "threshold dose" or "no-effect dose" for a carcinogen "is a highly relative level which applies only for the precise experimental conditions generated."[47]

One practical condition, not considered in experimental tests of pesticide toxicity, is biomagnification, or heightened concentration, of these chemicals as one moves higher in the food chain. Although one might believe that he is theoretically exposed to a subthreshold amount of pesticidal residue as a result of eating particular meats, fruits, vegetables and dairy products, because of food chain effects, this might not be the case. Since a person's intake of pesticides is a function of the level of the particular chemicals concentrated in his food, the amounts thus ingested could vary widely. In a California fresh water lake, for example, which showed no traces of a pesticide (in terms of *ppb,* parts per billion), birds and fish using the lake for their water supply had as much as 2500 ppm (parts per million) of this toxic substance.[48] Likewise, some crustaceans, exposed for 24 hours to 8 *ppb* of a chemical, concentrated it in their bodies at a level 23,000 times as great as the pesticide level present in their water.[49] Eventually, says one government scientist, "this constitutes a health hazard to man via contaminated water and food."[50] Thus another clear problem, with ascertaining when one is receiving subthreshold amounts of a pesticide, is determining wide-ranging effects of being part of a particular food chain which has members who have ingested varying concentrations of the toxic chemicals.

Besides food-chain concentrations, other factors also make it difficult to estimate when one has received a theshold dose of a particular pesticide. One of these variables is the delayed effect of toxic chemicals. (This point was addressed earlier.) Another parameter is transport of pesticides via vaporization, rainfall, runoff, dust, and wind. Since rainfall in some locales of the country often contains concentrations of these chemicals 1000 times greater than that in other areas, and because pesticides can be transported for thousands of miles by the elements,[51] it is not clear how one could ever know if and when he had received a threshold dose of a particular toxic substance. As is well known, for example, there are measurable levels of DDT in Antarctic penguins.[52] The fact that DDT has never been used in Antarctica illustrates well the fact of pesticide transport and the difficulty of predicting what and where measurable amounts of the chemical might be found.

A related reason why it is difficult to determine when a human might ingest a threshold dose of a particular pesticide is that many such chemicals are extremely persistent in the environment. With pesticide residues, as with radioactive isotopes, their toxic potential through time is described in terms of the notion of a "half-life." (The half-life designates the period of time necessary for one half of the residue or radioactivity to lose its toxic character. In two half-lives, three-fourths of the residue has become harmless. In three half-lives, seven-eighths of the residue has become harmless, and so on.)

For many pesticides, especially those in the chlorinated hydrocarbon family such as DDT, the half-life is quite long. Government tests of DDT, for example, revealed that in some instances, half of the toxic residue remained after 17 years.[53] Because there are so many pesticides in the environment, because a significant percentage of them are long-lived, and because their effects are cumulative (fat soluble residues accumulate to high levels in tissues of organisms, as was explained previously),[54] it is difficult to determine when a given individual has received a threshold dose of any one of them.

There is an obvious response to all these difficulties (e.g., food-chain, transport, and synergistic effects) with knowing when an individual has received a threshold dose of a certain pesticide. One might simply argue that these consequences involve practical problems of monitoring and that, in principle, it would be possible to know when a person had received a given amount of a chemical. Therefore, goes the response, it would be in principle possible to protect persons from carcinogenic risks borne as a consequence of pesticide exposures above threshold.

The problem with this answer is that, apart from whether it is in principle or in practice impossible to know how much pesticide a person ingests, anything short of monitoring each and every individual will not yield an accurate picture of who has or has not received a threshold dose. Obviously, it is impractical and perhaps even impossible to monitor each person. If so, then there will clearly be unknown people who, for one reason or another, will have received inordinate doses of pesticides and who therefore will not receive equal protection under the law.

At present, no laws exist which require pesticide manufacturers, distributors or users to report amounts of the chemicals which they have sold or used; likewise there are no federal statutes governing aerial applications of these toxins.[55] Less than 1 percent of the total number of interstate food shipments are checked by the Food and Drug Administration for pesticide content, and only 30 diets in five geographical regions are sampled yearly by the FDA to determine chemical residues.[56] Obviously, therefore, a great many improvements in monitoring could be made, and it would be possible to have a clearer idea of what pesticide doses people were actually receiving. As was mentioned before, however, nothing short of monitoring every person would provide a failproof guarantee that all citizens were receiving equal protection under the law. Since such extensive monitoring is impractical, this means we are left with statistical sampling procedures. But if we are left with statistical sampling procedures, then our analyses of "acceptable" doses commit the fallacy of assuming that what is safe for the "average" person in the sample is therefore safe for everyone.

This fallacy of presupposing that all individuals are average and according rights to equal protection (from pesticides) on the basis of statistical or average characteristics is a dangerous one. One need only think what might occur, if this assumption were used in other areas of reasoning, in order to see that such a presupposition is erroneous.

If one assumed that rights to equal protection were awarded only on the basis of statistical or average characteristics, then one might argue, for example, that handicapped people deserve no more protection than the average person. Obviously, however, such an argument would be erroneous; rights are accorded in part on the basis of individual characteristics and not merely because of the statistical characteristics of the group to which someone belongs. For example, on the average, women score lower than men on standardized tests (e.g., the Graduate Record Exam, the College Entrance Board) allegedly measuring mathematical ability. Even if one presupposes that such tests do in fact measure real *ability,* which is doubtful, the results clearly do not warrant the assumption that all individual women have less mathematical aptitude than all individual men. Nor do they warrant the assumption that the right to a mathematics-related job is determined on the basis of gender rather than on the grounds of individual characteristics. But if rights are accorded on the basis of individual, as well as group, characteristics, then it is at least questionable whether the right to equal protection (from pesticides) can be determined on the basis of "average" limits below which there is no pesticidal injury. To see why this is questionable, let us examine the consequences of applying a threshold law (assuming it is correct) for pesticidal residues.

Even if a threshold for such injury were established, this level would still carry with it all the disadvantages of being the measure of harm for the "average" person. This means that whoever did not have the average response to pesticidal injury would be likely to be harmed by these toxic chemicals. On the basis of threshold studies, it is now widely known that a number of classes of persons are more sensitive to (and more likely to be hurt by) pesticide ingestion. These groups include infants, children, women, people with allergies, and those leading sedentary lives.[57] Hence if one were to permit use of pesticides below some threshold level established on the basis of the average person, then such a policy would be likely to discriminate against the rights of infants, children, women, allergic persons, and others, to equal protection under the law.

Apart from the chemical sensitivities of these classes of people, it is well known that "the tolerance [of humans to pesticides] is extremely variable from person to person."[58] Because there are so many avenues of exposure to pesticides, there are also "an infinite number of different exposure levels" among persons and "wide variations of residues in food groups obtained in ... different ... cities."[59] Even if the infinity of tolerance levels and exposure levels did not clearly point to persons involuntarily placed in higher pesticide-risk categories, however, the mere existence and employment of a threshold in setting pesticide policy would be discriminatory to some individuals. It simply would not be as easy to identify them.

The discriminatory nature of threshold-setting is evident on the basis of the probabilistic properties of the conclusions drawn regarding pesticidal risks. These statistical problems, in fact, characterize most laboratory tests for carcinogens. For example, suppose one found an incidence of zero tumors in 100 animals during an appropriate time period and under appropriate experimental conditions when one was testing a particular carcinogenic threshold. If one assumed, on the basis of these test results, that the threshold was "safe," then there would very likely be some persons for whom the threshold was not safe. The test result of zero tumors in 100 subjects is statistically consistent with a true risk of 4.5 percent when an assurance level of 99 percent is used.[60] My point is this; no matter how many subjects are studied, there is still a small probability that the actual risk is greater than that indicated, on the surface, by the test results. Admittedly, this risk decreases as the sample size increases and no tumors are found. Nevertheless, some hazard always remains; some persons are therefore, in principle, in a higher risk category as a consequence of using statistically established thresholds for carcinogenesis.

Whether one considers specific classes of high-pesticide-risk persons, or the infinity of tolerance and exposure levels for toxic chemicals, or the statistical properties of test results on carcinogenesis, each of these considerations leads to the same conclusion. Employment of a threshold for determining what level of pesticide residue is acceptable inevitably discriminates against some persons' rights to equal protection. Perhaps this is why one scientist has argued that acceptance of any threshold for pesticidal residues amounts to playing a deadly game of Russian roulette.[61]

Rachel Carson makes much the same point when she notes that we have allowed poisonous chemicals to be used quite indiscriminately by those largely ignorant of their potential for harm. The problem, as Carson sees it, is that people have been subjected to contact with pesticides "without their consent and often without their knowledge." She views subjection to inordinate amounts of these chemicals as a violation of guarantees expressed in the Bill of Rights.[62]

Although she does not say so, Carson probably had in mind the Fifth and Fourteenth Amendments. They assure all citizens of the right to equal protection and to due process under the law. What is significant about these and other principles established in the Bill of Rights is that they establish an "egalitarian" basis for assignment of rights and duties, and they imply the existence in the US of an egalitarian social ethic. Such an ethic is based on "equality in the assignment of basic rights and duties," apart from the totality of good that might be achieved. In other words, the moral goal of an egalitarian ethic is not to maximize the good for the majority of mankind (which is thc goal of a utilitarian philosophy), but instead to maximize equality in the assignment of basic rights and duties.[63]

If in fact the Bill of Rights does provide for an egalitarian ethic, rather than a utilitarian one, then this has great consequences for public policy regarding pesticide use. To the extent that the threshold hypothesis dictates such policy, and to the extent that it fails to guarantee equal protection to all persons, then to that same degree will our pesticide policy fail to serve the ethical demands dictated by liberal interpretation of the Bill of Rights.

In response to these considerations regarding pesticide policy, the Bill of Rights, and egalitarian ethics, I suspect that many people will be tempted to argue as follows. "Economic, technological, agricultural, and medical benefits gained by employing a threshold hypothesis in pesticide policy are worth the violations of equity that might ensue and worth the possible increase in cancer and genetic damage." This is a strong argument as well as a realistic one. There are, however, problems with it.

If one assumes that the pesticide-injury tradeoff is indeed a moral one, then, unfortunately, the sole basic standard of right and wrong action is the "principle of utility." It posits, as the moral goal of all human actions, the greatest possible balance of good over bad for mankind as a whole.[64] Since the moral end of utilitarianism is to maximize the well-being of the majority, principles of individual rights and equal justice are recognized only to the extent that doing so will lead to the greatest good. As a consequence, employing the principles of utilitarianism can lead to individual violations of equity and justice, at the same time as the general welfare is optimized.[65] This means that any given minority could be deprived of rights in order to serve the interests of the majority. For example, says J.J.C. Smart, a utilitarian would be bound to accept as right the action of framing and executing an innocent man, if his death would prevent serious riots which would take the lives of many persons.[66] Even utilitarians admit that it is not difficult to show that their beliefs could "have some very horrible consequences."[67]

The pesticide-injury tradeoff and its utilitarian underpinnings, however, do not merely affront one's ethical sensibilities or rationality. As was discussed earlier, they may also be incompatible with the principles upon which the US Constitution rests. Marshall Cohen, Professor of Philosophy at the City University of New York, noted this point and said that utilitarianism is "incompatible with our moral judgment and with the principles on which our Constitution rests."[68] Thus when one argues that the benefits of pesticide use outweigh considerations of equity, one is really arguing against affirmation of the Constitutional rights of all persons. As was mentioned previously, equal protection is guaranteed by the Fifth and Fourteenth Amendments. This is an inalienable right, and its recognition is not contingent on whether it is useful, expedient, or practical to affirm it. When conditions (e.g., promoting the benefits of pesticide use) are put on the exercise of this right, then it can be said neither to be inalienable nor to be guaranteed by the Constitution. A utilitarian tradeoff, which exchanges the benefits of indiscriminate pesticide use for the right to equal protection, may be compatible neither with the Bill of Rights nor with rational considerations of justice. Since the threshold hypothesis leads to such a tradeoff, it too may be said to be incompatible both with equity and with one interpretation of the principles on which the Constitution rests.

Admittedly, not all those who support the threshold hypothesis for pesticide policy are thereby following an ethic contrary to principles of equity and Constitutional rights. The focus of this discussion has been whether one ought to follow a threshold hypothesis in using chemicals which are known to cause cancer, genetic damage, and other injuries in animals. If specific pesticides are not known to cause these consequences in humans or animals, then following some form of the principle is not necessarily unethical. In other words, what I have presented in the preceding discussion is not a condemnation of all employment of threshold hypotheses but merely a criticism of its use in policy affecting known carcinogens or known sources of injury. Hence my remarks constitute an ethical substantiation of current US policy, viz., the Delaney Amendment, which requires one to reject totally its employment in the case of carcinogens.[69] Hopefully this discussion also provides an analysis of why many scientists, who support using this hypothesis with known carcinogens, are making some questionable ethical assumptions.

Carcinogens and Cocarcinogens

A fourth argument in favor of widespread employment of pesticides is also based on some doubtful ethical assumptions. Some scientists argue that banning DDT in 1972 was not "clearly warranted" because there is strong evidence that it is not a carcinogen but is rather only a cocarcinogen, a chemical that acts synergistically with something else to produce cancer. In the case of DDT, it is not alone a known carcinogen but interacts metabolically to produce a chemical, DDE, which *is* a carcinogen. It is likely, therefore, that cancer is not a "primary effect" of DDT but rather that DDE and not DDT causes the cancer.[70] As a consequence, they say, quite a few pesticides (like DDT) have been "misclassified as carcinogens" and perhaps wrongly "excluded from the environment,"[71] since they are merely cocarcinogens.

At first glance, this point regarding DDE and DDT seems to be a sound one. Why ban DDT if DDE is the problem? The difficulty with this reasoning, however, is that DDE is a metabolite of its parent compound, DDT. Within one month after application of DDT, a majority of the dose of DDT metabolizes (in the body of the organism that has ingested it) to DDE.[72] In other words, DDT is stored in an organism as DDE, since DDE is a degradation product of DDT.[73] In fact, DDE is the most frequently detected pesticide residue in the US.[74] Hence if DDT always metabolizes to DDE, then one way to avoid the carcinogen DDE is to avoid the parent compound DDT. Obviously if DDE rather than DDT causes cancer, then DDE rather than DDT ought to be banned. But if DDE is an irrevocable consequence of DDT, and DDE ought to be banned, then DDT ought also to be banned, in order to avoid DDE.

The problem with the argument that DDT is not a carcinogen is that it assumes that one could ban DDE without banning the cause of DDE, which is DDT. Hence, although this distinction between a carcinogen and a cocarcinogen is an important one, it does not in this instance show why DDT is not harmful to health.

Pesticide Dosage in Cancer Experiments

A fifth argument also presents some interesting considerations regarding the methodology employed to determine whether pesticides are carcinogenic. According to proponents of this reasoning, data on pesticide toxicity incorrectly suggest that the chemicals are more dangerous than they really are, since test animals receive exorbitant doses of the substance being evaluated. One scientist criticizes the fact that doses of DDT given to mice, for example, were the equivalent of 853 times the dietary exposure and three times the workplace exposure for humans.[75] Therefore, he suggests, pesticides are probably not as harmful as believed, and there are not necessarily moral reasons for constraining their use, as has been required in the case of DDT.

The preceding point is well taken, since many scientists would probably agree that almost any substance, taken in large enough doses, would induce cancer in most organisms. Is this consideration sufficient, however, to render widespread use of DDT and other broad-spectrum pesticides morally acceptable?

As was argued earlier, a number of parameters make it impossible to determine the pesticide dose a given individual has received. Among these factors are synergistic effects, aerial spraying results, pesticide transport processes, and food chain effects of biomagnification and bioconcentration. Hence, as was indicated previously, there is an infinity of exposure and tolerance levels for chemicals throughout the US. When scientists question the fact that pesticide dosage under test conditions is 853 times the exposure levels to which humans (in a non-workplace situation) are subject, they are criticizing the fact that test dosage is so much higher than "average" exposure levels. If everyone were average, this argument might be sound. As it is, however, the wide variety of exposure levels to which average people are subject makes it necessary to have a margin of safety.

Consider the fact, for example, that organisms living in or near pesticide-contaminated water concentrate these chemicals in their tissues "to levels thousands of times greater than the water concentration."[76] Consider, too, that US rivers vary in their pesticide-concentration levels by a factor of 60 or more, and that two of the key routes of chemicals to man are through potable water and through consumption of crab, shrimp, fish, and waterfowl.[77]

So long as there is so much variation in pesticide intake as a result of multiplier effects in the food chain, and so long as some individuals receive chemical doses significantly above average, then the high exposures to which test animals are subject are needed to protect persons whose pesticide intake is above average. If doses given test animals were not somewhat greater than those to which the average human might be subject, countless people would be put in jeopardy. If they were, then one would have grounds for arguing that experimental information was not used to establish an egalitarian ethic for pesticide protection, but instead to support a discriminatory and utilitarian one. As I have argued earlier, there are numerous reasons for rejecting such a philosophy.

Government scientists are well aware of the ethical problems of legislating for pesticide safety on the basis of the "average" tolerance or exposure levels of citizens. In a recent classic work on US pesticide use, several researchers from the Environmental Protection Agency made precisely this point. They explained: "In an effort to make up for this deficiency [that some persons do not fit average exposure or tolerance levels for pesticides], a fraction of the highest 'no-effect dose' is taken as safe for human consumption."[78] If their points about the inadequacy of statistical testing of these substances are correct, and if my misgivings about utilitarian frameworks for pesticide policy are well founded, then current modes of testing chemical toxicity cannot necessarily be faulted on grounds that the animal doses are too large. And if these exposure levels are not necessarily too high, then they give little basis for believing that certain pesticides are actually safer than government reports indicate. But if specific chemicals (e.g., DDT) are in fact no safer than has been thought, then it might be ethically questionable to argue for more widespread use of them

Alternatives to Pesticide Use

In addition to the many arguments over safety, one of the most important focal points in the pesticide controversy is whether there are natural or biological methods of pest control. Many scientists argue for continued widespread use of chemicals on the grounds that such alternatives do not exist, and that pesticides are the only effective method of pest control at this time.[79] Kraybill writes: "pesticides are the only feasible control procedure for pests"; they are warranted "if mankind is to live in balance with its environment."[80]

Both of these arguments in favor of pesticides are important ones, since they address a pragmatic approach to the ethical question of whether or not it is acceptable to use the chemicals. Clearly both factors, namely, the existence of alternative means of control and the consequence of enabling man to live in balance with nature, are capable of influencing one's judgment as to the moral acceptability of increased and continuing pesticide use. If there are no other practical options to chemical management, and achieving pest control enables man to live in balance with the environment, then one is more likely to judge pesticide use morally acceptable than if there were alternatives to it. On the other hand, if there are feasible alternatives to chemical pest control, then one is more likely to judge pesticide use morally unacceptable than if there were no options besides it.[81] In the case of DDT, for example, the major public decision to ban it cited the availability of other means of pest control as part of the rationale for doing away with it.[82]

At least two questions need to be addressed in order to determine the plausibility of arguing for or against the moral acceptability of increased pesticide use on the basis of the existence or nonexistence of alternative forms of control. These are:

1. "Do alternatives to chemicals exist?"

2. "Does use of chemicals enable humanity to live in balance with its environment?"

Let us examine each of these issues. There is widespread belief that continuing, large-scale forms of chemical pest control are necessary and that there are no alternatives to pesticide use. According to government scientists, however, this is not the case. Instead what has happened is that, because of the easy availability of chemicals, many agriculturists have put off evaluating and using alternative forms of management. Now that serious ecological problems have arisen as a result of pesticides, more persons are utilizing natural or biological means of managing pests.[83] There are not only less hazardous chemicals available to take the place of DDT,[84] for example, but also a wide variety of nonchemical ways of controlling pests. In a recent report on pesticide use, one government scientist concluded: "In many cases, pest control, with minimum damage to the environment, can be achieved by alternative procedures."[85] Some of these alternative procedures include agricultural management (e.g., crop rotation and destroying crop remains after harvest); importation and release of natural pest enemies; use of insect hormones; release of sterile males to the insect population; employment of microbial insecticides; planting resistant varieties of crops; use of insect attractants and repellants; and various physical and mechanical means of inhibiting pests).[86]

"Many of the economic pests in the U.S. have come from other countries without their biological parasites or predators," say government scientists.[87] Hence the logical way to cope with these insect immigrants is to introduce their natural enemies. Approximately 100 major pest insect species have been "substantially or completely controlled by imported natural enemies."[88] *Polistes* wasps, for example, have been effective in reducing tobacco hornworm damage by 76 percent when they were used in North Carolina.[89] Wasps have also been effective in controlling gypsy moth damage.[90] In California the citrus industry has been protected from major pest populations of mealy bugs by importing Australian predators, *vedalia ladybirds*.[91] Likewise citrus diseases, such as Florida red scale and purple scale, have been controlled by using parasites introduced from California.[92]

Successful biological control is not only exemplified by cases of importing natural predators, but also by numerous other nonchemical means. Screw worms, for example, were completely eliminated in 1958 from southeastern US and from Curacao, simply by releasing 100 sterile males per square mile.[93] Aquatic weeds in Florida are successfully controlled by large fresh-water snails and by the flea beetle.[94] In areas of the South, where codling moth pest is common in apples, resistant varieties of the fruit can be grown so as to avoid the moth, which (with the cotton bollworm and cotton boll weevil) is one of the three major crop pests in the US; crop sanitation is also an effective means of reducing the numbers of the moth.[95] Although numerous other examples of biological control of pests could be cited, these few instances suffice to illustrate the point defended both by university scientists and by government entomologists, viz., in many cases widespread use of biological management is now feasible as an alternative to chemical control.[96]

The hallmark of newer programs adopted by agriculturists and environmentalists in the US is something called "integrated control." Whereas chemical methods are based on routine spray schedules and 100 percent pest eradication, integrated control is achieved primarily by bioenvironmental means (e.g., importing predators, releasing sterilized males) of pest management. These are built upon monitoring insect populations and employing lower-level controls based on economic-threshold densities for the pest populations. Government scientists indicate that if all economically feasible means of biological management were presently employed in programs of "integrated control," rather than "chemical eradication," pesticide use could be reduced to 50 percent "with little or no effect on crop production."[97] Presumably, as more bioenvironmental controls are tested and employed, pesticide usage could be cut even further, thus reducing the medical and ecological threats posed by utilization of chemicals.

Since more pesticides are employed in the US on cotton than on any other crop, "integrated control" of the cotton boll weevil and cotton bollworm would significantly reduce pesticide usage. In Peru integrated control has been successfully utilized for a number of years on their cotton crops;[98] likewise integrated control has long been used on the rice crops of China. Through this environmentally sound means of pest management, China has been able not only to feed its numerous citizens but also to be a major rice exporter.[99]

In the face of these obvious instances of bioenvironmental methods and the clear successes of "integrated control," two obvious questions arise. Why has integrated control, the current policy of the Department of Agriculture,[100] been so slow in taking hold in the US? Any why has the government, despite this policy, continued to maintain that US use of chemicals will increase indefinitely in the foreseeable future? In fact, pesticide use in the US is expected to increase by 25 percent between now and 1984.[101]

The answer to both these questions lies in the realm of politics. According to one famous entomologist (University of California at Berkeley) who did research in pest control for the last 30 years, the powerful agri-chemical industry has a vested interest in lobbying for continued widescale employment of pesticides; since most of the major banks in this country have invested billions of dollars in the chemical industry, there are far-reaching economic reasons for their supporting continuing policies of chemical, rather than integrated, pest control.[102] Although current knowledge of alternative means of pest management is adequate to "solve many of the problems associated with the use of conventional pesticides" and to enable these controls to "pay for themselves," safer means of dealing with pests probably will not be utilized on a large scale, scientists say. This is because "the structures that evolved among the government, industry, and farmers to implement the use of conventional pesticides appear to be unsuitable for the initiation of new control measures."[103]

What are the "structures" that are unsuited for implementing biological forms of control? Perhaps the most obvious of these is the Department of Agriculture (USDA), which is heavily staffed with persons sympathetic to the chemical mode of pest management. Scientists with the Environmental Protection Agency have charged that, because of the influence of the chemical industry, the USDA has failed to set adequate priorities for implementing "alternative methods of pest control" in the US.[104] They also claim that "although alternative pest control technology exists, conventional chemicals will remain the chief means of insect pest control for the foreseeable future." This is because the Department of Agriculture has not financially supported programs for "utilizing ... alternative pest control technology."[105] Such criticisms have been so pronounced that Ralph Nader has been led to call the USDA "the Department of Agribusiness."[106]

Business would surely suffer if the government implemented plans for biological control which (as was discussed earlier) could result in cutting annual pesticide use by 50 percent. If current annual US pesticide sales of five billion dollars were halved, this would constitute a serious problem for the chemical industry.[107] This suggests that economic and political considerations are largely responsible for the failure of the USDA to implement its avowed policy of integrated control, which could result in the 50 percent reduction in pesticide use. Even members of the chemical industry admit that government money is needed to implement programs of biological management.[108] Most agriculturists, for example, are not familiar with the techniques necessary to employ a program of integrated methods. This means that government funding for educational and developmental programs in biological control is necessary.[109]

The difficulty, posed by inadequate government funding for programs dealing with biological management, is heightened by the fact that the industry pays for most pesticide-related research in US universities. As a consequence, many scientists report that they have been either harassed by their university or corporate employers or fired for being "anti-chemical"; grants from pesticide industries represent a significant amount of income for many land-grant universities.[110] Because of this, a number of scientists have been led to speak of the "hire education" offered to academics by pesticide companies.[111]

Even, however, if programs in biological control were receiving adequate USDA funding, and even if much pesticide research were not sponsored by the chemical industry, there would still be one significant reason why pesticide use is predicted to increase substantially in the US in the future. This is because, at present, it is very difficult for biological controls to compete financially with existing, inexpensive chemical methods. The National Agricultural Chemicals Association said that this was one of three major reasons why "conventional" chemicals will continue to be used in increasing amounts. Members of the Association explained that they needed "economic incentives" to develop alternative forms of pest control. (Their other two reasons for supporting pesticides were a lack of funds for developing and implementing new forms of pest management and inadequate knowledge of plant biochemistry on the part of crop owners.)[112]

The problem, with failing to use medically and environmentally safer methods of biological control because pesticides are cheaper in the short term, is that employment of chemicals might not be economical in the long run. Their use might be both financially and ethically questionable. For these reasons, one ought to be certain that what Henry James called "the huge American rattle of gold" does not drown out our moral sensibilities. There is some danger, in fact, that this has already occurred. In speaking of government policy regarding science and technology, the President's Science Advisory Committee in 1965 warned that " the corporation's convenience has been allowed to rule national policy."[113] (More will be said later about this point and about how biological control can be used cost-effectively. At this time it is sufficient to note that politics, rather than science or ethics, seems to have influenced much recent pesticide promotion and regulation.)

Pesticides and the Balance of Nature

If there were no question, however, as to whether political and economic factors have influenced US policies encouraging chemical control of pests, there would still be another problem concerning the desirability of using pesticides. "Do chemical forms of pest management enable man to live in an ecologically sound environment?" According to a number of scientists, the toxins are needed not only because biological forms of control are not available, but also because mankind must "live in balance with its environment."[114] If this belief is true, then (as was discussed previously) pesticides obviously have a greater degree of moral acceptability than if it were not true. Let us examine the claim.

In two clear senses, chemical control contributes to man's living in harmony with his environment. First, use of pesticides allows humans to conserve land resources, since if greater losses to insects result from the absence of pest management, then more land probably will have to be put in cultivation to make up for insect damage. Secondly, employment of these chemicals also enables us to conserve energy resources, since energy output per unit of crop yield is lower when pest control is employed.[115] The critical question, however, is not whether pesticides enable one to conserve land and energy resources, since obviously they do, but whether use of biological management also conserves land and energy. If it does, then these two environmental benefits are not the consequence of pesticide use *per se,* but the result of pest control in general.

There are several important senses, however, in which pesticides do not enable mankind to live in balance with nature. First, because the most commonly used of these chemicals are broad-spectrum biocides, they also kill non-target organisms and predators of the pests. According to the US Department of Agriculture, this is one of the most serious consequences of chemical pest control.[116] Not only is the balance of nature upset by the killing of non-target animals, but also serious insect infestations often occur after extensive use of chemicals, because natural enemies of the pests have been destroyed. This sets up a vicious circle: spraying, reinfestation at greater densities (because natural enemies have been killed), spraying greater quantities, reinfestation at an even greater densities, spraying even greater amounts, and so on. This positive-feedback cycle is what Van den Bosch, the famous Berkeley entomologist, calls "the insecticide treadmill," since the spraying eventually becomes prolonged, magnified, and "addictive."[117]

In the course of the cycle of spraying and reinfestation at higher levels, there are numerous "alterations in predator-prey relationships, decreased floral and faunal fecundity and specific physiological alterations which reduce the ability" of both target and non-target organisms to compete for survival. This means that the balance of nature is upset in numerous ways. As a result of spraying citrus groves in California recently, for example, the mealy bug reached major pest populations because DDT had eliminated the bird species that were its predators.[118] In another classic and often replicated case, the wild robin had greater than 85 percent mortality and no reproduction in certain areas of Wisconsin where DDT was sprayed to control elm disease.[119]

Besides the significant amount of harm sustained by non-target organisms as a result of pesticide employment, there is another major consequence of repeated spraying. This is that eventually many insects become immune to the chemicals used. Because of their immunity, they cause even greater damage both to crops and to the balance of nature. In Guatemala, for example, 12 years after chemical programs were begun to control mosquitos, the pests became resistant to the pesticide used, DDT. At the same time, DDT had passed all through the food chains, causing Guatemalan women to have the highest concentrations of DDT in their milk in the world.[120]

More than 30 species of major pests are resistant to DDT alone. More than 120 species of insect and mite pests on earth are now "less susceptible to control chemicals than they were fifteen years ago."[121] Put in terms of percents, we know that 72 percent of all insects are resistant to one or more pesticides, and that 96 percent of all insect populations are actually increased because of use of pesticides.[122] According to the US Department of Agriculture, as a consequence of chemically induced pest resistance and predator destruction, pesticide use in the US is increasing, while crop losses to insects are also rising.[123]

In the past 30 years, pesticide use in the US has increased twelve-fold, while crop losses to insects have doubled, i.e., they have gone from 7 percent preharvest losses to 13 percent. In California alone, pest losses (in dollars) have gone from 254.6 million in 1970 to 643.3 million in 1974.[124]

Because of these disastrous effects of pesticides, even chemical industry spokespersons have admitted that the long-term economics of crop production requires a very low use of pesticides, if insect resistance and destruction of predator-prey relationships are to be avoided. In fact, one chemical-industry scientist wrote: "the growth of resistance [to pesticides] ... has reached a point where control of the principal crops can no longer be assured by pesticides alone."[125] This statement is especially true of cotton, the crop on which most US pesticides have been and are used. Numerous statistics, on reduction of cotton yield because of insect resistance to pesticides, have caused even chemical-company scientists to caution against heavy pesticide use in cotton production.[126]

The ecosystem is not merely upset, however, by pesticide-induced insect resistance and destruction of natural predator-prey relations. Besides these difficulties, chemical means of pest control also skew the balance of nature in more serious ways. Like radiation, pesticides are known to cause genetic damage in animals.[127] This means, according to government experts, that the chemicals can affect not only individual survival and reproducive success but also "the genetic constitution of future generations."[128] Scientists agree that pesticides cause mutations in test animals, and that human genetic material is not significantly different from that of the animals.[129] For these two reasons, it can be argued that the rights of future generations demand that current individuals not jeopardize our genetic endowment to posterity by employing certain chemicals which are known mutagens.[130] The Environmental Mutagen Society has agreed with these arguments; the society has stipulated that "the guiding principle in all cases should be that no risk whatsoever is acceptable when ... an alternative nonmutagenic compound is available."[131] Hence if my previous conclusions (regarding the availability of biological controls to reduce pesticide employment by 50 percent) are correct, then there are reasons for arguing that many uses of mutagenic pesticides are not morally acceptable.

On the basis of the preceding discussion, it is clear that there are numerous ways in which pesticides disrupt the balance of nature. Hence, there is reason to question the claim of scientists who maintain that use of the toxins enables mankind to live in balance with nature. Employment of many of these chemicals, in current high levels of application, causes numerous deaths among members of non-target animal populations, threatens the genetic health of humans and animals, and leads to even greater environmental problems because of pesticide-induced insect resistance.

The Contributions of Pesticides to Food, Fiber, and Survival

Despite the ecological calamities of pesticide use, what about the claim that use of these chemicals is necessary to provide humans with food and fiber and therefore essential to the survival of the human race?[132] According to this belief, although use of pesticides entails some very great risks, their hazards must be balanced with the benefits arising from their employment.

The argument that the benefits (e.g., increased food and fiber production, and therefore increased rates of survival among humans) of pesticide use outweigh its costs is a very persuasive one. Obviously, there are no simple responses to this line of reasoning; evaluating it depends upon careful weighing of advantages and disadvantages. The most moral course of action is probably to determine what is the lesser of two evils. Is pesticide injury to people and the environment a lesser evil than crop losses caused by failure to employ chemical means of pest control?

If the pesticide problem were indeed describable simply as the choice between use of chemicals versus crop losses, then it might not be difficult to sanction widespread pesticide employment if it meant saving the lives of persons who otherwise might starve to death. The problem, however, is not this simple. For one thing, as was argued earlier, alternatives to chemical control are workable; according to government scientists, if all economically feasible means of biological management were presently employed in programs of "integrated control," then pesticide use could be reduced to 50 percent "with little or no effect on crop production."[133] Hence, we are not faced with a simple "either-or" of using chemicals or losing our crops, as some persons have suggested.

Secondly, if my previous remarks are correct, then it is not necessarily plausible to assume that using chemicals prevents crop losses, while the failure to employ the toxins heightens them. As was pointed our earlier, both government and pesticide-industry scientists agree that, in many cases, current policies of intensive spraying actually cause greater crop damages because pests become resistant and their natural predators are killed.

Thirdly, as I said before, even if using pesticides always did prevent crop losses, this fact would not insure the *moral acceptability* of widespread and increasing use of these chemicals. This is because the ethics of pesticide use is determined not merely by the magnitude of costs and benefits, but also by the equity with which costs and benefits are distributed. Hence, to the degree that one's assessments of pesticide use ignore questions of justice (as Kraybill's arguments have), to that extent they might be said to sanction violations of rights guaranteed by the Fifth and Fourteenth Amendments.

In addition to these three considerations regarding the "pesticide use versus crop loss" argument, a number of other factors are essential to assessing the moral acceptability of increasing employment of chemical pest controls. For one thing, the ethics of pesticide use might be somewhat different for developed, as opposed to underdeveloped, nations. That is, the morality of using great amounts of chemicals might be a function of whether the failure to do so would result in serious food shortages or merely in a slight decrease in agricultural profits. A number of writers have argued that, since underdeveloped nations face grave risks of famine and malnutrition from pest losses, and since developed countries do not generally face hazards of the same magnitude, therefore use of pesticides in underdeveloped countries is more acceptable morally than in developed nations.[134]

Admittedly, some cases of employment of pesticides anywhere might be questionable on purely pragmatic grounds (e.g., whether they induce pest resistance and therefore a greater problem). However, to the extent that some uses of chemicals are effective in controlling health problems or crop losses, employment of them is more justifiable in underdeveloped than in developed countries. Not only are the consequences of pest destruction worse for poorer nations, but also they are usually those least able to provide the educational and economic foundations for alternative forms of biological control. In other words, the greater is one's freedom to choose among varying alternatives of pest management, the greater is one's obligation to choose the least harmful and most beneficial alternative. To the extent that poverty, lack of biological training, or some other factor limits a country's freedom to choose among available means of pest control, to that degree circumstances mitigate the responsibility of the country to choose the least harmful form of pest management. Hence, all other things being equal, underdeveloped countries are probably more often ethically justified in using great amounts of dangerous pesticides than are developed nations.

Paradoxically, however, underdeveloped nations taken together annually employ 8 percent of all pesticides produced in the world,[135] while US use alone accounts for 38 percent of total pesticides manufactured each year; in fact, the US uses more such chemicals than any area (e.g., all of Europe) on the planet.[136] As a consequence, for example, pesticide residue levels in US-processed seafood greatly exceed those of non-US-processed seafood.[137] Moreover, the proportion of US crop acres treated with pesticides has been increasing annually. It went, for instance, from 41 percent in 1971 to 56 percent in 1976. US farmers are also applying more pesticides per acre.[138] These facts, coupled with government estimates that the US could (without a loss in profits) cut pesticide use by 50 percent by means of biological controls, reveal that more pesticides are used in the one country that could most easily not do so, while fewer chemicals are used in nations that have the most ethical basis for using more. Hence to the degree that food, fiber, and survival justify large-scale pesticide employment, then to that extent the US is less able than underdeveloped nations to cite this ethical justification. For this reason, the "pesticide use versus crop loss" argument is misleading if one does not distinguish the varying ethical positions of developed and underdeveloped nations.

In response to the thesis that the US can more easily afford to reduce pesticide use than can underdeveloped countries, one might argue that this is not so, since the US feeds much of the world unable to feed itself. This is a good point; if true, it establishes another set of ethical considerations that ought to be taken into account in assessing the morality of chemical means of pest control.

Does high US employment of pesticides enable this country to feed the rest of the world? For one thing, most insectides used in the US are for the non-food crops of cotton and tobacco rather than for food crops.[139] Moreover, as was pointed out previously by chemical company spokesmen and US agriculturists, use of biological controls and very small amounts of pesticides is the only way for cotton growing to be cost-effective. This is because of boll-weevil and bollworm resistance to the toxins and because of chemical destruction of pest predators.[140] Hence eliminating the bulk of pesticides used in the US could, in principle, mean that there are no reductions at all in food-crop employment of these poisons. In fact, the US Department of Agriculture has predicted that the overall supply of food crops would not be diminished in the US if pesticide use were substantially reduced; it has also calculated that annual crop losses from pests would rise from 33.6 percent to 40.7 percent annually if no chemicals at all were used.[141] Since the Department of Agriculture also points out that the US produces a 10 percent surplus of crops annually,[142] excess production could easily offset losses incurred as a result of no pesticide use. However, if the US wished to continue to produce such surpluses (whether to help feed the rest of the world or for another reason), this would still be possible through use of integrated control of pests.[143]

According to US Department of Agriculture data, elimination of pesticides might even be cost-effective in the long run. Three quarters of a billion dollars are spent annually to apply pesticides, and three to four billion dollars are spent per year to divert nearly 60 million acres into a soil bank through federal farm subsidies. On the basis of these facts, David Pimental, a well-known entomologist and agriculturist, reasons that if chemical controls were eliminated completely and if the 60 million acres were not diverted to a soil bank but utilized instead to increase crop production (so as to offset losses to insects incurred as a result of not using pesticides), approximately four to five billion dollars could be saved yearly. Since the seven percent annual increase in crop damage from pests represents a 2.1 billion dollar loss, eliminating chemicals and increasing acres in cultivation could accomplish an annual savings of two to three billion dollars. Moreover, he reasons, this would result in a 28.4 percent rise in farm product value and only a nine percent increase in retail food prices. As a consequence, farmers would step up their output so as to take advantage of the new prices. Through their efforts to plant more, the seven percent increased annual crop loss would be compensated and retail food prices would be lowered.[144]

If the preceding considerations are correct, and if most pesticides are not used on food crops, then the US Department of Agriculture is correct in asserting that pesticide use could be reduced to 50 percent "with little or no effect on crop production."[145] But if this is correct, then there is reason to doubt that widespread, increasing use of pesticides is necessary either for food production or for survival. But if there is reason to doubt that chemical controls are essential to food production or to survival, then there is also reason to doubt that the benefits of pesticide use outweigh the ecological and health costs. Hence, it is questionable whether widespread and increasing use of pesticides is morally justified.

Pesticides and Cost-Benefit Analyses

Besides the fact that employment of pesticides is not essential to maintaining adequate levels of food production, there are several other reasons for believing that the benefits of widespread use of these chemicals are less than the dangers. Several pesticide-industry spokesmen have argued that accepting such risks is realistic, since if one performs a cost-benefit analysis, he will see that societal and economic benefits resulting from pesticide use are of greater magnitude than the hazards induced by it.[146] They also have claimed that, in "a general view," if carcinogenic pesticides are banned, it can only be because there have been no assessments of cost-effectiveness; if such studies were done, then they might not necessarily be prohibited. In other words, "some [carcinogenic] pesticides could be retained by virtue of their benefits."[147]

This particular argument is based on the ethical assumption that, so long as the desirable "social and economic" consequences of pesticide use are of greater magnitude than the costs, then employment of these chemicals is morally desirable. This view, however, has precisely the same ethical shortcomings as an assumption discussed earlier (viz., that it is morally acceptable to subject all persons to a particular threshold dose of a carcinogenic pesticide). Both of these presuppositions wrongly posit that the ethical permissibility of a particular pesticide policy is simply a function of the magnitude of total costs and benefits. Just as the threshold hypothesis fails to consider adequately the inequitable distribution of health effects resulting from a given pesticide dose, so also the "social and economic" consequences hypothesis fails to take into account the inequitable cost-benefit distribution resulting from employment of these chemicals. In other words, both of these assumptions implicitly sanction a utilitarian disavowal of some persons' rights to the equal protection guaranteed by the Fifth and Fourteenth Amendments. Since the ethical and political problems inherent in utilitarian ethics were discussed previously, I will not address this topic again here. However, let us examine some of the reasons why a "social and economic" cost-benefit analysis of pesticide use gives implicit assent to the failure to recognize equity.

The most basic reason why cost-effectiveness studies fail to provide for considerations of equity is that numerous social risks of pesticide use are typically not taken into account. In other words, economically speaking, there are differences between the purely *private* losses and benefits affecting only the producers and users of the chemicals, and the ecological, medical, and ethical costs and benefits of pesticide employment which are imposed on the *public*. The chief cause of these private-versus-social discrepancies is that existing systems of property rights imperfectly specify sets of ownership rights. This means that risks and benefits of pesticide use are involuntarily imposed on society, but they have not been included in the original contract between the parties (society and pesticide producers/users) to the exchange. These factors which are not accounted for are called "externalities," since they are external to the standard cost-benefit analysis used to justify the exchange.[148]

Let us examine some of the externalities not included in analyses of pesticide use. Although there are many, we will discuss five representative "social costs" typically omitted from government assessments of pesticide costs and benefits:

1. Unreported pesticide poisonings;
2. Water pollution;
3. Long-term carcinogenic and mutagenic health effects;
4. Wildlife kills;
5. Injury to the shellfish industry.

Although one of the liabilities of pesticide employment is unintended poisoning, especially among children, the social cost of this problem is not adequately accounted for in government risk-benefit analyses because most poisonings go unreported and most of those reported are not recognized as being effects of pesticides. A major US Environmental Protection Agency Report calculated that only 10 to 15 percent of all serious pesticide poisonings are ever reported, and that there are definitive accounts of approximately four thousand children per year unintentionally poisoned in this manner.[149] A noted California entomologist, basing his arguments on facts provided by the California Department of Public Health, claims that only one percent of all pesticide poisonings are reported and that there are approximately 1,300 such cases known per year in California alone.[150] What these data indicate is that there are thousands of persons, especially children, who are victims of toxic chemicals annually, but that the human and economic costs of their deaths and injuries are not included in assessments of pesticide use. The existence of unreported poisonings provides grounds for the claim that the risks of employing these chemicals have probably been underestimated.

A second significant social cost of pesticide use is water pollution. The principal source of such pollution is runoff from land.[151] Chronic, low-level concentrations of chemicals are difficult to remove by standard water-treatment practices;[152] as a consequence, treated drinking water in the US contains pesticide concentrations "comparable to the raw water."[153] Since water pollution is not included as a cost of pesticide use, two sorts of inequities result. First, persons who bear a higher carcinogenic or mutagenic risk as a result of drinking pesticide-laden water are not compensated for this hazard. This means that there are at least potential grounds for arguing that their rights to due process and to equal protection have been violated. Secondly, if the chemicals are not ingested, it is because water companies or individuals often must pay the extra costs associated with drinking bottled water or with using powdered activated charcoal to filter water. The latter alternative, however, is of limited effectiveness in removing chemicals. Such a situation is inherently inequitable since the contamination is induced by producers and users of the chemicals. In other words, those people receiving the greatest benefits from pesticides (their producers and users) are not those who are saddled with the pollution costs associated with them. This means, again, that there are at least potential grounds for arguing that persons who must bear the hazards of pesticide-contaminated water are not treated equitably.

Perhaps the most important social cost not included in economic assessments of pesticide use arises from the carcinogenic and mutagenic injury the chemicals cause to humans. In one of the latest government documents providing a methodology for cost-benefit analysis of pesticide use, the authors explicitly note that only *acute* effects of the chemicals are included in government analyses of environmental and health damages. Chronic, long-term consequences of pesticide exposure, such as genetic injury, cancer, and birth defects, all of which are "intergenerational health effects," are "excluded from consideration because they are less obvious and more difficult to detect."[154] Therefore, what are perhaps the most significant medical hazards of the toxic chemicals, the long-term injuries, are not even considered in cost-benefit analyses of pest control. This can only mean that those who receive such chronic injuries are not being granted equal protection and due process. It also indicates that the alleged costs of pesticide use have been greatly underestimated. Even government scientists admit that their failure to consider long-term effects of the chemicals begs the question of their total costs.[155]

Another externality excluded from risk-benefit studies of pesticide employment is damage to wildlife. For aesthetic, ecological, and humane reasons, these costs ought not to be ignored. Pesticide residues have caused many species of birds, for example, to be decimated because of chemically induced eggshell thinning.[156] These killings represent a loss not only for those who delight in nature, but also for all those who believe it is a moral duty to keep the ecosystem intact. Death of a number of members of the system results in a skewing of predator-prey relationships, food chains, and the balance of nature. Hence whether or not one believes that animals have rights,[157] failure to recognize the significance of wildlife "kills" constitutes neglect of their effects on both human and planetary well-being.

In more obvious economic ways, pesticide-induced fish and wildlife deaths can lead to violations of equity. It is well known that runoff from pesticide-treated lands has killed fish and other marine animals.[158] Therefore, to the extent that commercial fishing is endangered by pesticidal pollution, to that degree producers and users of these chemicals have jeopardized the interests of the fishing industry. A number of US government reports have made it clear, for instance, that the level and persistence of pesticides in Gulf of Mexico estuaries have endangered commercial species of shrimp and perhaps resulted in long-term contamination of these shellfish-growing waters.[159] Because government analyses of the effects of pesticide use ignore the social costs borne by fishermen, for example, they clearly have not provided a complete treatment of the costs and benefits of employing chemical pest controls.

There are a number of reasons why a majority of government assessments have not included the sorts of social costs or externalities just exemplified. First, the method of assessing risks and benefits has often been the "alternative cost approach."[160] According to this format, one simply determines the dollar cost of using pesticides as compared to the dollar cost of not using pesticides on the same crop areas. In other words, social risks (such as pesticidal pollution) and social benefits (such as cosmetically preferable fruits and vegetables) are ignored as consequences of chemical pest control. Instead crop production, with its attendant risks and benefits, is described as a purely private exchange involving primarily the pesticide manufacturer and the farmer. This method skews actual gains and losses, however, in ways other than its neglect of social consequences. It also is based on the false assumption that failure to use pesticides will not ultimately result in changing the type of crop produced or the location of the production. Moreover, its assumptions overstate the costs associated with restricted use of pesticides, because they are based on the premise that there will be no adjustments in the amount of the crop produced as a result of lessened employment of chemicals. In addition, the method has no way to account for adjustments that the economy might make "in response to changing production costs."[161] For all these reasons, a number of scholars have argued that "the very nature of synthetic organic pesticides," for example, their "persistence, wide spectrum toxicity, biological concentration" and the methods of applying them, provides for "spillover effects of such magnitude and distribution that the marketplace cannot adjust the values."[162]

Analysis of the costs and benefits of pesticide use is not questionable, however, only because various externalities are ignored. It is also problematic because many of the parameters which ought to be included in the assessment are not capable of being translated into monetary terms. Environmental, health, and aesthetic hazards of pesticides, for example, are three factors which "can't be incorporated into monetary terms."[163] These conversions, instead, can only be accomplished in part and by "subjective judgment techniques."[164] As a result of this incompleteness, more weight is given to quantifiable, and perhaps less important, consequences of the use of pesticides. With so many important social costs typically ignored in government assessments of chemical pest control, it is not surprising that many scientists believe that risk-benefit analyses vindicate current pest-control practices. But if these studies do not necessarily indicate that greater social good arises from using (rather than not using) pesticides, then there are no obvious grounds for maintaining that the current scope of pesticide employment is ethically desirable.

Conclusion

By criticizing some of the major tenets used to justify current levels of pesticide use, I have attempted to clarify the ethical questions involved in debate over pesticide policy. Although I have not argued that all chemical means of pest control ought to be banned, I have given reasons why I believe the more morally responsible course to follow, than that currently pursued, would be to reduce use of these chemicals by 50 percent, as suggested by the US Department of Agriculture.

Current public policy regarding pesticide use is questionable on a number of grounds. According to some interpreters, it ignores the constitutional requirement to provide equal protection and due process to all citizens, and instead presupposes the moral acceptability of a utilitarian ethic. Much current policy regarding "acceptable risks" caused by exposure to such chemicals is also questionable because it includes the assumptions that it is ethically desirable to engage in human experimentation, but logically justifiable to appeal to ignorance when evaluating the results of animal experimentation.[165] Moreover most current judgments that the benefits of pesticide use outweigh its costs are built upon erroneous economic methodology. Because the methodology does not account for a number of environmental, health, and aesthetic considerations, it is incomplete and therefore an unsound basis for establishing that pesticide use represents a greater good than nonuse. More sophisticated analyses of pesticide policy are needed. One ought to look, not only at dollar-per-acre yields with and without employment of chemical pest control, but also at the ethical questions raised by our attack against pests. What my considerations have shown, I think, is that in using chemicals against insects, we have unwittingly used them against ourselves and against the earth.

Notes

1. This saying is quoted by R. and M. Buchsbaum, *Basic Ecology*, Pittsburgh, The Boxwood Press, 1957, p. 1.

2. The roach-lizard-cat-rat episode was told by Professor L.C. Cole, an ecologist at Cornell University. The story was reprinted in N. de Nevers (ed.), *Technology and Society*, Reading, Massachusetts, Addison-Wesley, 1972, p. 233.

3. A number of philosophers have argued recently that present persons do have obligations to members of future generations. See, for example, R.M. Green, "Intergenerational Distributive Justice and Environmental Responsibility," *Bioscience*, 27, No. 4 (April 1977): 261; Joel Feinberg, "The Rights of Animals and Unborn Generations," in T.A. Mappes and J.S. Zembaty (eds.), *Social Ethics*, New York, McGraw-Hill, 1977, 358-59; and Daniel Callahan, "What Obligations Do We Have to Future Generations?" *The American Ecclesiastical Review*, 164, No. 4 (April 1971): 265-68.

4. For philosophical treatment of the question of animal rights, see Peter Singer, "Animal Liberation," *The New York Review of Books*, 20, No. 5 (April 5, 1973): 17-21; Michael Fox, "'Animal Liberation': A Critique," *Ethics*, 88, No. 2 (January 1978): 106-118; and Tom Regan and Peter Singer (eds.), *Animal Rights and Human Obligations*, Englewood Cliffs, New Jersey, Prentice-Hall, 1976.

5. Statistics on pesticide use and demand are taken from T.R. Eichers and P.A. Andrilenas, *Evaluation of Pesticide Supplies and Demand for 1978*, Agricultural Economic Report No. 399, Washington, D.C., US Department of Agriculture, March 1978, pp. v, vi, 7, 16; hereafter cited as: *Pesticide 1978*. According to this report, each year more acres are treated with chemicals and more pesticides are used per acre. See also G.W. Irving, "Agricultural Pest Control and the Environment," *Science*, 168, No. 3938 (19 June 1970): 1419; and S.C. Bloom and S.E. Degler, *Pesticides and Pollution*, Washington, D.C., Bureau of National Affairs, 1969, pp. 1-2; hereafter cited as: *P and P*.

6. Bloom and Degler, *P and P*, p. 1.

7. William Eblen, et al., *Pesticides: Mist of Death*, West Haven, Connecticut, Pendulum Press, 1971, p. 13; hereafter cited as: *Death*.

8. R.F. Glasser, "Pesticides: the Legal Environment," in D.L. Gunn and J.G. Stevens (eds.), *Pesticides and Human Welfare*, Oxford, Oxford University Press, 1976, pp. 228-39; Gunn and Stevens hereafter cited as: *Welfare*.

9. R.M. Harmer, *Unfit for Human Consumption*, Englewood Cliffs, New Jersey, Prentice-Hall, 1971, p. 11; hereafter cited as: *Human*.

10. Rachel Carson, *Silent Spring*, Boston, Houghton Mifflin, 1962, p. 12; hereafter cited as: *SS*.

11. Specifically I will analyze the faulty ethical assumptions underlying representative articles published recently in trade journals of the pesticide industry. One of the best of these is by H.F. Kraybill, a scientist-researcher at the National Cancer Institute. His article is "Pesticide Toxicity and the Potential for Cancer: A Proper Perspective," *Pest Control*, 43, No. 12 (December 1975): 10-16; hereafter cited as: Toxicity.

12. Kraybill, Toxicity, pp. 13-15. See also T. Jukes, *et al.*, *Effects of DDT on Man and Other Mammals: I*, New York, MSS Information Corporation, 1973.

13. W.W. Lowrance, *Of Acceptable Risk: Science and the Determination of Safety*, Los Altos, William Kaufman, 1976, p. 169 (hereafter cited as: *Risk*); and R.E. Train, *et al.*, *Development Document for Interim Final Effluent Limitations Guidelines for the Pesticide Chemicals Manufacturing Point Source Category*, Washington, D.C., US Environmental Protection Agency, November, 1976, p. 125; hereafter cited as: *Pesticide Chemicals*.

14. Robert Van Den Bosch, *The Pesticide Conspiracy*, Garden City, New York, Doubleday, 1978, p. 91; hereafter cited as: *Pesticide*.

15. Van Den Bosch, *Pesticide*, p. 91.

16. Kraybill, Toxicity, pp. 10, 16. W.R. Furtick, "Uncontrolled Pests or Adequate Food," in Gunn and Stevens, *Welfare,* p. 12, makes this same argument in favor of pesticides.

17. Kraybill, Toxicity, p. 10.

18. Kraybill, Toxicity, p 16.

19. Kraybill, Toxicity, p. 16.

20. R.A. Baker, *Pesticide Usage and its Impact on the Aquatic Environment in the Southeast,* Washington, D.C., US Environmental Protection Agency, September 1972, p. 9; hereafter cited as: *Pesticide Usage.*

21. C.D. Reese, et al., *Pesticides in the Aquatic Environment,* Washington, Environmental Protection Agency, 1972, p. 23; hereafter cited as: *Pesticides 1972.*

22. Reese, et al., *Pesticides 1972,* pp. 13, 8, 24, and Baker, *Pesticide Usage,* pp. 127-128, 70-73, 1-2.

23. Harmer, *Human,* p. 49.

24. Lowrance, *Risk,* pp. 86-94, discusses these and other considerations influencing safety judgments.

25. Kraybill, Toxicity, pp. 13, 15. J.M. Barnes, "Hazards to People," in Gunn and Stevens, *Welfare,* p. 183, uses this same argument in favor of pesticides.

26. Kraybill, Toxicity, pp. 12, 15. J.M. Barnes, "Hazards to People," p. 191, and W.R. Furtick, "Uncontrolled Pests or Adequate Food?" p. 12, both in Gunn and Stevens, *Welfare,* use this same argument.

27. Kraybill, Toxicity, pp. 13-14; J.C. Headley and J.N. Lewis, *The Pesticide Problem: An Economic Approach to Public Policy,* Washington, D.C., Resources for the Future, 1967, p. 96; hereafter cited as: *Pesticide Policy;* D.J. Epp, *et al., Identification and Specification of Inputs for Benefit-Cost Modeling of Pesticide Use,* EPA-600/5-77-012, Washington, US EPA, August 1977, pp. 51-56; hereafter cited as: *Pesticide Use.* See also Lowrance, *Risk,* p. 84; N.H. Crawford and A.S. Donigian, *Pesticide Transport and Runoff Model for Agricultural Lands,* EPA-660/2-74-013, Washington, Government Printing Office, December 1973, p. 8; hereafter cited as: *Pesticide Transport;* Carson, SS, pp. 50-51; and J.M. Barnes, "Hazards to People," in Gunn and Stevens, *Welfare,* p. 183.

28. Professor William Hoskins, quoted by Harmer, *Human,* pp. 38-39; see also p. 7.

29. Lowrance, *Risk,* p. 82; for further discussion of the Delaney Principle, see also Harmer, *Human,* p. 91; Epp, et al., *Pesticide Use,* pp. 55, 73, 78; and T. Jukes, *et al., Effects of DDT on Man and Other Mammals: I,* New York, MSS Information Corporation, 1973, pp. 22-23; hereafter cited as: *DDT.*

30. Lowrance, *Risk,* p. 67; see also pp. 17, 52-54, 65.

31. Lowrance, *Risk,* p. 52.

32. Harmer, *Human,* p. 41 and Lowrance, *Risk,* pp. 161-62.

33. Lowrance, *Risk,* p. 172.

34. Kraybill, Toxicity, p. 13.

35. Epp, et al., *Pesticide Use,* p. 55; see also p. 9.

36. Lowrance, *Risk,* p. 69. Epp et al., *Pesticide Use,* p. 56. See also Frank Graham, *Since Silent Spring,* Boston, Houghton Mifflin, 1970, p. 215; hereafter cited as: *SSS;* and Committee on Commerce, *Pesticide Oversight,* Hearings Before the Subcommittee on Environment of the Committee on Commerce, US Senate, 93rd Congress, Second Session, August 9 and 16, 1974, No. 93-109, Washington, Government Printing Office, 1974, p. 54; hereafter cited as: *Oversight.*

37. Carson, *SS,* pp. 222-27.

38. Baker, *Pesticide Usage,* pp. 174-75; see also Crawford and Donigian, *Pesticide Transport,* p. 6; and Reese, *et al., Pesticides 1972,* pp. 55-56.

39. Carson, *SS,* pp. 188-95, makes this same point. See also Baker, *Pesticide Usage,* p. 12 and R.L. Rudd, *Pesticides and the Living Landscape,* Madison, University of Wisconsin Press, 1964, pp. 248-67; hereafter cited as: *LL.*

40. Lowrance, *Risk,* p. 84.

41. Carson, *SS,* pp. 181-82.

42. See notes 20-22.

43. See, for example, T. Jukes, et al., *DDT,* pp. 103-106; Reese, et al., *Pesticides 1972,* p. 23, and Carson, *SS,* pp. 155-72; see also pp. 85-152.

44. Baker, *Pesticide Usage,* pp. 2-3 explains the synergistic effects of pesticides.

45. Carson, *SS,* p. 237; see also pp. 232-237.

46. Baker, *Pesticide Usage,* p. 191; see also pp. 187-91.

47. Epp, et al., *Pesticide Use,* p. 56.

48. Harmer, *Human,* p. 62, and Baker, *Pesticide Usage,* pp. 2-3.

49. Reese, et al., *Pesticides 1972,* pp. 31-32.

50. Baker, *Pesticide Usage,* p. 3.

51. Baker, *Pesticide Usage,* pp. 127-128 and Reese, et al., *Pesticides 1972,* pp. 13-24.

52. Reese, et al., *Pesticides 1972,* p. 8.

53. Baker, *Pesticide Usage,* p. 104; see also Rudd, *LL,* pp. 149-56, 160-61.

54. See note 38.

55. Baker, *Pesticide Usage,* pp. 16, 35.

56. Harmer, *Human,* p. 94, and Carson, *SS,* pp. 183-4.

57. J.E. Davis, *Pesticide Protection,* Washington, D.C. US Environmental Protection Agency, March 1977, p. 27; Carson, *SS,* p. 194; Harmer, *Human,* pp. 39, 49, 153; Graham, *SSS,* pp. 131-40; and Headley and Lewis, *Pesticide Policy,* p. 89.

58. Jukes, et al., *DDT,* p. 102.

59. Epp, et al., *Pesticide Use,* pp. 52, 79; see also Baker, *Pesticide Usage,* p. 3.

60. This example is used by Epp, et al., *Pesticide Use,* pp. 55-56.

61. Van den Bosch, *Pesticide,* p. 34.

62. Carson, *SS,* pp. 12-13.

63. John Rawls, *A Theory of Justice,* Cambridge, Harvard University Press, 1971, pp. 14-15. See also Charles Fried, *An Anatomy of Values,* Cambridge, Harvard University Press, 1970, pp. 42-43; Charles Fried, *Right and Wrong,* Cambridge, Harvard University Press, 1978, pp. 116-17, 126-27; and Alan Donagan, *The Theory of Morality,* Chicago, University of Chicago Press, 1977, pp. 221-39.

64. For a discussion of utilitarianism, see John Stuart Mill, *Utilitarianism, Liberty, and Representative Government,* New York, E.P. Dutton and Company, 1910, esp. pp. 6-24; hereafter cited as: *Utilitarianism.* See also Jeremy Bentham, *The Utilitarians: An Introduction to the Principles of Morals and Legislation,* Garden City, New York, Doubleday, 1961, esp. pp. 17-22; hereafter cited as: *Principles.* See P. Nowell-Smith, *Ethics,* Baltimore, Penguin, 1954, p. 34, and J.J.C. Smart, "An Outline of a System of Utilitarian Ethics," in J.J.C. Smart and B. Williams, *Utilitarianism: For And Against,* Cambridge, University Press, 1973, pp. 3-74; hereafter cited as: *Utilitarianism.*

65. Smart, *Utilitarianism,* pp. 69-71 and Mill, *Utilitarianism,* pp. 38-60.

66. Smart, *Utilitarianism,* pp. 69-70.

67. Smart, *Utilitarianism,* p. 69.

68. Marshall Cohen, "The Social Contract Explained and Defended," *The New York Times Book Review,* Section 7 (July 16, 1972), 1, 16-17.

69. See footnote 29. See also Epp, et al., *Pesticide Use,* pp. 52-55, 79; and Headley and Lewis, *Pesticide Policy,* p. 96.

70. Kraybill, Toxicity, p. 13. See also Jukes, et al., *DDT.*

71. Kraybill, Toxicity, p. 15. See also Jukes, et al., *DDT.*

72. Baker, *Pesticide Usage,* pp. 167, 216, 235.

73. G.L. Henderson, *et al., Effects of DDT on Man and Other Mammals II,* New York, MSS Information Corporation, 1973, pp. 123-63; hereafter cited as: *DDT II;* US Food and Drug Administration, *FY 1973 Pesticide/PCB in Foods Program: Evaluation Report,* Washington, FDA, 1973, pp. 6, 15; hereafter cited as: *Pesticide/PCB;* Baker, *Pesticide Usage,* pp. 195-96.

74. FDA, *Pesticide/PCB,* Part 3, Chart 1; and Reese, et al., *Pesticides 1972,* p. 8.

75. Kraybill, Toxicity, pp. 14-15.

76. Baker, *Pesticide Usage,* p. 194.

77. Reese, et al., *Pesticides 1972,* p. 53.

78. Epp, et al., *Pesticide Use,* p. 56.

79. Kraybill, Toxicity, pp. 10, 14, 15.

80. Kraybill, Toxicity, pp. 10, 14.

81. Lowrance, *Risk,* p. 88 discusses how the existence of available alternatives to a questionable scientific practice renders the practice less acceptable.

82. Lowrance, *Risk,* p. 88.

83. Baker, *Pesticide Usage,* pp. 295-397.

84. Lowrance, *Risk,* p. 171.

85. Baker, *Pesticide Usage,* p. 4.

86. See Graham, *SSS,* pp. 260-66; Carson, *SS,* pp. 277-97; Harmer, *Human,* pp. 232-41; Reese, et al., *Pesticides 1972,* pp. 75-141; Baker, *Pesticide Usage,* pp. 295-397; R.W. Holcomb, "Insect Control: Alternatives to the Use of Conventional Pesticides," *Science,* 168, No. 3930 (24 April 1970); 456-58; hereafter cited as: Control. See also Helmut van Emden, *Pest Control and Its Ecology,* London, Edward Arnold, 1974, pp. 19-33; hereafter cited as: *Pest Control;* and David Pimental, "Realities of a Pesticide Ban," *Environment,* 15, No. 2 (March 1973): 19-31; hereafter cited as: Realities.

87. Baker, *Pesticide Usage,* p. 338; see also Harmer, *Human,* p. 173.

88. Robert Van den Bosch, "The Cost of Poisons," *Environment,* 14, No. 7 (September 1972): 31; hereafter cited as: Cost.

89. Baker, *Pesticide Usage,* p. 374.

90. Reese, et al., *Pesticides 1972,* p. 109.

91. Baker, *Pesticide Usage,* p. 321 and Reese, et al., *Pesticides 1972,* pp. 94 ff.

92. Baker, *Pesticide Usage,* p. 331 and Reese, et al., *Pesticides 1972,* p. 108.

93. Baker, *Pesticide Usage,* pp. 4, 347.

94. Reese, et al., *Pesticides 1972,* pp. 111 ff. and Baker, *Pesticide Usage,* pp. 335-36.

95. Pimental, Realities, pp. 28-29.

96. See notes 83, 85, and 88.

97. Pimental, Realities, p. 29; see also Baker, *Pesticide Usage,* p. 375; Crawford and Donigian, *Pesticide Transport,* p. 6; Harmer, *Human,* pp. 188-214; Rudd, *LL,* pp. 30-31; Irving, "Agricultural Pest Control," pp. 1420-23.

98. Van Emden, *Pest Control,* pp. 51-59; see also pp. 35-50.

99. Van den Bosch, *Pesticide,* pp. 147-51; see also pp. 152-78.

100. Crawford and Donigian, *Pesticide Transport,* p. 7.

101. See note 5.

102. Van den Bosch, *Pesticide,* pp. 173-177.

103. Holcomb, Control, p. 458.

104. Baker, *Pesticide Usage,* p. 21.

105. Reese, et al., *Pesticides 1972,* p. 140.

106. Graham, *SSS,* p. 222.

107. Statistics on national annual pesticide sales are given in Van den Bosch, *Pesticide,* p. 60.

108. H.C. Chapman, "The Present Limitations of Biological Control of Public Health Insects," *Pest Control,* 45, No. 1 (January 1977): 14; hereafter cited as: Limitations.

109. Baker, *Pesticide Usage,* p. 18, makes this point.

110. Van den Bosch, *Pesticide,* pp. 62-70; see also Graham, *SSS,* pp. 167-75.

111. Harmer, *Human,* pp. 90-110, 111-127.

112. Cited in Baker, *Pesticide Usage,* p. 383. This point is also supported by Van den Bosch, *Pesticide,* 203-205.

113. Graham, *SSS,* p. 185.

114. See note 80.

115. D. Price Jones, "Pesticides and the Conservation of Land and Energy Resources," in Gunn and Stevens, *Welfare,* pp. 206-16, makes these two points.

116. Chapman, Limitations, p. 13.

117. Van den Bosch, *Pesticide,* pp. 23-25; see also Carson, *SS,* pp. 245-61; Barry Commoner, *The Closing Circle,* New York, Bantam, 1974, pp. 149, 245; hereafter cited as: *Circle.*

118. Baker, *Pesticide Usage,* pp. 2, 321.

119. Epp, et al., *Pesticide Use,* p. 91.

120. Commoner, *Circle,* pp. 149, 245.

121. Rudd, *LL,* pp. 141-148. See also Carson, *SS,* pp. 263-75.

122. Van den Bosch, *Pesticide,* p. 27.

123. Pimental, Realities, pp. 20, 25 cites this information from the USDA.

124. Van den Bosch, *Pesticide,* pp. 27-31.

125. J.R. Busvine, "Pest Resistance to Pesticides," in Gunn and Stevens, *Welfare,* p. 199.

126. G.A. Matthews, "Cotton," in Gunn and Stevens, *Welfare,* pp. 157-67. See also Van den Bosch, Cost, p. 20.

127. Carson, *SS,* pp. 119-216.

128. Epp, et al., *Pesticide Use,* p. 96. See also Carson, *SS,* p. 13.

129. Epp, et al., *Pesticide Use,* p. 48.

130. Arguments for this point may be found in the works cited in note 3.

131. Cited in Lowrance, *Risk,* p. 82.

132. Kraybill, Toxicity, pp. 14-15. This claim is also supported by Irving, Agricultural Pest Control, p. 1419; E.E. Cheeseman, "Cocoa," in Gunn and Stevens, *Welfare*, pp. 147-49; G. Schuhmann, "The Economic Impact of Pesticides on Advanced Countries," in Gunn and Stevens, *Welfare*, pp. 55-71; R.H. Gray, "Population and Food Production," in Gunn and Stevens, *Welfare*, pp. 13-27; and H.S. Stein, "We're Part of This Picture," *Pest Control*, 41, No. 1 (January 1973): 11-12.

133. See note 97.

134. See, for example, K. Mellanby, "Pesticides, the Environment, and the Balance of Nature," in Gunn and Stevens, *Welfare*, pp. 217-27. See also A.V. Adam, "The Importance of Pesticides in Developing Countries," in Gunn and Stevens, *Welfare*, pp. 115-30; J.A. Freeman, "Food in Store," in Gunn and Stevens, *Welfare*, pp. 169 ff.; and G. Schuhmann, "The Economic Impact of Pesticides on Advanced Countries," in Gunn and Stevens, *Welfare*, pp. 55-71.

135. A.V. Adam, "The Importance of Pesticides in Developing Countries," in Gunn and Stevens, *Welfare*, pp. 115-30.

136. Eichers and Andrilenas, *Pesticide 1978*, pp. 7, 16, 136.

137. Baker, *Pesticide Usage*, pp. 195, 55.

138. Eichers and Andrilenas, *Pesticide 1978*, p. 7.

139. Pimental, Realities, p. 19.

140. Van den Bosch, *Pesticide*, pp. 39 ff. See also notes 125 and 126.

141. Cited in Pimental, Realities, pp. 19, 20, 25.

142. Cited in Pimental, Realities, p. 26.

143. See Carson, *SS*, p. 9.

144. Pimental, Realities, pp. 19-31, esp. p. 26.

145. See note 133.

146. Kraybill, Toxicity, p. 15.

147. Kraybill, Toxicity, p. 15.

148. See Arthur C. Pigou, *The Economics of Welfare*, London, Macmillan, 1962. For futher discussion of environment-related externalities, see R.T. Roelofs, J.N. Crowley, and D.L. Hardesty, "Environment and the New Economics," in Roelofs, *et al.* (eds.), *Environment and Society*, Englewood Cliffs, N.J., Prentice-Hall, 1974, pp. 113-18, and Commoner, *Circle*, pp. 249-91. In theory, externalities may be either positive or negative, although it is easier to find examples of the latter type. Like negative externalities, positive ones are also involuntarily imposed; unlike negative externalities, positive ones are economically beneficial to those upon whom they are placed. Such an externality is enjoyed, for example, by a homeowner who lives next to a quiet, spacious park. Externalities, and especially negative externalities, are often referred to as "spillovers," or "diseconomies."

149. Baker, *Pesticide Usage*, pp. 16, 273.

150. Van den Bosch, *Pesticide*, pp. 29-35.

151. Baker, *Pesticide Usage*, p. 105.

152. Baker, *Pesticide Usage*, p. 193, and Reese, *et al.*, *Pesticides 1972*, pp. 10, 54.

153. Reese, et al., *Pesticides 1972*, p. 53.

154. A.D. Bernstein and R.A. Lowrey, *Methodology for Designing Cost-Effective Monitoring and Compliance Strategies for Pesticide Use*, EPA-600/5-78-020, Athens, Georgia, US Environmental Protection Agency, September 1978, p. 3; hereafter cited as: *Methodology*.

155. Epp, et al., *Pesticide Use*, p. 5.

156. Baker, *Pesticide Usage*, p. 185; Graham, *SSS*, pp. 117-130.

157. See footnote 4.

158. Baker, *Pesticide Usage,* pp. 130-33; Graham, *SSS,* pp. 117-30.

159. Baker, *Pesticide Usage,* pp. 132, 185-87, and Reese, et al., *Pesticides 1972,* p. 49.

160. For examples of use of this method by government scientists, see the Office of Water Planning and Standards, US Environmental Protection Agency, *Economic Analysis of Interim Final Effluent Guidelines for the Pesticides and Agricultural Chemicals Industry,* EPA-230/1-76-065 f, Washington, US EPA, September 1976, pp. 36-39; see also pp. 40-44; hereafter cited as: *Economic Analysis.*

161. Epp, et al., *Pesticide Use,* pp. 22-42, esp. p. 26. See also Headley and Lewis, *Pesticide Policy,* pp. 37, 59.

162. Headley and Lewis, *Pesticide Policy,* pp. 129-30.

163. Epp, et al., *Pesticide Use,* pp. 117-18; see also pp. iii, 86-126.

164. Bernstein and Lowrey, *Methodology,* pp. 32-33; 57; see also Office of Water Planning and Standards, US EPA, *Economic Analysis,* pp. 36, 51-52.

165. See note 160.

Further Reading

Adam, A.V. 1976. The Importance of Pesticides in Developing Countries. In *Pesticides and Human Welfare.* ed. D.L. Gunn and J.G.R. Stevens, pp. 115-130. Oxford: Oxford University Press.

Adams, J. 1991. ALAR. *NRDC Newslines* 8 (April): 2.

Baker, R.A. 1972. *Pesticide Usage and Its Impact on the Aquatic Environment in the Southeast.* Washington, D.C.: US Environmental Protection Agency (September).

Bernstein, A.D. and Lowrey, R.A. 1978. *Methodology for Designing Pesticide Cost-Effective Monitoring and Compliiance Strategies for Pesticide Use,* EPA-600/5-78-020. Athens, Georgia: US Environmental Protection Agency (September)..

Bloom, S.C., and Degler, S.E. 1969. *Pesticides and Pollution.* Bureau of National Affairs.

Brown, L., Flavin, C., and Postel, S. 1990. Picturing a Sustainable Society. In *State of the World 1990.* ed. L. Brown et al., pp. 173-190. New York: Norton.

Brown, L., and Wolf, E. 1985. *Reversing Africa's Decline.* Washington, D.C., Worldwatch Institute.

Carothers, A. 1990. Targeting Pesticide Exports. *Greenpeace.*(May/June): 23.

Carson, R. 1962. *Silent Spring.* Boston: Houghton Mifflin.

Chapman, H.C. 1977. The Present Limitations of Biological Control of Public Health Insects. *Pest Control* 45 (January): 14.

Committee on Commerce. 1974. *Pesticide Oversight.* Hearings Before the Subcommittee on Environment of the Committee on Commerce, U.S. Senate, 93rd Congress, Second Session, August 9 and 16, No. 93-109. Washington, D.C. Government Printing Office.

Crawford, N.H., and Donigian, A.S. 1973. *Pesticide Transport and Runoff Model for Agricultural Lands.* EPA-660/2-74-013. Washington, DC.: Government Printing Office (December).

Detwyler, T.R. 1971. *Man's Impact on the Environment*, pp. 549)581. New York, McGraw-Hill.

Devlin, R.M. 1974. DDT: A Renaissance? *Environmental Science and Technology* 88: 322-325.

Dover, M. *A Better Mousetrap.* Washington, D.C.: World Resources Institute.

Dubos, R., and Ward, B. 1972. *Only One Earth.* pp. 60-66. New York: W.W. Norton.

Eblen, W. Daller, J., and Schwartz, B. 1971. *Pesticides: Mist of Death.* West Haven, Connecticut: Pendulum.

Eichers, T.R., and Andrilenas, P.A. 1978. *Evaluation of Pesticide Supplies and Demand for 1978.* Agricultural Economic Report No. 399. Washington, D.C.: US Department of Agriculture (March).

Epp, D.J., et al. 1977. *Identification and Specification of Inputs for Benefit-Cost Modeling of Pesticide Use,* EPA-660/5-77-012, Washington, D.C.: U.S. Environmental Protection Agency (August).

French, H. 1990. *Green Revolutions.* Washington, D.C.: Worldwatch Institute.

Graham, F. 1970. *Since Silent Spring.* Boston: Houghton Mifflin.

Grayson, M.J., and Shepard, T.R. 1973. *The Disaster Lobby,* esp. pp. 27-43. Chicago, Follett.
Harmer, R.M. 1971. Pesticides and Living Things. In *Unfit for Human Consumption,* pp. 56-71. Englewood Cliffs, New Jersey: Prentics-Hall.
Headley, J.C., and Lewis, J.N. 1967. *The Pesticide Problem: An Economic Approach to Public Policy.* Baltimore: Johns Hopkins Press.
Holcomb, R.W. 1970.Insect Control: Alternatives to the Use of Conventional Pesticides. *Science* 168: 456-458.
Irving, C.W. 1970. Agricultural Pest Control and the Environment. *Science* 168: 1419-1424.
Jones, D. 1976. Pesticides and the Conservation of Land and Energy Resources. In *Pesticides and Human Welfare.* ed. D.L. Gunn and J.G.R. Stevens, pp. 206-216. Oxford University Press.
Jukes, T., et al. 1973. *Effects of DDT on Man and Other Mammals: I.* New York: MSS Information Corporation.
Millar, F.1991. Too Close for Comfort. *Friends of the Earth.*21, No. 1 (Winter): 10-13.
Miller, A., Mintzer, I., Hoagland, S. 1986. *Growing Power.* Washington, D.C.: World Resources Institute.
Moyers, B. 1990. *Global Dumping Ground.* Washington: Seven Locks Press.
Murphy, D. 1988. Challenges to Biological Diversity in Urban Areas. In *Biodiversity,* ed. E.O. Wilson, pp. 71-78. Washington, D.C.: National Academy Press.
Office of Water Planning and Standards, Environmental Protection Agency. 1976. *Economic Analysis of Interim Final Effluent Guidelines for the Pesticides and Agricultural Chemicals Industry.* EPA-230/1-76-065 f, Washington, D.C.: US Environmental Protection Agency (September).
Pimentel, D. 1973. Realities of a Pesticide Ban. *Environment* 15: 18-31.
Postel, S. 1986. *Altering the Earth's Chemistry.* Washington, D.C.
Reese, C.D., et al. 1972. *Pesticides in the Aquatic Environment,* Washington D.C.: US Environmental Agency.
Rudd, R.L. 1964. *Pesticides and the Living Landscape.* Madison, Wisconsin: University of Wisconsin.
Thornton, J. 1991. The Dioxin Deception.*Greenpeace* (May/June): 16-21.
Train, R.E., et al. 1976. *Development Document for Interim Final Effluent Limitations Guidelides for the Pesticide Chemicals Manufacturing Point Source Category.* Washington, D.C.: US Environmental Protection Agency (November).
Van den Bosch, R. 1972. The Cost of Poisons. *Environment* 14 (September): 18-31.
————. 1978. *The Pesticide Conspiracy.* Garden City: New York: Doubleday.
Van Emden, H.F. 1974. *Pest Control and Its Ecology.* Institute of Biology's Studies in Biology, No. 50, London: Arnold.
Wasserstrom, R., and Wiles, R.1985.*Field Duty: US Farm workers and Pesticide Safety.* Washington, D.C.: World Resources Institute.
Whiteside, T. 1970. *Defoliation.* New York: Ballantine.

12

NUCLEAR ENERGY AND RADIOACTIVE POLLUTION

Introduction

THE Global weapons budget is now at $1 trillion per year. (This is equivalent to spending approximately $32,000 every second for a year.) Less than one day of the annual global arms race could pay to wipe out cholera, malaria, polio, typhus, and yellow fever in developing nations. Several days of the cost of the global arms race could end famine on the planet, or save all the rainforests, or restore polluted rivers worldwide. 40 days worth of global military expenditures, for example, could halt Third World desertification. Moreover, the people saved by not spending so much money on nuclear weapons would not be merely those dying of disease or starvation in developing nations. Many persons in developed countries, including Americans and Soviets, would also benefit by slowing the arms race. 500,000 USA servicemen were exposed to dangerous levels of fallout after nuclear weapons testing in the Pacific and the Southwest; many have already died of radiation-induced leukemia and multiple myeloma. One fifth of the Soviets living near the Semipalatinsk nuclear test site have died of premature, radiation-induced cancer, and infant mortality has doubled since testing began.

Apart from the morality of developing atomic weapons that may kill us all, nuclear arms raise a key environmental issue. Do we have the right to jeopardize the health and continued existence of all our planetary ecosystems? If Carl Sagan is correct, even a few nuclear detonations would blanket the earth in impermeable smoke and dust, causing a "nuclear winter," destruction of most plants and animals, global crop failure, and world-wide famine.

Because commercial nuclear power plants create plutonium usable for nuclear weapons, a key ethical issue is whether we ought to use a "peaceful" technology whose side effects are necessary for the technology of nuclear war. Proliferation, however, is not the only problem facing commercial nuclear energy. Ever since the first fission plant generating electricity opened in 1957, the use of atomic energy has been ethically and environmentally controversial. The USA claimed that fission was the energy source of the future, and that by the year 2000, half of all USA electricity would be supplied by 1000 fission reactors. By the early 1990s, however, 112 nuclear plants supplied only 13 percent of USA electricity. Moreover, since 1974, there have been no new orders for commercial reactors in the USA. Worldwide, there are about 425 nuclear plants with another 80 under construction. Most new facilities are in third-world nations. Some environmentalists claim that, because nuclear energy is not cost-effective, countries build these new reactors in order to attain nuclear-weapons capability.

The possibility of catastrophic nuclear accidents also presents numerous scientific and ethical difficulties. Although the moral issue here is whether one ought to risk a major accident which (according to USA government estimates in WASH-740, "The Brookhaven Report") could cause 45,000 immediate casualties and 100,000 later deaths, there are no complete data available on such a catastrophe. WASH-1400, known as "The Rasmussen Report," was said by government to be the only complete study of nuclear power safety. It concluded that a major nuclear incident had a very low probability of occurrence, 1 in 17,000 per reactor year, or 1 in 4 for the 30-year lifetimes of all USA reactors. However, after this study was challenged by the US Environmental Protection Agency, the American Physical Society, and other groups in 1979, the US Nuclear Regulatory Commission withdrew its approval from WASH-1400.

At present, there is no government approved study of the risks of catastrophic nuclear accidents. The Soviet nuclear core melt at Chernobyl in April 1986 and the USA core melt at Three Mile Island in March 1979, however, have convinced the public that catastrophic nuclear accidents are possible. European scientists reporting to the World Information Service on Energy (WISE) have calculated that between 50,000 and 75,000 persons will die prematurely from cancers caused by Chernobyl radiation. Six years after the disaster, 1 in 4 children born in Soviet Mogilev—and 1 in 2 in Soviet Narodichi—suffer from thyroid disease. In Cherkassy, 1 in 5 babies are born with deformities—limbs, eyes, or ears missing. Blood and liver problems, immune disorders, and gastrointestinal diseases have also plagued thousands of victims. Outside the Soviet Union, Polish children have died from eating food contaminated by Chernobyl fallout; the number of genetically deformed German livestock has risen; and many farms in Scotland are still subject to government restrictions on movement and slaughter of contaminated animals.

The Chernobyl accident raises the ethical issue of whether government ought to jeopardize citizens' rights to life by a potentially catastrophic energy technology. Is it ethical for government to impose such risks on the population, even though the probability of catastrophe is allegedly low? Do government experts have the right to decide, for the people, how safe is safe enough? Do the ethical answers to these questions depend on whether alternative energy sources are available? Since the US Office of Technology Assessment claims that solar energy can now supply, cost-effectively, 40 percent of total USA energy needs, is this a more ethical option than nuclear power? Responding to the Chernobyl and Three-Mile-Island accidents, nuclear proponents have argued that future reactor designs can make nuclear power safe. With this hope, the Bush administration unveiled a new "National Energy Strategy" in 1991 that called for billions of dollars in new reactor research and more than a doubling of USA nuclear energy capacity by the year 2030.

Environmentalists have responded that, if commercial nuclear power is really safe, then the USA government ought to repeal the Price-Anderson Act, the USA law which limits nuclear-accident liability to less than one percent of total possible damages. The only reason for the liability limit demanded by the nuclear industry, say environmentalists, is that a massive accident could bankrupt the nuclear utility. Yet, they say, if nuclear power is safe, then accidents are unlikely, and the liability limit is not needed. Hence, environmentalists claim that the utilities cannot consistently claim both that nuclear power is safe and that they need the liability limit. The also say that such limits violate citizens' rights to due process of law and to equal protection.

Even if government removes the liability limit and prevents all catastrophic reactor accidents, commercial nuclear energy would still face the problem of "normal," low-level radioactive emissions. The USA government has calculated that, even if there are no accidents, normal commercial nuclear emissions in the USA will cause at least 30,000 cancer deaths, worldwide, up to the year 2000. The deaths amount to trading lives for electricity. Is such a tradeoff justifiable? If a zero-risk society is impossible, at which risk level might a tradeoff be unethical? Is government allowed to ignore nuclear-induced deaths, to achieve "the greatest good for the greatest number of persons"? Does nuclear power achieve a "greater economic good," despite its health costs?

The US Nuclear Regulatory Commission contends that tighter control of radioactive pollution is not economically advantageous to the nuclear utility industry. As a consequence, reactor licenses are not required to contain radioactive emissions if it would cost them more than $1,000 per person-rem to do so, and if the 0.5 rem annual standard is met. The chief ethical question, raised by public policy regarding low-level radiation emissions, is whether such a tradeoff between economic benefits and health losses is acceptable.

Another major environmental and ethical problem associated with nuclear energy is intergenerational equity. Because of the long half-lives of many of the radioactive isotopes, the nuclear fuel cycle of each plant generates waste that will remain lethal—capable of causing cancer, genetic damage, and injury—in perpetuity. Ought present generations to mortgage the health and safety of future persons? Can we contract a radioactive debt that must be paid by our descendants? Do any future benefits counterbalance this radioactive cost to the future? Have present individuals ignored their duties not to jeopardize future persons' free, informed consent to radioactive risks?

In the two essays which follow, each author takes a different view of the moral problems raised by the issues just outlined. Bernard Cohen, Director of the Scaife Nuclear Laboratories at the University of Pittsburgh and author of *Nuclear Science and Society,* argues that the questions surrounding the acceptability of nuclear fission generation of electricity are largely technical ones. He cites the probability of injury resulting from nuclear power plants, compares these hazards to those from coal-fired plants, and concludes that atomic energy provides an acceptable risk. Ralph Nader, consumer advocate and coauthor of *The Menace of Atomic Energy,* disagrees with Cohen on all three counts.

Nader shows why the question of nuclear fission is primarily a political, institutional, and moral issue, rather than a scientific or technical one. He argues that atomic power is only *thought* to be safer and cheaper than electricity generated from coal, because of massive government subsidies of nuclear energy and because many health effects of fission power have been ignored.

Unlike Cohen, Nader believes that the question of the acceptability of nuclear generation of electricity must be analyzed in the light of alternative technologies, such as solar and geothermal power. While Cohen ignores discussion of these alternatives, Nader stresses their importance and criticizes the US government for spending so much money on nuclear research rather than on studies related to "soft" energy paths. Whereas Cohen accepts the conclusions of WASH-1400, the Rasmussen Report, Nader sides with the US Environmental Protection Agency, the American Physical Society, and the Union of Concerned Scientists in criticizing it.

Although Cohen and Nader disagree on numerous technical questions, such as the acceptance of the Rasmussen Report, their ethical conclusions are even more widely divergent. Cohen assumes that because the alleged probability of nuclear catastrophe is low and because fission reactors allegedly cause fewer health effects than coal-fired plants, atomic energy is morally acceptable. Nader, on the other hand, says that the grave ethical uncertainties regarding nuclear fission cannot be "swept away" merely by probability estimates. Some of the uncertainties in this category include the fact that numerous infringements of civil liberties are likely to occur as a result of widespread employment of atomic power. Another problem mentioned by Nader is the violation of rights to due process and equal protection. The violations are allowed by virtue of the government-established liability limit for nuclear accidents. According to the Price-Anderson Act, total claims against a nuclear utility are prohibited from exceeding approximately $680 million, even though government estimates (WASH-740 update) of property damages, alone, from a major reactor accident are $18 billion. Nader criticizes the fact that the nuclear industry has been allowed this protection against possible claims.

Apart from the scientific disagreements among nuclear proponents and opponents, there is common ground on which they both agree, viz., that nuclear power is not a 100 percent safe technology. The question raised by their disagreement and by the controversy is an ethical one: "How safe is safe enough?"

Readings

1. Cohen, B. 1978. Health Risks of Nuclear Power. *The Physics Teacher* 16 (November): 526-32. Copyright 1978 by The American Association of Physics Teachers. Reprinted by permission of Bernard Cohen and *The Physics Teacher*.

2. Nader, Ralph. 1976. Nuclear Power: More Than a Technological Issue? *Mechanical Engineering* 98 (February): 32-37. Copyright 1976 by *Mechanical Engineering*. Reprinted by permission of the editor.

HEALTH RISKS OF NUCLEAR POWER

Bernard L. Cohen

IF A UTILITY wants to build a power plant to produce electricity, its choice is now generally restricted to burning coal in a "coal-fired" power plant, or burning uranium in a nuclear power plant. When these fuels are burned they don't simply disappear, but rather they are changed into other forms which may loosely be called "wastes." When these wastes get into the human environment they may cause damage to health and/or property; therein lie the principal environmental impacts of producing electricity. This paper will deal with the wastes generated in nuclear power, but for perspective let us begin with a brief outline of the wastes generated by a single large (one million kilowatt) coal-fired plant.

The pricipal waste product of such a plant is carbon dioxide, emitted at a rate of one ton every 5 sec. This is not basically a dangerous gas, but the vast quantities of it being emitted from burning coal, oil, and gas have increased its concentration in the atmosphere appreciably and there are serious concerns that this may have far-reaching long-term effects on world climate. The most important pollutant emitted is sulfur dioxide, discharged at a rate of about 10 pounds per sec. According to a recent high level study, this typically causes 25 fatalities, 60,000 cases of respiratory disease, and $12 million in property damage per year. Among the other gases emitted are nitrogen oxides, the principal pollutants emitted by automobiles; our one power plant discharges as much of these as 200,000 automobiles. Second only to sulfur dioxide as a health hazard is the smoke, made up of tiny solid particles; one sixth of all man-made smoke comes from these power plants. And finally there is the solid wastes or ashes, produced at a rate of about one ton per minute.

The waste products of a nuclear power plant consist of gases released to the atmosphere, contaminated water released to rivers, and solid wastes which will be buried deep underground. But these wastes differ from those of a coal-fired plant in two very spectacular ways: firstly, their quantity is 5 million times less—the wastes produced by a large nuclear power plant in one year could fit under a dining room table; and secondly, they are *radioactive.*

Somehow radioactivity has a very bad public image; it is generally viewed as something very new, highly mysterious, and extremely dangerous. Actually there is nothing *new* about radioactivity; it is, and always has been present in the earth (uranium, thorium, radium, etc.) and therefore also in the bricks and stones we use as building materials, in our bodies (potassium, an element vital to life is naturally radioactive), and in the air we breathe (radon gas) in far greater quantities than we will ever experience from nuclear power. Moreover, a new type of man-made radiation has been introduced into our environment in recent years in the form of medical x rays, almost equal to our exposure from natural radioactivity.

There is also nothing so *mysterious* about radioactivity and the radiations it emits. They are much simpler and consequently much better understood than nearly any other insult to our environment such as air or water pollution, insecticides, and food additives. Measuring techniques for radiation are cheap, accurate, and extremely sensitive—easily capable of detecting one-millionth of an amount that would do harm. Hundreds of millions of dollars have been spent studying the biological effects of radiation, and as a result we have rather reliable determinations of the worst effects a given amount of radiation can cause; it is these worst estimates that are always used in assessing environmental impacts (note that this is *not* done in assessing environmental impacts of coal-fired power plants; air pollution is assumed to be harmless unless there is direct evidence for the harm it does at the levels in question).

Radioactivity can be *dangerous* in large quantitites, but probably the main reason for the emotional reaction to this danger is the 100,000+ fatalities caused by the atomic bombs dropped on Japan during World War II. Blaming radioactivity for these deaths is unjustified: the great majority of these fatalities were caused by blast, heat, and fire, having nothing to do with radiation. Perhaps 20% were due to the direct radiation from the bomb but *none* were caused by the remnant radioactivity, which is the only source of potential danger from nuclear wastes. In fact, no trace of radioactivity from the bombs was ever found in Hiroshima or Nagasaki.

Before closing this discussion, it should perhaps be pointed out that radioactivity is saving thousands of lives each year through its medical applications, and billions of dollars from its applications in industry, agriculture, commerce, health care, and research.

Gaseous Waste

Returning now to the nuclear wastes, the effects of gaseous releases of radioactivity have been estimated by various investigators, and there is little disagreement in the scientific community about them. If all U.S. electric power were nuclear, the radiation these plants would administer to the average American would be less than 1% of what he already receives from cosmic rays (radiations from outer space), less than 1% of what he receives from the natural radioactivity in the ground, about 1% of what he receives from natural radioactivity in the bricks and stones from buildings (avoidable by living in a wooden house), 0.3% of what he receives from medical x rays, and 0.2% (in effectiveness) of what he receives from the natural radon in the air he breathes. It is also less radiation than he would get from one coast-to-coast airplane flight every ten years, or from spending one day per year in Colorado where the natural radiation is twice the national average (of course millions of people spend their whole lives in Colorado, and there has been no detectable effect on their health from this extra radiation; there are places in India and Brazil where the natural radiation is ten times higher than in Colorado, still with no detectable effects). While these comparisons may make the gaseous releases seem completely unimportant, by most reckonings they are responsible for nearly half of all the health effects of nuclear power.

Solid Wastes

The contaminated water released from nuclear facilities is less than one-tenth as effective as the gaseous releases, so we will ignore it here and go on to the question of the solid wastes which contain the great majority of the radioactivity and have been the center of so much discussion in recent years. These substances will be made part of a Pyrex glass-type of material and buried deep underground (about 1/3 mile deep) in some carefully selected rock formation. The principal danger from this waste after burial is that it will be contacted by ground water, leached into solution, and seep through the ground until it reaches the surface and gets into food and water supplies. To evaluate this hazard we must therefore consider the toxicity of the waste if ingested with food or water. This is given in the following table for various times after burial:

years after burial	0	100	200	300	500	1000	1 million
ounces ingested for 50% chance of health effect	0.001	0.01	0.1	1.0	6	16	160

From this it seems reasonable to conclude that the principle danger lasts only for a few hundred years. It is difficult to see how one can worry much about a rock-like material buried 1/3 mile underground of which several ounces ingested would be required to do harm; we have more dangerous materials in our homes.

We will develop a complete evaluation of this problem over the millions of years some like to consider, but first let us discuss the protections we have during the first few hundred years when the waste is so dangerous. When people hear that this waste must be isolated from mankind for hundreds of years, their immediate response is horror; few things on this earth can be counted on to remain unchanged for hundreds of years. However, one would have a very different reaction if he were a rock lying 1/3 mile below the surface fo the earth. The environment there is very different. If these rocks had a newspaper, it would only be published once in a million years, because there would be little news to report. Changes in rocks at these depths take place typically only once in 100 million years.

Let us list some of the protections we have against buried waste returning to our environment via the ground water route within the first few hundred years when it is highly toxic. *First,* the waste will be buried in a rock formation in which there is no ground water and in which geologists are quite certain there will be no ground water for at least a thousand years (a thousand years is a rather short time for geologists). *Second,* if ground water should somehow get into this rock formation, it would have to dissolve or leach much of it away before reaching the waste; this would be easiest if the waste were buried in salt, but if all the ground water in the region of the proposed burial site in New Mexico were diverted to flow through the salt formation it would take 50,000 years to dissolve away the salt in which one year's waste would be buried. *Third,* if ground water reaches the glass in which the waste is incorporated, it would leach very slowly; the glass form was chosen for its leach resistance, and if ongoing research indicates problems in this regard, other (ceramic) forms can readily be substituted. *Fourth,* the ground water in which the waste would be dissolved seeps through the ground very slowly; it typically takes ground water 1,000 years to reach the surface from 1/3 mile deep. *Fifth,* the radioactive material would be constantly filtered out of the ground water by the rock and soil ("ion exchange" processes), so it would travel 100 to 10,000 times slower than the water itself, thus typically taking 100,000 to 10 million years to reach the surface (If it seems hard to believe that material dissolved in water deep underground takes so long to reach the surface, one might consider petroleum, which is more mobile than this material but has been underground for tens of millions of years). *Sixth,* radioactivity is very easily detected, so if this material did get out and get into food or water supplies, it would almost surely be discovered before people could be harmed by it. With all of these protections, it seems virtually certain that once this waste is buried, it will not get back to our environment and do harm to mankind for at least a few hundred years.

We now turn to the hazard from the waste over the "millions of years" time period. This is best understood by comparing it with the natural radioactivity in the ground. The ground is full of natural radioactivity, and the wastes we bury add to it by only about one part per million per year. Moreover, the natural radioactivity from deep underground is doing virtually no harm. A quantitative evaluation of the hazard from buried waste may be derived if we assume that buried waste is no more likely to be contacted by ground water and leached than is *average* rock at that depth. The result of a calculation based on that assumption is that if we used nuclear power for a million years, the accumulated waste from all of this time would probably cause *less than one* fatality per year. Moreover, one should not forget that nuclear power plants are burning up uranium which is a major source of natural radiation; by doing so it would be *saving* about 50 lives per year, a hundred times more than the lives lost due to the waste. Thus, on any very long time scale, nuclear power must be viewed as a means for cleansing the earth of radioactivity.

Before closing our discussion of the solid wastes, some comments are in order on the often-raised question of the burden we are placing on future generations in watching our waste. In the first place it should be noted that our estimate above of less than one fatality per year from the wastes accumulated by a million years of all nuclear power in the U.S. is with *no* watching. It was derived by comparing the waste with "average rock," and clearly nobody is watching average rock; watching would serve only to reduce this already small effect. In the second place, once the waste is securely buried watching it would be a trivial burden; a thousand years of waste accumulation would be buried under an area only 10 miles square, and watching it (keeping signs maintained, checking against deep drilling or mining, collecting a few water samples once a month to measure their radioactivity) would be a part-time job for one person.

Clearly we are placing an infinitely greater burden on our progeny by consuming all the earth's rich mineral resources. Within a few generations we are using up all the copper, tin, lead, zinc, mercury, etc., leaving few alternatives for our progeny to use for metals. And we are burning up those very valuable hydrocarbons—coal, oil and gas—at a rate of millions of tons each per day, thereby depriving our descendants materials for producing plastics and organic chemicals. In doing this, we are placing a tremendous burden on our progeny, and the only way we can compensate them is by leaving them a technology that will allow them to live in reasonable comfort without these resources. The key to such a technology must be cheap and abundant energy. With it, they can develop substitutes for almost anything. Without it, their only recourse will be to return to a primitive life. Thus we owe our progeny a source of cheap and abundant energy, and the only such source we can now guarantee is nuclear energy.

Plutonium

There is one component of what we have been calling nuclear waste that will not be buried deep underground—namely the plutonium which will be separated out because it is valuable as a fuel. However, plutonium has also somehow received a very unfavorable public image. It has been called "the most toxic substance known to man" (a demonstrable falsehood) and Ralph Nader has claimed that "a pound of plutonium can cause 8 billion lung cancers." The basis for such statements has never been explained but clearly they depend on assuming that every bit of the plutonium finds its way into a human lung, and further on the validity of the "hot particle" theory of alpha particle toxicity. The first assumption is clearly unrealistic without collaboration from the victims, and with such collaboration the water in a swimming pool could drown a million people. (In a typical dispersal situation, only a few parts per million would be inhaled by people.) The "hot particle" theory hypothesizes that usual toxicity estimates, which are based on a uniform distribution of plutonium over the lung, underestimate the effects because the plutonium actually is concentrated in individual particles which therefore administer very high local radiation doses. This hypothesis has been studied and rejected by the following prestigious groups of radiobiomedical scientists: the U.S. National Council on Radiation Protection and Measurements (made up of about 70 of our nation's top scientists in that field), a special National Academy of Sciences Committee assembled to study the problem, the British Medical Research Council, the United Kingdom National Radiological Protection Board, the U.S. Atomic Energy Commission (in a thorough paper written by some of the world's top experts on effects of plutonium), and the U.S. Nuclear Regulatory Commission; and by the International Commission on Radiation Protection and the U.S. Environmental Protection Agency in deciding not to revise their standards on allowable plutonium intakes. No official body or prestigious scientific group anywhere in the world has supported the "hot particle" theory. Perhaps the simplest argument against it is that there are 25 workers who were exposed to plutonium in 1945-46 at Los Alamos; according to the "hot particle" theory, they should have had an average of 200 lung cancers each by now, whereas none of them has had lung cancer.

If one does not use the "hot particle" theory and assumes normal meteorology in the dispersal, it turns out that a pound of plutonium dispersed in the most effective way in a large city would cause about 20 fatalities.

It has been charged that in producing large quantities of plutonium we will be introducing a terrible new toxic material into our world in unprecedented amounts. Since plutonium is dangerous only if inhaled as a fine dust, let us compare it with other lethal inhalants we produce. The number of lethal doses (in trillions) produced per year in the U.S. is: chlorine—400, phosgene—20, ammonia—6, hydrogen cyanide—6, plutonium—1 if all U.S. power were produced by fast breeder reactors which are the type involving the most plutonium. It should be noted in this comparison that the other materials are gases which are readily dispersed whereas plutonium dust is very difficult to disperse effectively (it tends to agglomerate into large particles which are not dangerous to inhale, and to clog up dispersal devices). It may be argued that plutonium will last far longer than the gases with which we are comparing it, but it turns out that dispersed plutonium does more damage in the short time after dispersal when it is in a cloud of dust than it will accumulatively over the thousands of years it subsequently spends on the ground and in the soil. Finally, it should be pointed out that these effects of plutonium are most theoretical; there is no evidence that plutonium toxicity has ever harmed any human being, anywhere, ever.

One sometimes hears claims that we may "contaminate the earth" with plutonium-239, the isotope with a 25,000 year half life. However, it can be shown that if all the world's electric power were derived from fast breeder reactors, the total amount of plutonium-239 in the world at any one time (a constant amount since it would be burned up as fast as it would be produced) would be less than the radium in every eight inches of depth of the earth's crust. Moreover, radium is 40 times more dangerous than plutonium either as an ingestion or as an inhalation hazard, so all this plutonium would be five times less dangerous than the radium in each inch of depth. One hardly needs to be reminded that there are a lot of inches in the depth of the earth's crust.

As perspective on plutonium hazards, there have been 10,000 pounds of plutonium distributed through the earth's atmosphere in bomb tests, whereas the legally allowed total plutonium releases from an all fast breeder electric power industry would be about 1/1000 pound per year. Even a single largely unnoticed event in 1963, the reentry of a space satellite with a plutonium power source into the earth's atmosphere and its consequent burn-up, released more plutonium than a full nuclear power industry would release in a thousand years.

A much more credible problem than plutonium toxicity is the possibility of its use by terrorists to make crude nuclear bombs. There are several problems in such a scenario. In the first place, stealing plutonium would be very difficult and dangerous under present safeguards. Dr. Theodore Taylor, the one who has raised most of the fuss over the terrorist weapon threat, estimated in answer to a question from a congressional committee that a group of plutonium thieves would have much less than a 50-50 chance of escaping with their lives. In the second place, fabrication of a bomb from stolen plutonium would be a very difficult, time consuming, expensive, and dangerous job. Much of the public impression on this was derived from the NOVA television program about an MIT student figuring out how to make a bomb. However, his design was on paper, which is the easiest part of the job, and it was judged to have only a small chance of working. There were also newspaper stories about a Princeton student, but he only assembled information and never produced a design.

A third problem facing the potential terrorist-bomb maker is that his bombs can blow up a large building, not destroy a city. Taylor's scenario was that it could blow up the New York World Trade Center (that building sometimes contains 50,000 people, whence the oft-quoted remark that such a bomb could kill that many—a generally false impression). Most people equate all nuclear bombs to "city destroyers," but this is not valid here. There are many much easier ways to kill as many people (e.g., introducing readily available poison gases into the ventilation system of the World Trade Center). Terrorists have always had the option of killing thousands but have almost never intentionally killed more than a dozen or so.

Accidents

Up to this point we have been considering releases of radioactive wastes from normal operations of a nuclear industry, but there is always the possibility of accidents which will release it in much more dangerous ways. For example, worries have been expressed over transportation accidents resulting in radioactivity releases. However, these are much less of a hazard than one might think. Firstly, the waste is always shipped as a ceramic or glass, and nearly all release scenarios require that it be melted or vaporized; however, temperatures from fires that may be generated in transport accidents (e.g., gasoline fires) are not hot enough to melt this material. Secondly, these wastes are shipped in very elaborate casks; they cost \$2 million each, and one can well imagine that for that price a great deal of crash protection can be achieved. There have been several elaborately detailed estimates of radioactivity release probabilities in transport accidents, and of their consequences. The results have always come out to be less than *one fatality per century* if all U.S. power were nuclear. Of course there would be many more casualties from the ordinary traffic aspects of these accidents (a man is much more easily broken up than a specially designed $2 million cask), but coal-fired power would require 100 times as much transport and thus would cause 100 times as many fatalities.

Of much greater importance than these matters is power plant accidents. There is truly a vast quantity of radioactivity in a reactor so if it were released to the environment in a power plant accident, the consequences could indeed be grave. What is generally considered to be the most important potential accident is one in which the system (made of very thick, highest quality steel) suddenly bursts open, allowing the water inside, which is at very high temperature and pressure, to come pouring out, leaving the reactor core without cooling. In such a situation, the radioactivity would generate enough heat to melt the fuel, releasing the radioactivity within it partially as a radioactive dust. If this radioactive dust should escape to the environment, a great deal of very serious radiation exposure could occur.

Since this scenario is potentially so dangerous, a great deal of effort is expended to minimize the probability of the system bursting open, to prevent such an event from leading to a fuel melt, and to mitigate the consequences of a fuel melt if it should occur. First, fabrication standards are of the highest quality, exceeding those in any other industry. Second, the system is subjected to an inspection routine of unprecedented thoroughness, utilizing x-ray, magnetic particle, and ultrasonic techniques during the construction period, and after the plant is in operation it is periodically put through renewed ultrasonic and elaborate visual inspections (including removal of insulation) looking for imperfections or tiny cracks. Third, the system is highly instrumented with warning systems for any type of abnormality; for example, large leaks ordinarily develop from small leaks so the reactor is equipped with at least four different types of leak detection systems capable of detecting leaks hundreds of times smaller than what might cause problems. Fourth, if all of these measures should fail and the system does suddenly burst open leading to a loss of the water coolant, there is an elaborate emergency cooling system to reflood the reactor core and thus keep the fuel from melting. Fifth, if in spite of all this, somehow the fuel should melt, the system is enclosed within a powerfully built *containment* structure made of very thick, heavily reinforced concrete lined with steel plate. This containment protects the reactor against forces from the outside; for example, it would stop a truck load of explosives, or anything a tornado could throw against it (like a house or an automobile). But in a fuel melt-down situation, its function is to keep the radioactive dust contained for at least several hours; if it can do so, there are systems within to clean the radioactive dust out of the air, thus eliminating the danger. And *sixth*, even if the containment should be breached before this can happen, under ordinary weather conditions the escaping radioactive dust would be sufficiently dispersed to avoid a tragedy; only if there is a temperature inversion keeping the radioactive cloud close to the ground would large exposures result.

This series of protections is an example of "defense in depth" which is the basic philosophy of reactor design. It is sometimes said that if anything goes wrong in a reactor due to equipment failure, human error, or an "act of God," a tragedy would result. This is exactly *wrong.* In reactor design it is assumed that all sorts of equipment failures will occur, that the operator will make all sorts of mistakes, and that the worst credible tornadoes, earthquakes, and floods will occur causing systems to fail. But if one system fails there is another system behind to protect the reactor, if that system fails there is another system backing it up, etc. Of course you eventually run out of back-up systems, but it is only in the highly unlikely event that several independent systems fail one after the other that tragedy can occur.

It is also worth noting here that airplane safety is much less stringent on the first three of the above defenses, and has nothing corresponding to the last three. Materials used in airplanes are severely constrained by weight limitations, and inspections are much less elaborate than in reactors. There are no warning systems on mechanical failures—if a wing breaks off, everyone aboard is as good as dead. These first three levels of defense have failed in airplanes, but they have never failed in reactors. And if they should, there are still three levels of defense that have never been challenged.

Of course it is not *impossible* to have an accident, and most public discussion seems to concentrate on the worst accident that can be envisioned, expected less than once in a million years if all U.S. power were nuclear, in which there would eventually be tens of thousands of fatalities. Even this, it should be noted, is far less tragic than the worst accident from a hydroelectric dam failure, and the latter, whose risk we have long been willing to accept, is far more probable. Transport and storage of liquifed natural gas, another segment of the energy industry, could also lead to far greater tragedies than the worst nuclear accident.

The rational approach would seem to be considering the *average* effects of these accidents. If one accepts the estimates developed by the Rasmussen Study (a $4 million study financed by the AEC and directed by Norman Rasmussen, an MIT professor), if all U.S. power were nuclear, the average American's risk of dying as a result of a nuclear accident would be one in 20 million per year (98% of this risk is in the form of a slight increase in one's probability of dying of cancer many years after the accident).

Critics of nuclear power have raised many objections to the Rasmussen study (it is not difficult to criticize a process as complicated as estimating the probability of an accident). They have claimed that the uncertainties in the Rasmussen study are so large that it may be underestimating the average effects of a nuclear accident by as much as a factor of one hundred. One could argue about their claims, but it would be almost impossible to prove them wrong.

Summary and Perspectives

We are now in a position to summarize our discussion by considering the total annual death risk from an all nuclear U.S. power system to the average American. If one accepts the Rasmussen study, this is one in 10 million per year (50% from accidents, 45% from gaseous emissions, 5% from others), whereas according to the critics it is one in 200,000 per year. By comparison, an all-coal electric power system is estimated to give an average annual death risk of one in 20,000 with current technology, and perhaps one in 100,000 with all air pollution control equipment that might be feasible.

To put the nuclear risks into perspective, let us compare them with some of the other risks we take, like smoking cigarettes, being overweight, riding in automobiles, and using small cars rather than standard size cars. If we accept the Rasmussen results, all nuclear power would be as dangerous to the average American as smoking one cigarette every ten years, or as increasing his weight by a tenth of an ounce, or as riding in automobiles an extra three miles per year, or as riding in automobiles the same amount as now but doing 0.1% of it in a small car. Even if we accept the critics' estimates, it is only as dangerous as smoking one cigarette every 10 weeks, gaining one ounce of weight, or doing 5% of one's driving in a small car rather than in a standard size car.

Of course the risks are much larger if one lives very close to a nuclear power plant (the same is true for a coal-fired power plant), but if moving away from a nuclear power plant increases one's automobile travel by more than 250 yards per day (four miles per day according to the critics), it is safer to live next to the nuclear plant.

Surely these risks of nuclear power are acceptable, and in any case they are many times smaller than the risks from alternative energy sources. And surely they are infinitely less than the risks of doing without energy, which include large scale unemployment and a deteriorating standard of living. The bases for numerical estimates here are given in an article by this author in the September 1976 issue of *The American Scientist.* Further details and answers to questions are available from the author.

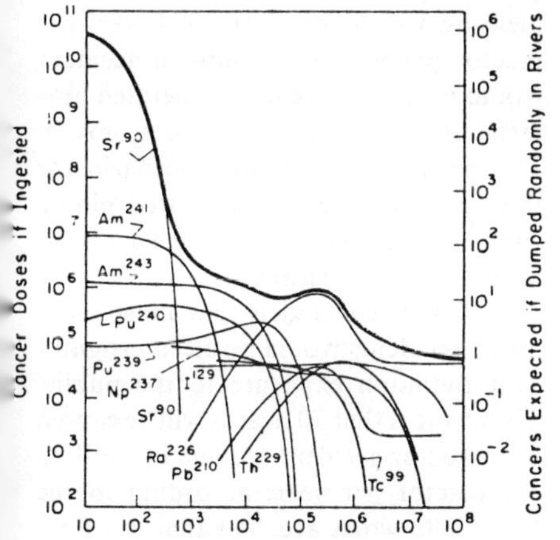

Hazard in the radioactive waste produced by one year of all-nuclear power in the U.S. Various curves are contributions of individual radioactive isotopes, and the thick curve is their sum. The scale at the left indicates the number of fatalities expected if the material were converted to digestible form and fed to people (assuming there were enough people) and the scale at the right shows the expected number of fatalities if it were converted to digestible form and dumped randomly into rivers. Clearly neither of these procedures would be acceptable.

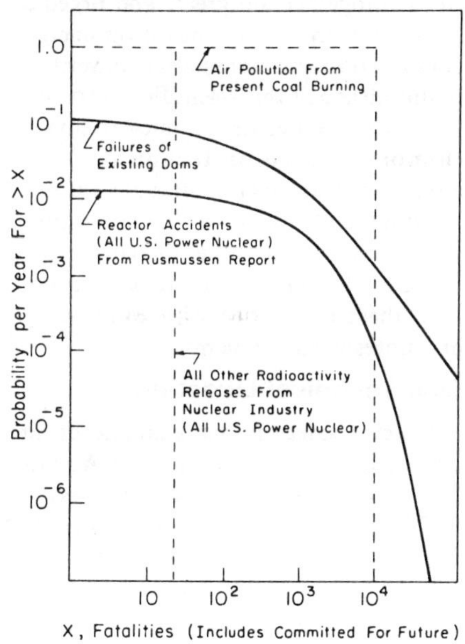

Probability per year for a number of fatalities is about 10^{-3} for dam failures and 10^{-4} for reactor accidents if all U.S. power were nuclear. The dashed lines represent large numbers of individual fatalities occurring frequently so the total number per year does not vary greatly.

NUCLEAR POWER: MORE THAN A TECHNOLOGICAL ISSUE?

Ralph Nader

ALTHOUGH the national debate on nuclear power is intensifying, some engineers and scientists view the nuclear issue as merely a technical one. Technologists such as Ralph Lapp and Norman Rasmussen condescendingly dismiss all doubts. If ordinary people will just leave nuclear power to the "experts," they feel, all the problems will be solved or will disappear.

Of course, this technical elitism has been severely undercut by more than 2,000 scientists and engineers who recently stated that the risks of nuclear power were "altogether too great" and urged a national energy policy based on energy conservation and nonnuclear power.[1]

But although this scientific controversy exists, the debate on nuclear power cannot be based on technical issues alone. Nuclear power is fraught with, if not dominated by, issues that are institutional and political rather than technical. There are technical issues, to be sure; but even these are ridden with controversy and unresolved problems

One Technical Question

Reactor safety is one example of an unsettled technical question. In August 1974, the Atomic Energy Commission (AEC) released a draft report which concluded that the chances of a major nuclear reactor accident were equivalent to the chances of a meteor striking a large city. From the fanfare with which the report was released, it is evident that the AEC and the nuclear industry hoped the report would lay to rest all concerns about reactor safety. But these hopes were not realized. The severe weaknesses of the report have contributed to greater skepticism over nuclear power.

At the outset, it should be emphasized that the report, the *Reactor Safety Study* (RSS),[2] only presumed to cover nuclear reactors themselves. The report did not cover the transportation of radioactive materials by truck, rail, and barge; the disposal of radioactive wastes; the risks of sabotage, theft, or terrorism; fuel reprocessing plants; or uranium mining processes and wastes.

The RSS attempted to predict probabilities and consequences of major reactor accidents. It assessed the chances of a meltdown accident as one in 17,000 per reactor per year. A meltdown accident would occur if the reactor generated heat faster than that heat could be removed. The most widely accepted scenario for this accident would begin with a piping rupture, allowing the water that cools the uranium fuel to escape.

The RSS concluded that most meltdown accidents would result in insignificant radiation exposure to the public. The chances that 70 deaths will be caused by a reactor accident are one in a millon per reactor per year, according to the RSS. The worst accident that the RSS considered was predicted to occur once every billion years per reactor. This accident would cause 2,300 immediate deaths, 5,600 immediate injuries, and 6.2 billion dollars in property damage.

Reactors are Unsafe

The most extensive and severe criticism of the RSS comes from a joint review by the Sierra Club (SC) and the Union of Concerned Scientists (UCS), a group of professionals in the Boston area. The SC-UCS critique[3] was organized by Henry Kendall, professor of nuclear physics at MIT.

The SC-UCS review found significant weaknesses in the conclusions of the RSS, with regard to both probabilities and consequences of reactor accidents. To predict reactor accident probabilities, the RSS used reliability estimating techniques developed by the aerospace industry. The SC-UCS pointed out that the aerospace industry later abandoned these techniques as a means of providing exact reliability estimates. As many as 35 percent of the failures in the Apollo program, for example, were deemed "incredible" before they happened.

The SC-UCS tested the validity of the RSS estimating techniques by applying them to a reactor accident that has already occurred. The result was an accident probability prediction of one in a billion-billion. The fact that this accident has already occurred casts a doubt on the RSS methods, to say the least. It seems that an accurate assessment of reactor accident probability might not be available until after an accident has occurred, which will be too late.

The SC-UCS review found that the RSS underestimates accident health consequences by at least a factor of 16. This finding is generally supported by the Environmental Protection Agency[4] and the AEC Regulatory Staff.[5] With the SC-UCS corrections, the most severe accident considered by the RSS could cause 36,800 *immediate* deaths and 90,000 *immediate* injuries. These totals are substantially equivalent to the results of accident consequence studies performed by the AEC in 1957 and 1965, which the RSS claimed to refute.[6]

When the SC-UCS corrects the RSS for errors in probabilities and consequences, they find that the risks from nuclear power plants are similar to nonnuclear risks. Where the RSS found nuclear risks similar to the dangers of being stuck by a meteor, SC-UCS finds nuclear power as dangerous as other technological hazards. With the correction factors, the probability of a nuclear plant catastrophe becomes similar to the probability of catastrophes from air crashes, fires, explosions, dam failures, and toxic chemical accidents.

Zero Credibility

A more recent review of the RSS by the American Physical Society (APS)[7] has destroyed whatever credibility the RSS might have retained after the critiques of EPA, SC-UCS, and the AEC Regulatory Staff. This latest review was conducted by a panel of 12 scientists for APS. Like the SC-UCS review, the APS study also had no confidence in the RSS accident probabilities. The APS study assessed the calculated probabilities of accident "branches," or hypothetical accident sequences, in these words: "[B]ased upon our experience with problems of this nature, involving very low probabilities, we do not now have confidence in the presently calculated absolute values of the probabilities of the various branches."[8]

The APS panel chose not to review the consequences of the most catastrophic RSS accident, as this had already been covered by other critiques. Instead, the study looked at a "small" accident, which the RSS predicted would cause 62 immediate deaths, 310 genetic defects, and 310 lethal cancers over the long term. The APS study identified impressive faults in the RSS analysis of long-term health effects. The RSS assumed that persons downwind of a reactor accident would receive radiation only through the first day following the accident. In reality ground-deposited radioactivity would irradiate people over a large area for an extended period. Correction of this factor alone increased lethal cancers and genetic defects by a factor of 25.

The APS study also found that RSS had neglected the selective and much larger irradiation of special tissues, notably the lungs and thyroid. APS corrects for these errors and others and concludes that the "small" accident would cause 10,000-15,000 lethal cancers and 3,000-20,000 genetic defects over the long term.[9]

Until the RSS, reactor safety researchers had held that the consequences of a meltdown accident would be severe, but that the probability of such an accident was very low—one in a million or less. The RSS concluded that a meltdown was much more likely to occur, but the consequences of an accident might not be severe. Professor Frank von Hippel, a member of the APS panel, assessed the RSS in this manner: "... two omissions in the calculations made there [the RSS] resulted in an underestimate of the average number of deaths from the reference accident by one or two orders of magnitude. Thus it appears that the APS group may have had the dubious honor of restoring the *status quo:* A reactor meltdown accident with containment failure would indeed be very serious.[10]

The critiques by the APS and other independent groups highlight the many uncertainties in nuclear reactor safety. These uncertainties cannot be swept away by the probability estimates of the thoroughly discredited *Reactor Safety Study*.

Too Little Insurance

The strongest business refutation of the RSS is the Price-Anderson Act, which provides for limited liability nuclear insurance. This Act requires each utility operating a nuclear plant to obtain the maximum available private insurance, which at present is $125 million. The federal government provides $435 million more in liability insurance. Were a nuclear accident to occur, liability of the reactor operator and the federal government would end at $560 million. This figure is less than one tenth the property damage alone of the worst RSS accident. More serious accidents are possible. A 1965 AEC study found that a reactor accident could cause 45,000 immediate deaths, 100,000 injuries, and $17 billion in property damage.

The Price-Anderson Act provides insufficient compensation for nuclear accidents, by any measure. But, if the chances of a reactor accident are really as remote as the RSS states, there will be no hardship for the nuclear industry to assume full liability for accidents it claims will never happen. If, on the other hand, reactor accidents are more likely than the RSS predicts, the public must be protected by unlimited liability. The question which must be put to the nuclear industry is: If nuclear reactors are so safe, why is limited liability necessary? To this, the industry has no satisfactory answer.

Plutonium Theft and Misuse

The Price-Anderson Act is by no means the only institutional issue facing nuclear power. If the nuclear establishment has its way, plutonium will soon be used as reactor fuel. This will subject an imperfect and unstable world to the dangers of a substance which is both extremely toxic and weapons grade material. Plutonium will subject workers and the environment to radioactive exposure from an industry that has not yet demonstrated it can handle the substance. But worse, plutonium will provide an attractive target for any group bent on sabotage or terrorism.

Experts within the nuclear establishment are concerned about plutonium theft and manufacture of an illicit nuclear weapon. An April 1974 AEC task force assessed measures to prevent plutonium theft and concluded that: "Even though safeguard regulations have just been reviewed and strengthened, we feel that the new regulations are inadequate and that immediate steps should be taken to greatly strengthen the protection of special nuclear materials."[11]

The AEC, in an environmental statement, admitted that present safeguards against theft would be inadequate if plutonium were introduced as reactor fuel.[12] The threat of an illicit nuclear weapon was reinforced by a public television program in which a 20-year-old student designed, on paper, a workable nuclear weapon. A member of the Swedish Defense Ministry advanced the opinion that the weapon had a good chance of exploding. The only missing ingredient would have been plutonium. It should be noted that were a terrorist group unable to make a nuclear weapon, the group could still create havoc by releasing plutonium in a populated area.

Don't Export Nuclear Power

Developing nations which may import reactor technology lack the technical infrastructure to adequately manage nuclear power, which is nowhere near safe in the U.S. itself. Moreover, there is probably no single item that developing nations need *less* than nuclear power. These nations cannot afford to tie large amounts of power production to a single central power station. When the plants shut down, a significant part of the country's power grid will be inoperative. In addition, many developing nations have substantial amounts of natural and renewable power that would be more appropriate. Clearly, a developing nation would want nuclear power largely for prestige and possible weapons manufacture.

The example of the nuclear weapon exploded by India indicates the ability of a country to divert plutonium from reactors to weapons production. The assurances of the State Department that U.S. reactors abroad will be subject to international safeguards are not at all convincing. International Atomic Energy Agency (IAEA) safeguards are little more than bookkeeping methods. The agency could detect a diversion of nuclear material after it occurred, but it requires minimal security measures. Even if diversion were detected, sanctions would have to be approved by the United Nations Security Council. If the Security Council could agree to act on a diversion, it might be hard-pressed to devise adequate sanctions.

Exporting nuclear power also means exporting sabotage targets. A General Accounting Office investigation in October 1974 concluded that nuclear plants in this country were susceptible to sabotage.[13] The same would be at least as true for reactors abroad. Unfortunately, reactor exports to areas of strife such as the Middle East, Pakistan, Korea, and Argentina are either in progress or planned. The sabotage threat makes reactor export to these areas even more senseless.

A Threat to Freedom

Civil liberties will also be affected. The question here is: If adequate safeguards against plutonium theft can be devised, can they be implemented without unacceptable losses of liberties? The AEC had proposed, for example, to establish a federal plutonium police force and widescale background security clearances as part of a safeguard system.[14] One thoughtful observer has evaluated what such measures might mean to society:

> A basic objection to theft-preventive safeguards is that they would require individuals to distort their assessment of their own role in society. For example, civilian employees of the nuclear power industry who would have to comply with stringent new security regulations might come to believe that they were more like soldiers than civilians in light of the back-

ground checks that they would undergo to secure employment and in light of the limitations on their off-the-job activities that they would have to observe to retain employment.

Post-theft recovery measures would create a situation approaching civil war, with the government arrayed against the perpetrators of the nuclear threat and with innocent citizens caught in the middle.[15]

Nuclear power threats to civil liberties have already surfaced, even without the use of plutonium fuel. The Virginia Electric Power Company (VEPCO), for example, has requested its state legislature to allow VEPCO to establish its own police force, with power to arrest persons anywhere in the state and to obtain access to confidential citizen records. VEPCO claimed such authority was required by AEC regulations, but the AEC denied that such sweeping private police authority was necessary.

There is a double danger for a society that chooses to use plutonium as an energy source. The first is that civil liberty infringements which undermine society will be necessary. The second is that a thief saboteur will slip through any security system, no matter how carefully it is enforced, thus rendering the defense of the nation impossible.

Dr. Donald Geesaman, biophysicist and plutonium specialist at the University of Minnesota, would remind us that:

> Reality is more inclusive than the *remove numerata* of the systems analysts. There are Klaus Fuchs and Lee Harvey Oswalds. There are heroin thefts at the New York Police Department and gambling hanky-panky in the highest echelons of AEC security personnel ... . There is terrorist violence at Khartoum and Munich and Tel Aviv; the attack on Princess Anne and the Hearst kidnapping. There are resurgent nationalism and alienated minorities, the hopeless, the poor, the vicious, the pathological, and most important of all, the brilliant, for raw human intelligence spreads across all human classifications; and "where there's a will there's a way."[16]

A Litany of Problems

There are other nuclear power institutional issues which because of limited space can only be mentioned:

• Nuclear power is an economic issue. Predictions of nuclear electricity so cheap it would not have to be metered have been shattered by skyrocketing nuclear plant costs and unreliable plant operation.

• Nuclear power is a subsidy issue. Even if utilities could produce electricity more cheaply for buyers they could only do so through subsidies and economic distortions throughout the nuclear fuel cycle. There is Price-Anderson, which reduces drastically the industry's insurance premiums. There is uranium enrichment, provided at reduced cost by government plants. There is fuel reprocessing, which will be subsidized by government floors for reused nuclear fuel. And, there are future costs which can only be estimated. The nuclear industry has practically no idea of the costs to decommission large plants. Add subsidies for safeguards and waste disposal and research—to name a few—to the rising demands for direct financial aid and tax reductions to utilities and the picture of nuclear costs becomes clearer.

• Nuclear power is an environmental and health issue. The power plant is supported by the nuclear fuel cycle of mining, milling, enrichment, fabrication, reprocessing, waste storage, and transportation in between. These steps impose the dangers of occupational radiation hazard, environmental degradation through radiation releases, and catastrophic accidents. Some steps pose all three dangers.

• Nuclear power is a moral issue. If no solution can be found for the disposal of radioactive wastes, guardianship may be required for the quarter-million years for which the wastes can cause cancer, leukemia, or genetic damage. Thus the power we consume now may leave lethal gar bage for thousands of generations to come.

• Nuclear power is an appropriations issue. For over 20 years, federal energy research has been almost totally devoted to nuclear power projects. If the breeder reactor and other nuclear projects continue to gobble up their present 60-70 percent of the energy research budget, development of new nonnuclear energy sources will be significantly forestalled.

There Are Alternatives

The last point raises the alternatives to a nuclear future. Nuclear power proponents would have us believe that the future requires continuing exponential growth in energy demand, with the only energy sources being imported oil, dirty coal, and nuclear power. These underlying assumptions must be challenged.

In the first place, there is an enormous potential for reducing energy use by cutting out waste. The Federal Energy Administration (FEA) estimates that per capita energy consumption could be cut 30 percent with no significant effect on lifestyle or industrial output.[17] Other experts believe that at least 40 percent of U.S. energy use is waste.

On a per capita basis, the U.S. consumes two to three times the energy of the industrial Western European nations. Hysterical nuclear proponents who claim that energy conservation will lead to economic chaos are refuted by Sweden, Denmark, and Switzerland. These countries get along on 50 percent or less of the per capita energy consumption of the U.S., but they all have higher per captia gross national products.[18]

The country should begin to think of conservation as an energy source. A recent American Institute of Architects study estimates that energy-efficient buildings alone could save more energy in 1990 than nuclear energy could supply; and the capital requirements to save that energy would be less than to provide it with nuclear power.[19]

On the supply side, one should not ignore oil as an alternative. Much publicity has been given to recent pessimistic reports on oil reserves, but little attention has been focused on the qualifications that accompany the pessimism. The latest U.S. Geological Survey's (USGS) estimates are that only a 40- to 60-year supply of domestic crude oil remains, at current production rates. But the USGS also admitted that its conclusions relied on data from the American petroleum Institute (API) and the American Gas Association (AGA); that its analysis was done for pre-1974 oil prices; and that significant discovered and potential resources, including oil shale, had been excluded.[20] One should realize that the API and AGA have a vested interest, along with a historical record, of drastically underreporting reserves in order to justify higher prices. Also, as prices at 1974 levels and beyond are taken into account, much more oil will become economical to recover.

With all factors considered, a more realistic conclusion is that at least 100 years of oil remain at present consumption levels. This conclusion should certainly not be an excuse for wasteful energy consumption. But it can keep the country from being stampeded into nuclear power reliance.

Persons who still remain convinced that oil reserves are scanty should refer to professional geologist M. King Hubbert. Dr. Hubbert's reserve estimates are even more pessimistic than those of the USGS.

Of nuclear power, he says, "I've gradually come around to look at the hazardous aspects and it scares the hell out of me."[21] Despite his low oil reserve estimates, Hubbert believes that dramatic conservation and development of solar energy are necessary to "phase out nuclear as fast as we can."[21]

Other sources include geothermal power and all the solar technologies—heating and cooling; conversion of urban, plant and animal waste to fuel[22]; wind power; and solar electricity. Even the FEA, which is becoming the Administration's nuclear power promoter, believes that solar power could replace the equivalent of 20 percent of our present energy needs by 1990.[23] Some of these sources are in everyday use right now, in this country and abroad. If nuclear power can be prevented from usurping the federal energy research budget, these sources will provide more and more of our energy in the future.

What If ...

So the choices for even the immediate future clearly are not limited to nuclear vs. coal. But even if they were, nuclear would lose, for two major reasons. The first reason is that the problems of coal power are largely technical in nature—safer mines, reclamation of mined land, and removal of pollutants from the coal before or after burning. The problems of nuclear power, on the other hand, are institutionally serious as well as technically unresolved. Nuclear power tempts saboteurs, terrorists, and hostile nations; it will strain civil liberties; it will require massive measures of government-financed corporate socialism; it will burden future generations; and it will require incredible stability in human nature and human institutions if its present radioactive inventory and future garbage are to be controlled.

The second reason nuclear power must be rejected is its social ramifications as a technology. Even proponents of nuclear power acknowledge that to be a viable energy alternative, it must be free of catastrophic accident. Were a catastrophe to occur, from any plant or in any step in the nuclear fuel cycle, what would be the reaction of the citizenry? After all, they have been told for years "it will never happen" by government officials and an industry grown smug from its own propaganda. If a catastrophe occurred, there would be such an outcry that nuclear power plants all over the country might be shut down forever. Clearly this country cannot let such a frail technology become a major energy source, now or in the future.

It is clear, then, that nuclear power is not merely a technical issue. Responsible technologists recognize the human ramifications of their technologies. Engineers and scientists must become sensitive to the myriad human issues in the nuclear power question and to the overwhelming legal problems and burdens that are developing under the stresses this technology is placing on our society and the world. Only when all these issues are recognized can one engage in a rational and reasoned policy deliberation. When only some of these issues are conceded, one realizes that the decision on nuclear power must be made by an entire enlightened population, not by delegated authority. Just as war is too important to be left to the generals, so nuclear power is too important to be left to the so-called nuclear "experts."

Notes

1. Union of Concerned Scientists, News Release, Aug. 6, 1975.

2. WASH-1400, *Reactor Safety Study Draft*, U.S. Atomic Energy Commission, Washington, Aug. 1974.

3. *Preliminary Review of the AEC Reactor Safety Study*, Sierra Club-Union of Concerned Scientists, San Francisco-Cambridge, Nov. 1974.

4. *Comments by the Environmental Protection Agency on Reactor Safety Study*, EPA, Washington, Nov. 1974.

5. *Review of the Reactor Safety Study (WASH-1400) Draft of August 1974*, U.S. Atomic Energy Commission Regulatory Staff, Dec. 2, 1974.

6. The 1957 study was WASH-740, *Theoretical Possibilities and Consequences of Major Accidents in Large Nuclear Power Plants*, U.S. Atomic Energy Commission, Mar. 1957.
The 1965 study updated the 1957 study by using for its model a more powerful reactor, which reflected the size of reactors being built in 1965.

7. *Report to the American Physical Society by the Study Group on Light-Water Reactor Safety*, Apr. 28, 1975. Referred to below as APS.

8. APS, pp. I-3 and I-4.

9. APS, p. V-68.

10. "Reactor Safety," invited Talk at the Spring Meeting of the American Physical Society, Apr. 30, 1975, Frank von Hippel, Center for Environmental Studies, Princeton Univ., p. 1.

11. "A Special Safeguards Study," AEC, reprinted in *Congressional Record*, S 6622, Apr. 30, 1974.

12. WASH-1327, *Generic Environmental Statement Mixed Oxide Fuel-Draft*, U.S. AEC, Aug. 1974, pp. V-6. Referred to below as GESMO.

13. Letter from U.S. General Accounting Office to Dixie Lee Ray, Chmn., U.S. AEC, Oct. 16, 1974.

14. GESMO, p. V-7 and elsewhere.

15. Ayres, R.W., "Policing Plutonium: The Civil Liberties Fallout," *Harvard Civil Rights—Civil Liberties Law Review*, Vol. 10, No. 2, Spring 1975, pp. 441-442.

16. Geesaman, D. "Statement in Support of a Nuclear Moratorium for the State of Minnesota," before the Minnesota Pollution Control Agency, Dec. 12, 1974.

17. U.S. Federal Energy Administration, brief filed with the State of California Public Utilities Commission, Apr. 2, 1975, p. 2.

18. Holdren, J.P., "Too Much Energy, Too Soon," University of California, Berkeley, *New York Times*, July 23, 1975, p. 31, and *Energy Reporter*, Federal Energy Administration, May 1975.

19. "Energy and the Built Environment: A Gap in Current Strategies" and "A Nation of Energy Efficient Buildings by 1990," American Institute of Architects, Washington, D.C., 1975.

20. "Geological Estimates of Undiscovered Recoverable Oil and Gas Resources in the United States," *Geological Survey Circular 725*, U.S. Department of the Interior, Washington, D.C. pp. 1, 2.

21. Prouty, D., " 'Convert Stresses N-Power Hazards," *Denver Post*, Feb. 4, 1975.

22. For an evaluation of the potential energy supply from plant waste, see "Energy Potential from Agriculture Field Residues," Farno L. Green, General Motors Corp., Detroit, June 1975.

23. *Project Independence Final Task Force Report: Solar Energy*, U.S. Federal Energy Administration, Washington, D.C., Nov. 1974, p. I-9.

Further Reading

Ayres, R.W. 1975. Policing Plutonium: The Civil Liberties Fallout. *Harvard Civil Rights—Civil Liberties Law Review* 10 (Spring): 369-443.

Berger, J.J. 1977. *Nuclear Power: The Unviable Option.* Palo Alto, California: Ramparts Press.

Bethe, H.A. 1976. The Necessity of Fission Power. *Scientific American* 234 (January): 21-31.

Caldwell, W.S., *et al.* 1974. "The ... Price-Anderson Act," *Rutgers-Camden Law Journal* 6 (Fall): 360-386.

Code of Federal Regulations. 1978. Energy. 10, Part 20, pp. 181-214. Washington, D.C.: Government Printing Office.

Canfield, C. 1990. *Multiple Exposures: Chronicles of the Radiation Age.* New York: Harper and Row.

Committee on Commerce, Science, and Transportation. 1977. *Radiation Health and Safety,* US Senate, 95th Congress, First Session (June 16, 17, 27-29), No. 95-49. Washington, D.C.: Government Printing Office.

Cooke, R.M. 1991. *Experts in Uncertainty: Subjective Probability and Expert Opinion.* New York: Oxford.

Elder, J. 1975. Nuclear Torts: The Price-Anderson Act and the Potential for Uncompensated Injury. *New England Law Review* 11 (Fall): 111-135.

Flavin, C. 1983. *Nuclear Power: The Market Test.* Washington, D.C.: Worldwatch Institute.

Flavin, C. 1984. Reassessing the Economics of Nuclear Power. In *State of the World 1984,* ed. L. Brown et al., pp. 115-135. New York: Norton.

Flavin, C., and Durning, A. 1988. *Building on Success: the Age of Energy Efficiency.* Washington, D.C.: Worldwatch Institute.

Flavin, C., and Lenssen, N. 1990. *Beyond the Petroleum Age: Designing a Solar Economy.* Washington, D.C.: Worldwatch Institute.

Flavin, C., and Polluck, C. 1985. Harnessing Renewable Energy. In *State of the World 1985,* ed. L. Brown et al., pp. 172-199. New York: Norton.

Fuller, J.G. 1975. *We Almost Lost Detroit.* New York: Ballantine.

Gilinsky, B. 1975. The Need for Nuclear Safeguards. *Atomic Energy Law Journal* 17 (Summer): 149-50.

Gofman, J., and Tamplin, A. 1970. *"Population Control" through Nuclear Pollution.* Chicago: Nelson-Hall.

______. 1971. *Poisoned Power.* Emmaus, Pennsylvania: Rodale Press.

Hohenemser, C., Kasperson, R., and Kates, R. 1977. The Distrust of Nuclear Power. *Science* 196 (April 1): 25-34.

Kirshner, D. 1991. Some Utilities Now Support Aalternative Energy. *Environmental Defense Fund Letter* 22, No. 2 (Aporil): 4.

Krauss, S., ed. 1991. The National Energy Tragedy. *Nucleus* 13, No. 1 (Spring): 1-8. (Union of Concerned Scientists, 26 Church Street, Cambridge, MA 02238).

Lewis, H.W., *et al.* 1975. Report to the American Physical Society by the Study Group on Light-Water Reactor Safety. *Reviews of Modern Physics* 47 (Summer): S1-S124.

Lovins, A. 1975. *Non-Nuclear Futures: The Case for An Ethical Energy Strategy.* San Francisco: Friends of the Earth International.

Lowenstein, R. 1977. The Price-Anderson Act. *Forum* 12 (Winter): 605-611.

Lowrance, W.W. 1976. *Of Acceptable Risk: Science and the Determination of Safety.* Los Altos, California: William Kaufmann.

Muchnicki, E.D. 1973. The Proper Role of the Public in Nuclear Power Plant Licensing Decisions. *Atomic Energy Law Journal* 15 (Spring): 34-59.

Mulvihill, R.J., Arnold, D.R., Bloomquist, C.E., and Epstein, B. 1975. Analysis of the United States Power Reactor Accident Probability, PRC R-695. Unpublished draft. From the file "WASH 740 Update," Public Documents Room, Nuclear Regulatory Commission. Los Angeles: Planning Research Corporation.

Munson, R., ed. 1976. *Countdown to a Nuclear Moratorium.* Washington, D.C.,: Environmental Action Foundation.

In *Values in the Electric Power Industry,* ed. Kenneth Sayre. Notre Dame, Indiana: University of Notre Dame Press.

Muchnicki, E.D. 1973. The Proper Role of the Public in Nuclear Power Plant Licensing Decisions. *Atomic Energy Law Journal* 15 (Spring): 34-59.

Mulvihill, R.J., Arnold, D.R., Bloomquist, C.E., and Epstein, B. 1965. Analysis of United States Power Reactor Accident Probability, PRC R-695. Unpublished draft. From the file "WASH 740 Update," Public Documents Room, Nuclear Regulatory Commission. Los Angeles: Planning Research Corporation.

Munson, R., ed. 1976. *Countdown to a Nuclear Moratorium*. Washington, D.C.: Environmental Action Foundation.

Nader, R., and Abbotts, J. 1977. *The Menace of Atomic Energy*. New York: W.W. Norton.

Novick, S./ 1976. *The Electric War: The Fight over Nuclear Power*. San Francisco: Sierra Club Books.

Office of Technology Assessment (OTA). 1985. *Managing the Nation's Commercial High-Level Radioactive Waste*. Washington, D.C.: US Government Printing Office.

Olson, M.C. 1976. *Unacceptable Risk: The Nuclear Power Controversy*. New York: Bantam.

Polluck, C. 1986. *Decommissioning: Nuclear Power's Missing Link*. Washington, D.C.: Worldwatch Institute.

Renner, M. 1990. Swords into Plowshares. Washington, D.C.: Worldwatch Institute.

Schmidt, F.H., and Bodansky, D. 1976. *The Energy Controversy: The Fight over Nuclear Power*. San Francisco: Albion.

Shrader-Frechette, K.S., ed. 1991. Nuclear Energy and Ethics. Geneva: World Council of Churches.

Shrader-Frechette, K.S. 1983. *Nuclear Power and Public Policy: The Social and Ethical Problems of Fission Technology*. Boston: D. Reidel.

Shrader-Frechette, K.S., 1985a. Science Policy, Ethics, and Economic Methodology. Boston, Massachusetts: Reidel.

Shrader-Frechette, K.S., 1985b. *Risk Analysis and Scientific Method*. Boston, Massachusetts: Reidel.

Shrader-Frechette, K.S., 1991. *Risk and Rationality*. Berkeley, California: University of California Press.

Shrader-Frechette, K.S. 1988. Values and Hydrogeological Method: How Not to Site the World's Largest Nuclear Dump. *Energy Policy Studies* 4: 101 ff.

Study Group on light-Water Reactor Safety. 1975. Nuclear Reactor Safety—the APS Submits Its Report. *Physics Today* 128 (July): 38-43.

Subcommittee on Oversight and Investigations of the Committee on Interstate and Foreign Commerce, House of Representatives. 1977. *Nuclear Waste Management and Disposal*, 95th Congress, First Session (July 29 and August 1), No. 95-67. Washington, D.C.: Government Printing Office.

US Atomic Energy Commission. 1957. *Theoretical Possibilities and Consequences of Major Accidents in Large Nuclear Power Plants*. US AEC Report WASH-740. Washington, D.C.: Government Printing Office.

______. 1974. Reactor Safety Study: An Assessment of Accident Risks in US Commercial Nuclear Power Plants. *Atomic Energy Law Journal* 16 (Fall): 177-204.

US Nuclear Regulatory Commission. 1975. *Reactor Safety Study—An Assessment of Accident Risks in US Commercial Nuclear Power Plants*, Report No. WASH-1400, NUREG-75/014. Washington, D.C.: Government Printing Office.

Willrich, M. 1971. *Global Politics of Nuclear Energy*. New York: Praeger.

World Health Organization (WHO). 1985. Health Implications of Nuclear Power Production. Paris: OECD.

World Information Service on Energy (WISE). 1991. Cancers on the Rise as Soviet Testing Continues. *Wise News Communique* 345 (18 January): 1-2.

WISE. 1990. Consequences of Chernobyl Still Worse. *Wise News Communique* 330 (6 April): 1-2.

WISE. Environmental Monitoring Since Chernobyl. *Wise News Communique* 3490350 (5 April): 18-19.

WISE. 1990. New Fears of Leaks from French N-Tests. *Wise News Communique* 338 (14 September): 1-2.

WISE. 1991. Number of Operational N-Plants Declines Worldwide. *Wise News Communique* 347 (22 February): 9.

WISE. 1991. Proliferation. *Wise News Communique* 348 (8 March): 1-2.

WISE. 1990. Proliferation. *Wise News Communique* 337 (31 August): 1-2.

WISE. 1990. Rongelap Islanders: Five Years After Evacuation. *Wise News Communique* 335 (6 July): 1-2.

Wilson, M.S. 1977. Nuclear Liability and the Price-Anderson Act. *Forum* 7 (Winter): 612-21.

Young, M.S. 1976. A Survey of the Governmental Regulation of Nuclear Power Generation. *Marquette Law Review* 59: 849-50.

INDEX